TensorFlow 2

人工神经网络学习手册

ANN架构机器学习项目

［印度］P．萨朗 （Poornachandra Sarang） 编著

周悦 曹旭阳 译

内容简介

随着深度学习理论的不断发展以及 TensorFlow 的广泛应用,基于深度学习的信号分析模型在不同领域产生了深远的影响。本书为 TensorFlow 2 的使用指南,从软件安装、数据下载、文件管理等方面入手为初级开发者提供细致而全面的介绍。在此基础上,本书系统地介绍了 TensorFlow 2 在人工神经网络实战项目中的应用,全面覆盖了各种深度学习架构,内容涉及:入门级的二分类模型、回归模型等;进阶级的文本生成模型、图像生成模型、机器翻译模型、时序预测模型等;以及最新的 Transformer 模型等。在每个项目中,本书完整地展示了模型设计、网络搭建、模型训练、模型保存、结果预测与显示的全过程,并提供了详细的实现代码。本书将深度学习理论与实际项目结合,为初学者搭建了进入人工智能领域的学习平台,为深度学习算法开发者提供了较为全面的应用范例,充分满足了不同群体的学习需求。

First published in English under the title
Artificial Neural Networks with TensorFlow 2: ANN Architecture Machine Learning Projects
by Poornachandra Sarang
Copyright © Poornachandra Sarang, 2021
This edition has been translated and published under licence from APress Media, LLC, part of Springer Nature.
APress Media, LLC, part of Springer Nature takes no responsibility and shall not be made liable for the accuracy of the translation.

本书中文简体字版由 APress Media, LLC 授权化学工业出版社独家出版发行。

本书仅限在中国内地(大陆)销售,不得销往中国香港、澳门和台湾地区。未经许可,不得以任何方式复制或抄袭本书的任何部分,违者必究。

北京市版权局著作权合同登记号:01-2022-0206

图书在版编目(CIP)数据

TensorFlow 2 人工神经网络学习手册 /(印)P. 萨朗 (Poornachandra Sarang)编著;周悦,曹旭阳译.
—北京:化学工业出版社,2022.3
书名原文:Artificial Neural Networks with TensorFlow 2
ISBN 978-7-122-40759-7

Ⅰ.①T… Ⅱ.①P… ②周… ③曹… Ⅲ.①人工神经网络-手册 Ⅳ.① TP183-62

中国版本图书馆 CIP 数据核字(2022)第 021334 号

责任编辑:周　红　张兴辉
责任校对:杜杏然
装帧设计:王晓宇

出版发行:化学工业出版社
　　　　　(北京市东城区青年湖南街 13 号　邮政编码 100011)
印　　装:三河市延风印装有限公司
787mm×1092mm　1/16　印张 27½　字数 681 千字
2022 年 4 月北京第 1 版第 1 次印刷

购书咨询:010-64518888
售后服务:010-64518899
网　　址:http://www.cip.com.cn

凡购买本书,如有缺损质量问题,本社销售中心负责调换。

定　　价:168.00 元　　　　　　　　版权所有　违者必究

Artificial Neural Networks with
TensorFlow 2

作者简介

  Poornachandra Sarang 博士在 IT 行业的职业生涯始于 20 世纪 80 年代，在这段漫长的职业生涯中，他广泛研究了各种技术。他曾在圣母大学和孟买大学教授计算机科学与工程。他曾是计算机科学学科的博士生导师，目前是计算机工程学科博士学位论文咨询委员会成员。他目前的研究兴趣是机器／深度学习，发表了多篇期刊文章，并在多个会议上发表过演讲。

审稿人简介

Artificial Neural Networks with TensorFlow 2

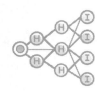

  Vishwesh Ravi Shrimali 于 2018 年毕业于 BITS Pilani，专业为机械工程。从那时起，他一直与 Big Vision LLC 合作研究深度学习和计算机视觉，并参与创建官方 OpenCV AI 课程。目前，他在梅赛德斯奔驰印度研发部工作。他对编程和人工智能有着浓厚的兴趣，并将这种兴趣应用于机械工程项目。他在 Learn OpenCV 网站上撰写了多篇关于 OpenCV 和深度学习的技术博客，该网站是计算机视觉领域的知名博客。他撰写了 Machine Learning for OpenCV（第 2 版）一书，由 Packt 出版。当他不写博客或做项目时，喜欢长距离散步或弹奏吉他。

致谢

　　首先要感谢我过去的学生 Abhijit Gole 教授，他目前是孟买大学附属自治学院计算机科学系的教员。Abhijit 让我为他的学生开设了机器学习的研究生课程，这引发了我对指导硕士和博士生进行机器/深度学习项目的兴趣。

　　衷心感谢我的四位实习生为编写本书做出的贡献。Mukul Rawat 今年（2020 年）毕业，Karan Aryan、Chandrakant Sharma 和 Udit Dashore 将于明年（2021年）毕业。这四位同学都是优秀的学生，如果没有他们帮助开发本书中的项目，我将会花更长的时间来完成这本书。他们都煞费苦心地审阅了本书初稿，以找出技术上的错误。

　　感谢 Apress 出版社的高级编辑 Aaron Black 先生，他给了我撰写本书的机会，他对本书的大纲做出了快速决定，并在撰写内容的过程中为我提供了最终确定大纲所需的灵活性。感谢助理编辑 Jessica Vakili 在各个方面的出色协调，尤其是在技术审查方面。我还要感谢整个 Apress 团队将这本书迅速推向市场。

　　最后，感谢 Vijay Jadhav 在运行所有项目源代码方面付出的很多努力，以找出写作和代码中的差异。需要特别提及他在按照 Apress 创作指南格式化手稿方面所做的努力。

前言

Artificial Neural Networks with TensorFlow 2

随着深度学习技术的不断发展，使用 TensorFlow 搭建的机器学习模型在多个领域产生了巨大影响。对于开发人员来说，这是一个很好的学习机会，可以将软件开发技能应用于更多实际项目中。了解时下热点的深度学习模型十分有必要，因此，本书以项目为导向，提供了基于深度神经网络和最新版本 TensorFlow 的实战项目，全面覆盖了各种深度学习模型。

第 1 章首先介绍了 TensorFlow 2，深度剖析了 TensorFlow 被广泛使用的原因，描述了 TensorFlow 的各种功能，包括使用 tf.keras 搭建模型、分布式训练、模型部署及数据通道的使用，并指导了本书开发环境的搭建。

第 2 章更加深入地介绍了 TensorFlow 2。按照惯例，第 2 章从一个简单的"Hello World"类应用程序开始，介绍了完整的机器学习开发过程。接下来使用 TensorFlow 完成机器学习的入门项目，即二分类问题。通过此项目的学习，有助于读者了解数据预处理、使用 tf.keras 的预留接口定义神经网络、模型训练、使用 TensorBoard 评估分类模型性能及学习使用混淆矩阵。

第 3 章详细介绍了 tf.keras 模块及带有模型构建功能的 API。本章向读者介绍如何使用 TensorFlow 面向对象的编程模式搭建网络模型及如何通过子类创建自定义的层。此外，读者将学习如何使用 TensorFlow 的 SavedModel 接口及几种部分或完整保存模型的方法。学习了上述基本方法后，读者将使用 tf.keras 中预定义的卷积神经网络层进行多分类项目实战，本书将向读者展示如何测试不同的网络架构并优化模型性能。

由于训练深度神经网络需要充足的计算资源，因此在第 4 章将向读者展示如何使用迁移学习技术加载预训练的网络模型。本章涉及两个项目：第 1 个项目展示如何使用在 ImageNet 数据集中预训练的分类器对其他图像集进行分类；第 2 个项目介绍如何基于预训练模型构建自己的分类器。

学习使用深度学习技术都是从分类和回归问题入手。在前面的章节中，我们深入探讨了分类问题，接下来，是否可以使用深度神经网络进行回归分析？针对这一问题，已有几种基于统计的深度学习模型可供我们分析。因此，第 5 章将基于 3 种不同的神经网络架构展示如何使用深度神经网络完成回归问题。

尽管使用预训练模型可以提高自定义模型的开发效率，但在许多情况下，很难找到合适的预训练模型。从头开始进行模型开发需要大量组件，TensorFlow 中的 Estimators 可以帮助开发过程中处理复杂的组件。第 6 章介绍 Estimators 的使用，共包含 4 个项目。前两个项目描述如何使用预定义的 Estimators 进行分类与回归，第 3 个项目讨论如何从头开始创建自定义的 Estimators，最后一个项目讨论为预训练模型创建 Estimators。

第 7 章讨论文本生成技术，介绍了用于文本生成的循环神经网络 (RNN) 和长短期记忆力模块 (LSTM)。第 1 个项目展示了如何将 LSTM 用于为新生儿起名的简单应用。第 2 个项目使用更加高级的文本生成技术，创建与列夫·托尔斯泰著名小说《战争与和平》的语言风格相匹配的文本。由于训练文本生成模型耗时较长，本章最后介绍如何在中断训练后继续进行模型训练。

第 8 章将进一步介绍文本生成技术，讲解了 seq2seq 模型、编码器和解码器架构及用于文本翻译的注意力模型。本章深入介绍了使用 GloVe 词嵌入模块的英语到西班牙语翻译器，向读者展示了如何使用编码器、解码器和注意力层自定义机器学习模型。

第 9 章介绍了自然语言处理中最先进的文本生成技术，即 Transformer。通过一个完整的实战项目教读者如何用 Python 语言构建 Transformer，并在没有预训练模型的前提下，在 Transformer 中使用双向编码器。

了解了文本生成技术后，本书第 10 章将讲解使用深度学习技术进行图像处理。第 10 章使用之前学到的长短期记忆力模块为图像生成标题。该项目使用预训练的 InceptionV3 模型进行图像处理，这是第 4 章介绍的迁移学习的实际应用。通过此项目开发，读者将学会如何使用 Bahdanau Attention 创建解码器，以及使用此模型为任何图像添加描述。

第 11 章进入机器学习的另一个领域，即时间序列分析和预测。本章提供了两个完整的项目，包括单变量时间序列预测及多变量时间序列预测。

第 12 章讲解图像处理领域中基于深度学习的风格迁移技术，可以将著名画家的风格应用于用相机拍摄的照片中，使照片看起来像著名画家的作品。本章包含两个项目：第 1 个项目使用来自 TensorFlow 中自带的预训练模型快速完成风格迁移；第 2 个项目深入讲解如何使用经典的 VGG16 架构提取图像和名画的特征，以及如何定义评价标准用于比较风格转换前后的内容和样式损失，并创建最终的风格化图片。

第 13 章介绍一个重要的深度神经网络架构，即生成对抗网络。本章共包含 3 个项目，第 1 个项目和第 2 个项目分别展示如何使用 GAN 生成手写数字图像和手写字母图像，第 3 个项目使用 GAN 生成复杂的动漫彩色角色图像。

第 14 章涉及一个重要的图像处理技术，即如何对黑白图像进行着色。本章使用 AutoEncoder 为图像着色，共包含两个项目。第 1 个项目构建了自定义的 AutoEncoder。第 2 个项目使用预训练的 VGG16 模型进行特征编码，用于提高模型性能。

本书详尽地收集了若干种深度神经网络架构，重点介绍了不同的深度学习模型在现实生活场景中的应用。

目录 Contents

第 1 章　TensorFlow 快速入门　001

- 1.1 什么是 TensorFlow 2.0　002
  - 1.1.1 TensorFlow 2.x 平台　002
  - 1.1.2 训练　003
  - 1.1.3 模型保存　005
  - 1.1.4 部署　005
- 1.2 TensorFlow 2.x 提供什么　006
  - 1.2.1 TensorFlow 中的 tf.keras　006
  - 1.2.2 Eager 执行　006
  - 1.2.3 分布式计算　007
  - 1.2.4 TensorBoard　007
  - 1.2.5 视觉套件（Vision Kit）　008
  - 1.2.6 语音套件（Voice Kit）　008
  - 1.2.7 边缘套件（Edge TPU）　008
  - 1.2.8 AIY 套件的预训练模型　009
  - 1.2.9 数据管道　009
- 1.3 安装　009
  - 1.3.1 安装步骤　009
  - 1.3.2 Docker 安装　010
  - 1.3.3 无安装　010
- 1.4 测试　010
- 总结　012

第 2 章　深入研究 TensorFlow　013

- 2.1 一个简单的机器学习应用程序　013
  - 2.1.1 创建 Colab 笔记本　014
  - 2.1.2 导入　015
  - 2.1.3 创建数据　016
  - 2.1.4 定义神经网络　018
  - 2.1.5 编译模型　018
  - 2.1.6 训练网络　018
  - 2.1.7 检查训练结果　019
  - 2.1.8 预测　021
  - 2.1.9 完整源码　022
- 2.2 使用 TensorFlow 解决二分类问题　024
  - 2.2.1 创建项目　024
  - 2.2.2 导入　024
  - 2.2.3 挂载 Google 云盘　025
  - 2.2.4 加载数据　026
  - 2.2.5 数据处理　027
  - 2.2.6 定义 ANN　030
  - 2.2.7 模型训练　032
  - 2.2.8 完整源码　036
- 总结　039

第 3 章　深入了解 tf.keras　040

- 3.1 开始　040
- 3.2 用于模型构建的函数式 API　041
  - 3.2.1 序列化模型　041
  - 3.2.2 模型子类　043
  - 3.2.3 预定义层　044
  - 3.2.4 自定义层　044
- 3.3 保存模型　046
- 3.4 卷积神经网络　049
- 3.5 使用 CNN 做图像分类　050
  - 3.5.1 创建项目　051
  - 3.5.2 图像数据　051
  - 3.5.3 加载数据　052

3.5.4	创建训练、测试数据集	052	3.5.8	保存模型	073
3.5.5	准备模型训练数据	053	3.5.9	预测未知图像	073
3.5.6	模型开发	055	总结		075
3.5.7	定义模型	060			

第 4 章　迁移学习　　　　　　　　　　　　　　　　　　　　076

4.1	知识迁移	076	4.4.6	处理图像	091
4.2	TensorFlow Hub	077	4.4.7	关联图像与标签	092
4.2.1	预训练模型	078	4.4.8	创建数据批次	093
4.2.2	模型的使用	079	4.4.9	显示图像函数	094
4.3	ImageNet 分类器	080	4.4.10	选择预训练模型	095
4.3.1	创建项目	080	4.4.11	定义模型	095
4.3.2	分类器 URL	080	4.4.12	创建数据集	097
4.3.3	创建模型	081	4.4.13	设置 TensorBoard	099
4.3.4	准备图像	082	4.4.14	训练模型	100
4.3.5	加载标签映射	083	4.4.15	训练日志	100
4.3.6	显示预测结果	084	4.4.16	验证模型性能	101
4.3.7	列出所有类别	085	4.4.17	预测测试图像	101
4.3.8	结果讨论	085	4.4.18	可视化测试结果	103
4.4	犬种分类器	085	4.4.19	预测未知图像	105
4.4.1	项目简介	086	4.4.20	使用小数据集训练	106
4.4.2	创建项目	086	4.4.21	保存、加载模型	107
4.4.3	加载数据	086	4.5	提交你的工作	108
4.4.4	设置图像和标签	088	4.6	进一步工作	108
4.4.5	图像预处理	091	总结		109

第 5 章　使用神经网络处理回归问题　　　　　　　　　　　　110

5.1	回归	110	5.2	神经网络中的回归问题	112
5.1.1	定义	110	5.2.1	创建项目	112
5.1.2	应用	111	5.2.2	提取特征和标签	113
5.1.3	回归问题	111	5.2.3	定义、训练模型	113
5.1.4	回归问题的类型	111	5.2.4	预测	114

目 录 Contents

5.3 分析葡萄酒质量 114
    5.3.1 创建项目 114
    5.3.2 数据准备 114
    5.3.3 下载数据 115
    5.3.4 准备数据集 115
    5.3.5 创建数据集 115
    5.3.6 数据归一化 116
    5.3.7 创建模型 119
    5.3.8 可视化评价函数 119
    5.3.9 小模型 120
    5.3.10 中模型 122
    5.3.11 大模型 124
    5.3.12 解决过拟合 126
    5.3.13 结果讨论 129
5.4 损失函数 130
    5.4.1 均方误差 130
    5.4.2 平均绝对误差 131
    5.4.3 Huber 损失 131
    5.4.4 Log Cosh 损失 131
    5.4.5 分位数损失 131
5.5 优化器 132
总结 132

第 6 章 Estimators（估算器） 134

6.1 Estimators 概述 134
    6.1.1 API 接口 135
    6.1.2 Estimators 的优点 135
    6.1.3 Estimators 的类型 136
    6.1.4 基于 Estimators 的项目
          开发流程 137
6.2 设置 Estimators 139
6.3 用于分类的 DNN 分类器 139
    6.3.1 加载数据 140
    6.3.2 准备数据 140
    6.3.3 Estimators 输入函数 141
    6.3.4 创建 Estimators 实例 142
    6.3.5 模型训练 142
    6.3.6 模型评价 143
    6.3.7 预测未知数据 144
    6.3.8 实验不同的 ANN 结构 144
    6.3.9 项目源码 145
6.4 用于回归的 LinearRegressor 147
    6.4.1 项目描述 147
    6.4.2 创建项目 147
    6.4.3 加载数据 148
    6.4.4 特征选择 148
    6.4.5 数据清洗 149
    6.4.6 创建数据集 151
    6.4.7 建立特征列 152
    6.4.8 定义输入函数 154
    6.4.9 创建 Estimators 实例对象 154
    6.4.10 模型训练 155
    6.4.11 模型评估 155
    6.4.12 项目源码 156
6.5 自定义 Estimators 158
    6.5.1 创建项目 159
    6.5.2 加载数据 159
    6.5.3 创建数据集 159
    6.5.4 定义模型 159
    6.5.5 定义输入函数 160
    6.5.6 将模型转换为 Estimator 160
    6.5.7 模型训练 161
    6.5.8 模型评价 161
    6.5.9 项目源码 161
6.6 为预训练模型定义 Estimators 163
    6.6.1 创建项目 163

6.6.2	导入 VGG16	163	6.6.7	训练、评价	166
6.6.3	创建自定义模型	163	6.6.8	项目源码	166
6.6.4	编译模型	165	总结		167
6.6.5	创建 Estimator	165			
6.6.6	处理数据	165			

第 7 章　文本生成　169

7.1	循环神经网络	170	7.3.9	项目源码 -TextGeneration BabyNames	184
	7.1.1 朴素 RNN	170			
	7.1.2 梯度消失和梯度爆炸	171	7.3.10	保存、重用模型	188
	7.1.3 LSTM（一个特例）	171	7.4	高级文本生成	188
7.2	文本生成	174	7.4.1	创建项目	189
	7.2.1 模型训练	174	7.4.2	加载文本	189
	7.2.2 预测	175	7.4.3	处理数据	190
	7.2.3 模型定义	176	7.4.4	定义模型	191
7.3	生成新生儿名字	176	7.4.5	创建 checkpoints	191
	7.3.1 创建项目	176	7.4.6	自定义回调类	192
	7.3.2 下载文本	177	7.4.7	模型训练	193
	7.3.3 处理文本	177	7.4.8	结果	193
	7.3.4 定义模型	180	7.4.9	断点续训练	194
	7.3.5 编译	181	7.4.10	过程观察	195
	7.3.6 创建 checkpoints	182	7.4.11	项目源码	196
	7.3.7 训练	182	7.5	进一步工作	199
	7.3.8 预测	182	总结		199

第 8 章　语言翻译　200

8.1	sequence-to-sequence 模型	200	8.3.3	创建数据集	205
	8.1.1 编码器、解码器	201	8.3.4	数据预处理	207
	8.1.2 Seq2seq 模型的缺点	203	8.3.5	GloVe 词嵌入	212
8.2	注意力模型	203	8.3.6	定义编码器	214
8.3	英语翻译为西班牙语	204	8.3.7	定义解码器	215
	8.3.1 创建项目	204	8.3.8	注意力网络	216
	8.3.2 下载数据集	205	8.3.9	定义模型	221

目 录 Contents

8.3.10 模型训练	222	8.3.12 项目源码	229
8.3.11 预测	222	总结	237

第 9 章 自然语言理解   238

9.1 Transformer 简介	238	9.2.11 解码器结构	254
9.2 Transformer 详解	239	9.2.12 定义解码器	257
9.2.1 下载原始数据	240	9.2.13 Transformer 模型	259
9.2.2 创建数据集	240	9.2.14 创建训练模型	261
9.2.3 数据预处理	240	9.2.15 损失函数	261
9.2.4 构建语料库	240	9.2.16 优化器	262
9.2.5 准备训练集数据	243	9.2.17 编译	262
9.2.6 Transformer 模型	244	9.2.18 训练	262
9.2.7 多头注意力（机制）	245	9.2.19 预测	263
9.2.8 Scaled Dot-Product 注意力模块	248	9.2.20 测试	263
		9.2.21 项目源码	264
9.2.9 编码器结构	249	9.3 下一步是什么	276
9.2.10 编码器	252	总结	276

第 10 章 图像描述   278

10.1 项目简介	280	10.13 创建解码器	288
10.2 创建项目	280	10.13.1 Bahdanau 注意力机制	289
10.3 下载数据	280	10.13.2 解码器功能	289
10.4 解析 Token 文件	282	10.13.3 解码器初始化	289
10.4.1 加载数据	282	10.13.4 解码器调用方法	290
10.4.2 创建列表	283	10.13.5 注意力得分	290
10.5 加载 InceptionV3 模型	284	10.13.6 注意力权重	290
10.6 准备数据集	285	10.13.7 上下文向量	291
10.7 提取特征	285	10.13.8 解码器实现	291
10.8 创建词汇表	286	10.14 编码器、解码器实例化	294
10.9 创建输入序列	286	10.15 定义优化器和损失函数	294
10.10 创建训练数据集	287	10.16 创建 checkpoints	296
10.11 创建模型	288	10.17 训练函数	297
10.12 创建编码器	288		

| 10.18 | 模型训练 | 298 | 10.20 | 项目源码 | 301 |
| 10.19 | 模型预测 | 298 | 总结 | | 310 |

第 11 章　时间序列预测　311

11.1	时间序列预测简介	311		11.2.10	项目源码	325
	11.1.1 什么是时间序列预测	311	11.3	多变量时间序列分析		330
	11.1.2 预测中的问题	312		11.3.1	创建项目	330
	11.1.3 时间序列组成	312		11.3.2	准备数据	331
	11.1.4 单变量与多变量	312		11.3.3	检查平稳性	331
11.2	单变量时间序列分析	313		11.3.4	探索数据	332
	11.2.1 创建项目	313		11.3.5	准备数据	333
	11.2.2 准备数据	313		11.3.6	创建模型	335
	11.2.3 创建训练集和测试集	316		11.3.7	训练	335
	11.2.4 创建输入张量	319		11.3.8	评估	335
	11.2.5 构建模型	320		11.3.9	预测未来点	336
	11.2.6 编译和训练	320		11.3.10	预测数据点区间	337
	11.2.7 评估	320		11.3.11	项目源码	339
	11.2.8 预测下一个数据点	322	总结		343
	11.2.9 预测数据点区间	323			

第 12 章　风格迁移　344

12.1	快速风格迁移	345		12.2.4	显示图像	354
	12.1.1 创建项目	345		12.2.5	图像预处理	354
	12.1.2 下载图像	345		12.2.6	构建模型	355
	12.1.3 准备模型输入图像	347		12.2.7	内容损失	357
	12.1.4 执行风格迁移	348		12.2.8	风格损失	357
	12.1.5 显示输出	348		12.2.9	全变分损失	357
	12.1.6 更多结果	348		12.2.10	计算内容和风格损失	358
	12.1.7 项目源码	350		12.2.11	Evaluator 类	359
12.2	自定义风格迁移	351		12.2.12	生成输出图像	359
	12.2.1 VGG16 结构	352		12.2.13	显示图像	360
	12.2.2 创建项目	352		12.2.14	项目源码	361
	12.2.3 下载图像	353	总结		365

目录 Contents

第 13 章　图像生成

- 13.1　GAN（生成对抗网络）　366
- 13.2　GAN 如何工作　366
- 13.3　生成器　367
- 13.4　判别器　367
- 13.5　数学公式　368
- 13.6　数字生成　369
  - 13.6.1　创建项目　369
  - 13.6.2　加载数据集　369
  - 13.6.3　准备数据集　370
  - 13.6.4　定义生成器模型　370
  - 13.6.5　测试生成器　372
  - 13.6.6　定义判别器模型　373
  - 13.6.7　测试判别器　374
  - 13.6.8　定义损失函数　375
  - 13.6.9　定义新训练函数　376
  - 13.6.10　项目源码　380
- 13.7　字母生成　385
  - 13.7.1　下载数据　385
  - 13.7.2　创建单字母数据集　385
  - 13.7.3　输出结果　386
  - 13.7.4　项目源码　387
- 13.8　印刷体到手写体　392
- 13.9　生成彩色卡通图像　392
  - 13.9.1　下载数据　392
  - 13.9.2　创建数据集　392
  - 13.9.3　显示图像　393
  - 13.9.4　输出结果　394
  - 13.9.5　项目源码　394
- 总结　400

第 14 章　图像转换　401

- 14.1　自动编码器　401
- 14.2　色彩空间　402
- 14.3　网络配置　402
  - 14.3.1　Vanilla 模型　403
  - 14.3.2　Merged 模型　403
  - 14.3.3　使用预训练的 Merged 模型　403
- 14.4　自动编码器　404
  - 14.4.1　加载数据　405
  - 14.4.2　创建训练、测试数据集　406
  - 14.4.3　准备训练数据　406
  - 14.4.4　定义模型　407
  - 14.4.5　模型训练　410
  - 14.4.6　测试　410
  - 14.4.7　未知图像预测　412
  - 14.4.8　项目源码　413
- 14.5　编码器的预训练模型　418
  - 14.5.1　项目简介　418
  - 14.5.2　定义模型　418
  - 14.5.3　提取特征　418
  - 14.5.4　定义网络　419
  - 14.5.5　模型训练　420
  - 14.5.6　预测　421
  - 14.5.7　未知图像预测　421
  - 14.5.8　项目源码　422
- 总结　426

第 1 章
TensorFlow 快速入门

　　TensorFlow 是一个开发和部署机器学习应用程序的端到端的开源平台,是完整的机器学习生态系统。很多人都在 Facebook 上的照片中见过"面部标记"功能,这就是一种基于机器学习算法开发的应用程序。除上述应用外,自动驾驶汽车使用物体检测来避免道路上的碰撞;机器翻译程序可以将西班牙语翻译为英语;语音转换算法将语音翻译为文本,用于创建数字文档。上述所有案例都是机器学习的应用程序。即使经常使用的微不足道的光学字符阅读器(optical character reader,OCR)应用程序也使用了机器学习。如今很多被开发出来的高级应用程序同样适用了机器学习算法,如为图像添加字幕、图像生成、图像翻译、时间序列预测、理解人类语言等。所有此类应用程序及更多未举例的应用程序都可以在 TensorFlow 平台上开发和部署,本书将详细介绍这部分内容。

　　无论是初学者还是领域专家,都可以利用 TensorFlow 轻松构建自己的机器学习模型。在 TensorFlow 中,可以定义自己的神经网络架构,并对它们进行试验、训练,最后将训练后的网络模型部署到生产服务器上。不仅如此,经过充分训练的模型可以部署在移动设备、嵌入式设备及支持 JavaScript 的 Web 上。

　　除了 TensorFlow 外还有其他机器学习开发软件,如 Keras、Torch、Theano 和 Pytorch。KDnuggets (www.kdnuggets.com) 最近展开了一项关于"哪个深度学习框架用户量增长最快?"的研究,结果如图 1-1 所示。

　　图 1-2 显示了在 2018 年对深度学习框架的性能得分调查情况。

　　显然,TensorFlow 是所有接受调查的深度学习框架中的赢家,因此,学习和使用 TensorFlow 2.x 进行深度学习应用程序开发是正确的选择。现在让我们开始学习 TensorFlow。

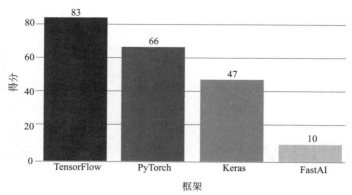

图 1-1　2019 年深度学习框架增长趋势

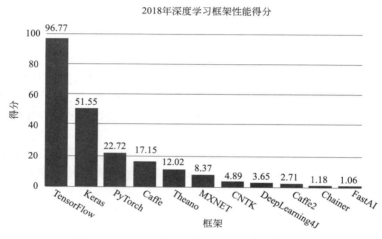

图 1-2　2018 年深度学习框架的性能得分

1.1　什么是 TensorFlow 2.0

一图胜千言，本节将展示一幅简化的 TensorFlow 平台整体概念图。

1.1.1　TensorFlow 2.x 平台

平台的整体概念如图 1-3 所示。

与典型的机器学习开发软件相同，TensorFlow 2.x 平台由三个不同的阶段组成。第一阶段称为训练阶段，即定义人工神经网络模型并在给定数据上对模型进行训练。此外，使用测试数据测试模型并重新训练，直到对模型性能满意为止。第二阶段，将模型保存到一个文件中，使得以后可以将其部署在生产服务器上。第三阶段，将保存的模型部署在生产服务器上，并准备对未见过的数据进行预测。

接下来，分别描述三个阶段中的不同组成部分。

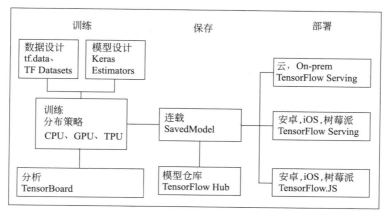

图 1-3　TensorFlow 2.x 平台

1.1.2　训练

训练包括读取数据、以模型所需的特定格式准备数据、创建模型及运行多个训练轮次 (Epochs) 训练模型。TensorFlow 2.x 提供了大量函数、库和工具以加速训练。一般地，训练机器学习模型在整个开发过程中需花费大量时间。与传统训练方法相比，TensorFlow 2.x 中提供的工具和设施使得模型能在更短的时间内训练。整个训练模块如图 1-4 所示。

接下来将解释训练模块中的各个组成部分。

1. 数据设计

数据设计模块如图 1-5 所示。

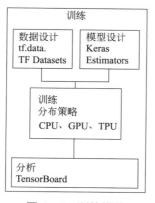

图 1-4　训练模块

图 1-5　数据设计模块

数据设计模块展示了两个子模块——tf.data 和 TF Datasets。本节讨论两者包含的内容，首先描述 tf.data。

模型训练需要根据模型设计将数据准备为特定格式。数据准备的步骤如下。

① 从外部源加载数据并对其进行清洗。清洗过程包括删除包含空字段的行，将分类字段映射到列，以及将数值缩放到 –1 到 +1 的范围（通常情况下）。

② 需要决定哪些列是特征（Features）及标签是什么。

③ 去除与模型训练无关的列。例如，数据集中的姓名和客户 ID 在机器训练中是冗余的。

④ 将数据拆分为训练集和测试集。

tf.data 模块提供了上述数据操作所需的各种功能。

TensorFlow 还提供了很多机器学习库中的内置数据集（TF Dataset），以在 tf.data 包中使用。其中包含了 100 多个即用型数据集，分为音频、图像、视频、文本、翻译等多个类别。用户可以根据自己的需求从其中一个类别中加载数据并快速进行模型开发。未来将会有更多的数据集加入此模块中。通过一句程序声明即可从 tf.data.dataset 模块中加载数据，该声明还能够创建训练集与测试集。上述操作节省了大量数据准备的工作，使用户快速专注于模型训练。尽管如此，在多数情况下可能无法将原始数据集直接输入到特定的机器学习算法中。例如，数值领域可能需要数值放缩，分类领域可能需要数据转换，像 IMDB（互联网电影资料库）电影评论数据集可能需要编码为不同的数据格式，还可能需要更改数据的维度等。因此，几乎总是需要对这些内置的数据集进行某种预处理。但这些内置的数据集依然为机器学习从业者提供了诸多便利。此外，这些数据集还支持高性能数据管道，以促进训练过程中的快速数据传输，从而加快训练速度。

2. 模型设计

Keras 应用程序接口（application programming interface, API）现已集成到 TensorFlow 库中，可以使用 tf.keras 访问整个 API。tf.keras 是一个高级 API，为 TensorFlow 1.x 中使用的许多 API 提供了标准化，使用户能够使用 TensorFlow 2.x 中引入的一些新功能。例如，模型开发可以利用 Eager 执行模式，使用 tf.keras 模块的函数式 API 可以设计复杂度极高的模型。

TensorFlow 2.x 还提供了 Estimator，用于模型之间的快速比较。Estimator 模块如图 1-6 所示。

Esitimator 由 tf.estimator 提供，是 TensorFlow 的一个高级 API。Estimator 封装了机器学习的各个阶段，如模型训练、评价、预测和导出模型，以利用生产服务器提供服务。库中提供了几个预制的 Estimator，如 LinearClassifier 和 DNNClassifier。除了使用预制的 Estimator 之外，还可以构建自定义的 Estimator。此外，该库还提供了一个名为 model_to_estimator 的函数，将现有模型转换为 Estimator，以进行 TensorFlow 的分布式训练。

图 1-6 模型设计模块

3. 分布策略

在整个机器学习过程中，最耗时的部分就是训练。即使在性能非常高的设备上，也可能需要几分钟到几天的训练时间。因此，需要强大的处理能力和内存来训练模型。幸运的是，TensorFlow 2.x 在此可以发挥作用。图 1-7 展示了分布策略模块。

模型训练可以在 CPU（central processing unit，中央处理器）、GPU（graphics processing unit，图形处理器）或 TPU（tensor processing unit，张量处理器）上完成。不仅如此，还可以在多个硬件单元之间进行分布式训练，极大地减少了训练时间。

4. 分析

在模型训练阶段，需要分析不同训练阶段的结果。基于这些结果，需要重新配置网络结构、修改损失函数、尝试不同的优化器等。为此，TensorFlow 提供了一个很好的分析工具：TensorBoard，如图 1-8 所示。

图 1-7　分布策略模块　　　　　图 1-8　分析模块

TensorBoard 提供了模型训练过程中各种指标的绘图，如正确率（Accuracy）和损失函数（Loss），这些都是训练过程中广泛使用的指标。TensorBoard 还提供了很多其他功能，读者将在进一步阅读本书时继续学习。

1.1.3　模型保存

通用架构中的模型保存模块如图 1-9 所示。

模型保存包括两部分：将开发完成的模型保存到磁盘和重用存储库中的预训练模型。

当模型训练到可接受的正确率后，将其保存到磁盘。TensorFlow 1.x 提供了多种保存模型的方法。TensorFlow 2.x 将模型保存标准化为一个名为 SavedModel 的抽象类。保存的模型可以直接加载到机器学习应用程序中，也可以上传到产品服务器上进行服务。TensorFlow 2.x 将模型保存为标准格式，以便能够将模型部署在移动设备、嵌入式设备和 Web 上。TensorFlow 提供了将已开发模型部署到不同平台的 API，包括支持 JavaScript 和 Node.js 的 Web。

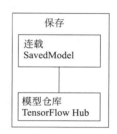

图 1-9　模型保存模块

TensorFlow Hub 是含有大量预训练模型的存储库，用户可以使用迁移学习（主要形式之一为加载预训练模型）来重用和扩展这些预训练模型以满足实际需求。使用预训练模型能够供开发者使用较小的数据集训练自己的模型，同时以更快的速度完成训练。现有的预训练模型库包括用于文本和图像识别的模型，在 Google 新闻数据集上训练的模型，以及用于渐进式生成对抗网络（progressive GAN）和谷歌标志点深度局部特征（Google landmarks deep local features）的模型。在撰写本书时，上述大多数模型都由 TensorFlow 1.x 所写，需要迁移至较新的版本。

接下来将介绍不同的模型部署选项。

1.1.4　部署

训练完成的模型可以部署到不同的平台上，如图 1-10 所示。

TensorFlow 2.x 最优秀的特性是能够在云端或本地部署训练完成的模型。不仅如此，TensorFlow 2.0 甚至可以在 Android 和 iOS 等移动设备及树莓派等嵌入式设备上部署模型。此外，还可以使用 Node.js 在 Web 端部署模型，以便在浏览器中使用该模型。一般来说，部署可以分为以下几类。

☑ TensorFlow Serving：一个允许模型通过 HTTP/REST 或 gRPC/Protocol 缓冲区提供服务的库。

☑ TensorFlow Lite：在 Android、iOS 和嵌入式系统（如树

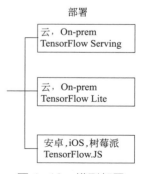

图 1-10　模型部署

莓派和 Edge TPU）上部署模型的轻量级解决方案。
- ☑ TensorFlow.js：使模型能够在 JavaScript 环境中部署，如通过 Node.js 在 Web 浏览器或服务器端部署。使用 TensorFlow.js，可以在 JavaScript 中定义模型，并使用类似 Keras 的 API 在 Web 浏览器中直接训练这些模型。

对 TensorFlow 进行简要介绍之后，接下来将简要描述 TensorFlow 2.x 的一些主要突出特性。

1.2 TensorFlow 2.x 提供什么

与早期版本相比，TensorFlow 2.x 引入了许多新特性。本节将简要总结 TensorFlow 2.x 的显著特性。

1.2.1 TensorFlow 中的 tf.keras

Keras API 现在可通过 TensorFlow 的 tf.keras 获得。这是一个高级 API，可为 TensorFlow 特定功能提供支持，如 Eager 执行、数据管道和 Estimator。用户可以按照 Keras 的方式使用 tf.keras 构建和训练模型，而不会损失灵活性和性能。

要在程序中使用 tf.keras，需要执行如下代码。

```
import tensorflow as tf
from tensorflow import keras
```

加载 TensorFlow 库后，即可自定义神经网络架构、创建模型及训练和测试模型。在第 2 章讨论 TensorFlow 2.x 的入门程序时将详细介绍上述内容。

1.2.2 Eager 执行

在 TensorFlow 2.x 之前，机器学习代码分为如下两部分。
① 构建计算图，即 computational graph。
② 创建一个会话来执行计算图，即 Session。
上述步骤可由如下代码表示。

```
import tensorflow as tf
a = 2
b = 3
c = tf.add(a, b, name='Add')
print(c)
```

上述代码输出如下内容。

```
Tensor("Add:0", shape=(), dtype=int32)
```

上述操作实际上构建了一个计算图，如图 1-11 所示。

要运行计算图，需按如下方式创建一个会话并运行该会话，以执行 Add 函数。

```
sess = tf.Session()
```

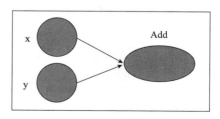

图 1-11　计算图

```
print(sess.run(c))
sess.close()
```

运行上述代码，将打印出结果"5"。

使用 TensorFlow 2.x，可以在不创建会话的情况下执行相同的操作，如下面的代码所示。

```
import tensorflow as tf
a = 2
b = 3
c = tf.add(a, b, name='Add')
print(c)
```

上述代码的执行结果为：

```
tf.Tensor(5, shape=(), dtype=int32)
```

可以看到，输出张量（Tensor）的值为"5"。

因此，创建会话的过程被完全取消了，这对构建大规模模型有很大帮助。通常，在开发过程中，如果模型开头的某个地方出现一个小错误，则需要重新构建整个计算图。每次修复错误时，都需要重新构建完整的计算图，这会造成诸多不便，且是一个非常耗时的过程。在 TensorFlow 2.x 中，Eager 执行模式允许程序运行部分代码，而无须构建完整的计算图。Eager 执行是默认的执行模式，因此在定义模型时不必有任何特殊考虑。此外，可以使用 tf.executing_eagerly() 命令来显式地声明 Eeger 执行模式。

随着创建会话过程的取缔，TensorFlow 代码现在可以像 Python 代码一样执行。TensorFlow 2.x 创建了所谓的动态计算图，而不是 TensorFlow 1.x 中的静态计算图。

1.2.3 分布式计算

机器学习模型构建过程中最耗时的操作是训练模型。TensorFlow 2.x 提供了一个名为 tf.distribute.Strategy 的 API，用于在多个 GPU 和 TPU 之间进行分布式训练。使用此 API，只需改动很少的代码即可分布已有模型和训练代码。该 API 提供六种分布策略：MirroredStrategy、CentralStorageStrategy、MultiWorkerMirroredStrategy、TPUStrategy、ParameterServerStrategy、OneDeviceStrategy。

读者可通过阅读官方文档深入研究上述策略的详细内容。

1.2.4 TensorBoard

TensorBoard 是一个可视化工具，可辅助开发者进行机器学习模型研发。TensorBoard 可以实现如下功能。

① 可视化损失和正确率评价指标。
② 可视化模型计算图。
③ 可视化权重、偏置等参数的直方图。
④ 显示图像。
⑤ 分析程序。

图 1-12 显示了一个 TensorBoard 的屏幕截图，包括正确率和损失指标。

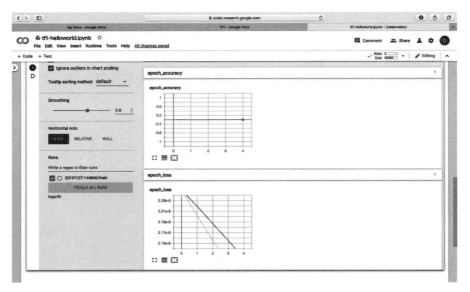

图 1-12　TensorBoard 指标显示

通过 TensorBoard，可以直接在 Jupyter 环境中可视化模型的训练过程。它提供了很多实用且令人兴奋的功能，如内存分析、查看混淆矩阵和模型概念图等。综上所述，TensorBoard 是一个可在机器学习工作流程中进行评价并查看结果的工具。

1.2.5　视觉套件（Vision Kit）

借助 TensorFlow 2.x，用户现在可以设计具有图像识别功能的物联网（IoT）设备，由 TensorFlow 的机器学习模型提供支持。使用谷歌 AIY 视觉套件（Google AIY Vision Kit），可以构建自定义的智能相机，使其看到和识别物体，可以创建自定义识别模型或使用针对此智能相机的预训练模型。整个套件可以装在一个小巧的纸板箱中，并由树莓派提供支持。该套件提供了构建自定义智能相机所需的一切元素。

1.2.6　语音套件（Voice Kit）

与为物联网设备提供视觉的智能相机一样，语音套件能够为物联网设备提供聆听和应答功能。使用谷歌 AIY 语音套件（Google AIY Voice Kit）能够创建自定义的自然语言处理器，该处理器可以连接到 Google Assistant 或云端的 peech-to-Text 服务，可以向物联网设备发出语音命令，还可以提出问题并获取答案。与视觉套件一样，语音套件也可以装在一个方便的纸板箱中，并由树莓派提供支持。语音套件包含构建具有音频功能的物联网设备（包括树莓派）所需的一切元素。

1.2.7　边缘套件（Edge TPU）

物联网设备制造商很乐意在其设备上构建新的机器学习模型，Coral 为此创建了 Edge TPU 开发板。Edge TPU 开发板可用于快速构建设备端机器学习产品原型，是具有可拆卸模块化系统(system-on-module, SOM)的单板计算机。SOM 包含 eMMC、SOC、无线电（wireless radios）和 Edge TPU。这也非常适合物联网设备和其他嵌入式系统所需的设备

1.2.8　AIY 套件的预训练模型

AIY 套件包含可供使用的多种预训练模型，部分预训练模型如下。
① 人脸检测器。
② 犬、猫、人检测器。
③ 用于识别食物种类的食品分类器。
④ 图像分类器。
⑤ 识别鸟类、昆虫和植物的自然检测器。
欢迎读者将自己开发的模型提交到谷歌，以在谷歌站点上的预训练模型列表中显示。

1.2.9　数据管道

如前文所述，TensorFlow 2 使得模型训练能够分布在 GPU 和 TPU 上，从而大大减少了执行单个训练步骤所需的时间。这要求在两个训练步骤之间提供有效的数据传输。最新的 tf.data API 有助于跨越不同模型和加速器构建灵活、高效的输入管道。第 2 章将介绍数据管道的使用。

1.3　安装

TensorFlow 2.x 可在如下平台上安装。
① macOS 10.12.6 或更高版本。
② Ubuntu 16.04 或更高版本。
③ Windows 7 或更高版本。
④ Raspbian 9.0 或更高版本。
本书使用 Mac 进行开发，教程中给出的所有程序都在 Mac 上开发和测试。
TensorFlow 的安装很简单，在控制台中运行如下命令以确保计算机上的 pip 为最新版本。

```
pip install --upgrade pip
```

使用如下命令安装仅支持 CPU 的 TensorFlow 版本。

```
pip install tensorflow
```

使用如下命令安装支持 CPU、GPU 的 TensorFlow 版本。

```
pip install tensorflow-gpu
```

1.3.1　安装步骤

在 Mac 上安装 TensorFlow，需保证 Xcode 为 9.2 或更高版本——机器上的命令行工具可用。在命令行中使用以下命令安装 pip 需要的依赖项。

```
pip install -U --user pip six numpy wheel setuptools mock 'future>=0.17.1'
pip install -U --user keras_applications --no-deps
```

```
pip install -U --user keras_preprocessing --no-deps
```

安装完上述依赖项后，可运行 pip install 命令来安装所需的任意版本 TensorFlow。

1.3.2 Docker 安装

如果不想自己安装 TensorFlow，可以使用 Docker 容器中的现成镜像。使用如下命令下载 Docker 镜像。

```
docker pull tensorflow/tensorflow
```

Docker 容器下载成功后，运行如下命令启动 Jupyter Notebook 服务器。

```
docker run -it -p 8888:8888 tensorflow/tensorflow
```

Jupyter 环境启动后，打开 Jupyter notebook 即可使用 TensorFlow。接下来，"测试"一节将对此做进一步解释。

1.3.3 无安装

至此，本书已展示了在几个平台上安装 TensorFlow 的方法。使用 Docker 镜像可以使读者免于研究依赖项。还有另一种学习和使用 TensorFlow 的简单方法——使用 Google Colab。这种方式下，用户无须安装即可使用 TensorFlow，只需在浏览器中启动 Google Colab 即可。Google Colab 是一个谷歌的研究项目，其实质是在浏览器中为开发者提供了一个 Jupyter notebook 环境。在 Google Colab 中使用 TensorFlow 无须配置，整个代码都在云中运行。本书将使用 Google Colab 来运行教程中的程序。

1.4 测试

由于本书中的项目使用了 Google Colab，因此本节介绍如何在 Google Colab 中测试 TensorFlow 的安装情况。

打开网址"http://colab.research.google.com"以启动 Google Colab。假设已登录谷歌账户，读者将看到如图 1-13 所示的屏幕截图。

选择 NEW PYTHON3 NOTEBOOK 菜单，在浏览器中打开一个空白的 notebook。然后，在代码窗口中输入以下两条语句。

```
%tensorflow_version 2.x
import tensorflow as tf
```

%tensorflow_version 命令被称为 Colab 魔法，用于加载 TensorFlow 2.x 而不是默认的 TensorFlow 1.x。本书在第 2 章讨论 TensorFlow 入门程序时将解释这种用法。

在当前版本的 Colab 中，不再需要使用 %tensorflow_version。因此这个声明是多余的，后续所有章节中已删除。

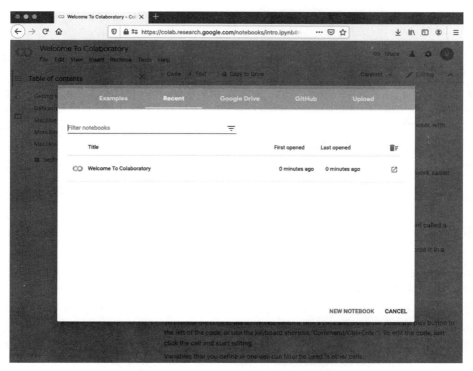

图 1-13　Colab 打开一个新的 Notebook

运行程序单元，可得到如图 1-14 所示的输出。

图 1-14　测试 Colab 设置

输出结果表明 TensorFlow 2.x 已被选中，因此可以使用。
在项目中写入如下代码。

```
c = tf.constant([[2.0, 3.0], [1.0, 4.0]])
d = tf.constant([[1.0, 2.0], [0.0, 1.0]])
e = tf.matmul(c, d)
print (e)
```

运行上述代码，可得到以下输出。

```
tf.Tensor(
[[2. 7.]
```

```
       [1. 6.]], shape=(2, 2), dtype=float32)
```

如果得到上述输出，则表示 Colab 环境中的 TensorFlow 2.x 已准备完成。

> 与 TensorFlow 1.x 不同，上述代码中没有创建会话。

至此，本章完成了 TensorFlow 的安装与设置。

总结

TensorFlow 2.x 为开发深度机器学习应用程序提供了一个非常强大的平台。该平台从数据准备、模型构建到生产服务器上的最终部署，都为开发者提供了便利，就像在使用一种端到端的开发工具。流行的机器开发库 Keras 现已完全集成到 TensorFlow 中。利用 TensorFlow 中的新功能，如 Eager 执行、分布训练及跨多个 CPU、GPU、TPU 预测和高效数据管道，用户能够在 Keras 中开发非常高效的机器学习程序。在开发过程中，TensorBoard 为用户提供了使用的分析工具来优化模型。经过充分训练的模型被保存为一种可以部署在移动设备和嵌入式设备上的存储格式。TensorFlow 还提供视觉套件和语音套件，供用户在嵌入式设备上部署图像/视频识别和语音控制的机器学习模型。开发者社区提供了多种预训练模型供用户使用。Edge TPU 的使用允许用户在设备端进行预测。综上所述，TensorFlow 2.x 可以被视为从开发到部署的机器学习一体化平台。

本章的结尾介绍了在不同平台上安装 TensorFlow 的方法，还讨论了 Google Colab 的使用。Google Colab 提供了一个基于云的项目开发环境，用于开发 TensorFlow 应用程序。如果读者拥有良好的互联网连接，那么可以依靠 Google Colab 来运行所有机器学习应用程序，这正是本书所做的。

下一章，将使用 TensorFlow 2.x 进行代码实战。

第 2 章
深入研究 TensorFlow

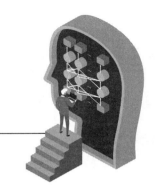

上一章介绍了 TensorFlow 的基本功能,这一章开始学习如何将 TensorFlow 运用到自己的应用程序中。

接下来,本章从一个简单的应用程序入手,该程序将带读者了解开发一个机器学习应用程序所面临的挑战。

2.1 一个简单的机器学习应用程序

为开始使用 TensorFlow 进行项目开发,本节将从一个简单的"Hello World"应用程序入手。在这个简单的应用程序中,使用 TensorFlow 开发一个基于统计回归进行结果预测的机器学习模型。

在本节的应用程序中,首先声明一组固定数据,由 (x, y) 坐标值组成,接下来计算程序输出值 z,z 与 x 和 y 有某种线性关系。例如,对于给定的 x 和 y 值,可以使用如下数学方程式计算 z 值。

$$z=7 \times x+6 \times y+5$$

本节任务是让机器自行学习映射关系,即在给定足够多的 x 和 y 值及相应的输出 z 值的情况下,找到最适合上述关系的映射模型。完成模型训练后,可以使用该模型预测任何新的 x 和 y 值对应的 z 值。例如,给定 $x = 2$ 且 $y = 3$ 时,该模型应该预测输出 $z = 37$。如果该模型成功预测输出值为 37,并且如果该模型能够 100% 成功预测任何未知的 x 和 y 对应的输出值 z,则可以定义该模型被完全训练,且正确率为 100%。实际上,永远不可能开发出正确率为 100% 的预测模型,开发者需要不断尝试优化模型性能以达到理想的 100% 正确率。

从上述讨论中可以得出,本节试图解决一个经典的线性回归案例。为简单起见,本节创建一个仅包含一个神经元的单层网络,该网络经过训练可以解决线性回归问题。实际应用中的神经网络通常由具有多个节点的多个层组成。使用 TensorFlow 定义复杂的网络结构需要对 Keras API 有足够深入的了解，因此，本节的样例避免使用过深的神经网络。在本章后面的小节中，使用 Keras API 进行更加复杂的神经网络搭建。

2.1.1 创建 Colab 笔记本

下面完成在 Colab 中创建、测试和推演第一个机器学习模型。为了方便刚接触机器学习开发的读者，本文对开发过程进行详细介绍。

通过在浏览器中输入以下网址启动 Google Colab "http://colab.research.google.com"，如图 2-1 所示。

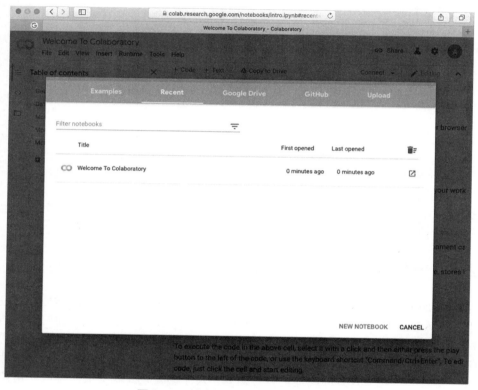

图 2-1 创建新的 Colab notebook

选择 NEW PYTHON3 NOTEBOOK 选项打开一个新的 Python3 notebook。登录 Google 账户后，将看到如图 2-2 所示的界面。

图 2-2 新的 Colab notebook

Colab notebook 的默认文件名为"Untitledxxx.ipnyb",可以将该文件名改为"Hello world"或任何自定义名称。接下来,通过编写代码在 Python 程序中导入 TensorFlow 依赖库。

2.1.2 导入

本节的应用程序需要三个附加依赖库,包括 TensorFlow 2.x、NumPy、Matplotlib。NumPy 用于数据处理,Matplotlab 用于画图。

1. 导入 TensorFlow 2.x

使用如下代码在 Python notebook 中导入 TensorFlow 2.x。

```
import tensorflow as tf
```

执行上述命令将导入默认版本的 TensorFlow,当前为 1.x(撰写本书时)。执行上述命令后的界面如图 2-3 所示。

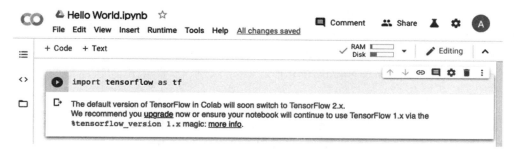

图 2-3　导入默认的 TensorFlow 依赖库

由于本书使用 TensorFlow 2.x 进行软件开发,需要导入 TensorFlow 2.x 版本。使用 Colab 中的魔法语句进行版本选择,如下所示。

```
%tensorflow_version 2.x
```

运行上述语句,完成 TensorFlow 版本选择后,得到如图 2-4 所示的界面。

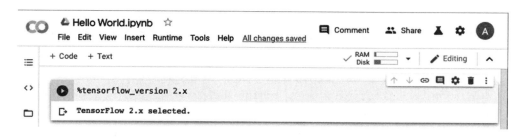

图 2-4　加载 TensorFlow 2.x

选择 TensorFlow 2.x 版本后,使用通用的 Import 语句加载 TensorFlow 2.x 依赖库,如下所示。

```
import tensorflow as tf
```

> 当前版本的 Colab 不再需要使用魔法函数。

Keras 依赖库是 TensorFlow 的一部分。要在本节应用程序中使用 Keras 库，需要从 TensorFlow 库导入，通过如下语句实现。

```
from tensorflow import keras
```

为了使用 Keras 模块，需要使用 tf.keras 语句。

接下来，将导入本节应用程序使用的其他依赖库。

2. 导入 NumPy 和 Matplotlib

在 Python 程序中，NumPy 库支持大型多维数组运算，通过高级数学函数操作数组。任何机器学习模型开发都依赖于数组的运算，并通过 NumPy 数组存储神经网络所需的输入数据。通过如下语句导入 NumPy 依赖库。

```
import numpy as np
```

Matplotlib 是一个用于创建高质量二维绘图的 Python 依赖库。本节应用程序使用 Matplotlib 库绘制正确率和误差指标。通过如下语句导入 Matplotlib 依赖库。

```
import matplotlib.pyplot as plt
```

上述过程完成了本节应用程序附加依赖库的导入。接下来，进行数据库创建。

2.1.3 创建数据

创建一组由 100 个 x 和 y 坐标组成的数据点。使用以下语句在 Python 变量中声明数据点的个数。

```
number_of_datapoints = 100
```

使用 NumPy 库中的 random 模块，生成 x 和 y 坐标。以 x 坐标为例，使用如下语句进行数据库创建。

```
# generate random x values in the range -5 to +5
x = np.random.uniform(low = -5 , high = 5 ,
    size = (number_of_datapoints, 1))
```

上述语句中，uniform 函数的 low 参数和 high 参数分别定义了随机数生成器的下限和上限，size 参数指定数组的维度，即要生成多少个值。uniform 函数返回一个由随机数组成的数组，行数为 100，列数为 1。可以使用以下语句打印生成数组的前 5 个值。

```
x[:5,:].round(2)
```

在输出数组中，调用 round 函数将每个值截断为两位十进制数。round 函数的输出如下。

```
array([[ 4.57],
       [-0.68],
       [ 2.64],
       [-3.17],
       [-4.86]])
```

> **注意** 上述函数每次运行的输出都不同。

上述过程为 x 变量的生成过程，y 值使用类似的语句生成，如下所示。

y = np.random.uniform(-5 , 5 , size = (number_of_datapoints , 1))

使用线性方程建立 x 和 y 之间的关系：

$$z=7\times x+6\times y+5$$

在机器学习术语中，变量 x 和 y 表示特征，z 表示标签，即真实值。完成预测模型的训练后，预测模型可以为给定的 x 和 y 预测对应的 z。之前提到过，网络训练学习了 x 和 y 之间的关系。为了得到更加稳定的预测模型，需要对由前面等式计算的每个 z 值引入一些噪声。可以使用前面的随机函数生成噪声，范围为 $-1 \sim 1$ 之间。噪声生成语句如下。

```
noise = np.random.uniform(low =-1 , high =1,
        size = (number_of_datapoints, 1))
```

现在，通过线性方程创建 z 数组并向其添加噪声，如以下语句所示。

$$z=7\times x+6\times y+5+noise$$

本节神经网络的输入是 100 行的单维数组，每行由一个以按列排布的 x 值和 y 值单维数组组成。为了创建上述输入数组，使用 column_stack 函数，如下所示。

input = np.column_stack((x,y))

打印输入数组的前 5 个值如下所示。

```
array([[-1.9 ,  2.91],
       [-2.14, -0.81],
       [ 4.18,  1.79],
       [-0.93, -4.41],
       [-1.8 , -1.31]])
```

上述过程创建了训练网络所需的数据集，下一个任务是构建网络模型。

2.1.4 定义神经网络

之前讲到本节的神经网络由一个接收一维向量并输出单个值的神经元组成，网络结构如图2-5所示。

Keras 库提供了一个 Sequential API 用于定义网络模型，Sequential API 能够构建多级复杂的网络架构。本小节使用 Sequential API 创建一个由单层和单个神经元组成的 ANN（人工神经网络）架构，通过如下语句实现。

图 2-5　单层、单节点网络结构

```
model = tf.keras.Sequential([keras.layers.Dense(units=1, input_shape=[1])])
```

其中，units 参数定义了输出空间的维度。上述语句中 units 指定值为 1，定义了一个单层网络，其中单个神经元输出单个值。Dense 函数包含多个参数，可以创建复杂的人工神经网络架构。本书中后续章节将使用 keras.Sequential API 创建复杂的神经网络架构。

完成模型初始化后，需要进行模型编译，并使用上一小节定义的数据集进行模型训练。

2.1.5 编译模型

为了训练模型，首先需要定义一个学习过程。模型编译是设置模型学习过程的一种方式。学习过程由以下几个部分组成：目标损失函数、优化器、评价指标。

首先，目标损失函数确定模型输出的预测值与目标真实值的距离。Keras 库提供了预定义的损失函数，如 categorical_crossentropy、mean_squared_error、huber_loss、poisson。其次，使用优化器进行损失函数最小化。优化器为一类改变神经网络属性的算法。神经网络属性包括权重和学习率。通过改变这些属性，减小网络损失函数。Keras 库提供了预定义的优化器，如 SGD 优化器（随机梯度下降）、RMSprop 优化器、Adagrad 优化器、Adam 优化器。最后，使用评价指标判断模型的性能，常用的评价指标包括：均方误差 (mean square error, MSE)、均方根误差 (root mean squared error, RMSE)、平均绝对误差 (mean absolute error, MAE)、平均绝对百分比误差 (mean absolute percentage error, MAPE)。

调用编译函数对上述学习过程进行设置，使用 compile 函数实现：

```
model.compile(optimizer = 'sgd' ,
              loss = 'mean_squared_error' ,
              metrics = ['mse'] )
```

上述代码使用随机梯度下降优化器，均方误差为损失函数和评价指标。到目前为止，编译了全部模型学习过程，接下来使用训练数据完成模型训练过程。

2.1.6 训练网络

经过多次迭代完成模型的训练过程。初始迭代时，为网络的各个节点分配预设权重。第一次迭代训练后，查看损失函数值，然后在接下来的迭代过程中调整网络权重，将损失最小化。上述每次迭代称为一个 epoch。每个 epoch 结束时，保存并监控损失函数值，以确保在正确的方向上优化网络。为了保存模型在每个 epoch 结束后的状态，需要使用 Keras 库创建一个历史对象，使用如下语句完成。

```
from tensorflow.keras.callbacks import History
history = History()
```

将上述 history 对象作为参数传递给模型的训练方法。模型训练通过调用模型的 fit 方法完成，如下所示。

```
model.fit(input, z , epochs = 15 , verbose = 1,
          validation_split = 0.2, callbacks = [history])
```

其中，input 参数指定之前创建的数据输入，z 参数指定目标值，epochs 参数定义迭代次数，verbose 参数指定是否观察训练进度，validation_split 参数指定使用输入数据的 20% 用于验证训练模型，使用 callbacks 参数指定中间监控数据的存储位置。模型训练过程的部分输出如图 2-6 所示。

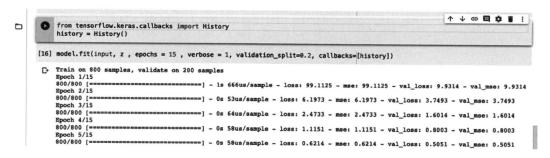

图 2-6　模型训练过程的部分输出

当全部 epoch 训练完成，需要验证模型是否经过充分训练。因此，对模型学习期间的评价指标进行分析，以检查训练过程的输出结果。

2.1.7　检查训练结果

上一小节已经将模型每个 epoch 的状态保存在历史变量中，可以借助以下语句检查历史记录。

```
print(history.history.keys())
```

上述语句的输出如下。

```
dict_keys(['loss', 'mse', 'val_loss', 'val_mse'])
```

历史记录保存了每个 epoch 的输出模型在训练集数据和验证集数据上的损失函数值和评价指标。验证集数据的损失函数值和评价指标使用前缀 val_ 表示。接下来，将损失函数值和评价指标以图像的方式显示，实现代码如下。

```
plt.plot(history.history['loss'])
plt.plot(history.history['val_loss'])
plt.title(Accuracy')
plt.ylabel('loss')
```

```
plt.xlabel('epoch')
plt.legend(['train', 'validation'], loc='upper right')
plt.show()
```

历史记录中每轮模型训练集和验证集的损失函数随 epoch 的变化如图 2-7 所示。

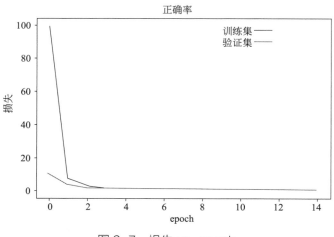

图 2-7　损失 vs. epoch

观察图 2-7 得出，损失函数在第 3 个 epoch 结束时完成了最小化过程，预定义模型在第 15 个 epoch 结束时被完全训练。15 是在 fit 方法中预定义的数值。为了进一步验证上述结论，绘制了训练集与验证集的 MSE 指标随 epoch 的变化曲线，实现代码如下。

```
plt.plot(history.history['mse'])
plt.plot(history.history['val_mse'])
plt.title('mean squared error')
plt.ylabel('mse')
plt.xlabel('epochs')
plt.legend(['train' , 'validation'] , loc = 'upper right')
plt.show()
```

上述代码的输出如图 2-8 所示。

最后，使用如下代码绘制训练集数据的预测值与真实值关系图。

```
plt.plot(np.squeeze(model.predict_on_batch(input)),
np.squeeze(z))
plt.xlabel('predicted output')
plt.ylabel('real output')
plt.show()
```

绘制结果如图 2-9 所示。

观察图 2-9 可知，模型对训练集数据的预测值与期望得到的真实值非常接近，即该模型被充分训练。接下来，使用网络未见过的测试集数据进行模型测试。

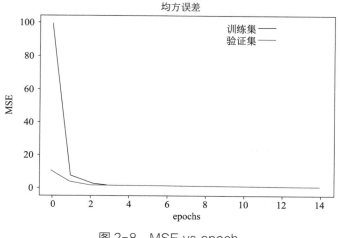

图 2-8　MSE vs.epoch

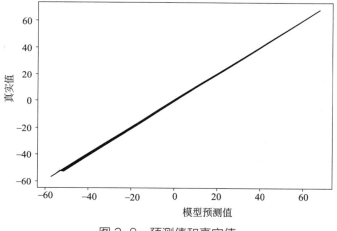

图 2-9　预测值和真实值

2.1.8　预测

利用模型对未知的 x 和 y 值进行预测，需要在模型上使用 predict 函数。使用如下语句实现。

```
print("Predicted z for x=2, y=3 ---> ",
        model.predict([[2,3]]).round(2))
```

上述语句指定 $x = 2$，$y = 3$，输出结果四舍五入为两位十进制数字，执行结果为：

```
Predicted z for x=2, y=3 ---> [[36.99]]
```

下面检查预测结果是否足够接近预期输出，使用如下代码查看预期输出。

```
# Checking from equation
# z = 7*x + 6*y + 5
print("Expected output: ", 7*2 + 6*3 + 5)
```

上述语句的执行结果为 37，模型的输出结果为 36.99，可以看出预测结果非常接近期望值。由于模型的预测结果每次都不同，因此为了达到满意的预测结果，可以使用更多 x 和 y 值测试模型。

2.1.9 完整源码

清单 2-1 给出了本节介绍的 "Hello World" 应用程序的完整源码，仅供参考（全书源码在 https://github.com/Apress/artificial-neural-networks-with-tensorflow-2 下载，余同）。

清单 2-1 一个简单的线性回归应用程序代码

```
# Load TensorFlow 2.x in a Colab project.
%tensorflow_version 2.x

# Import required libraries
import tensorflow as tf
import numpy as np
import matplotlib.pyplot as plt

# Set up data
number_of_datapoints = 1000
# generate random x values in the range -5 to +5
x = np.random.uniform(low = -5 , high = 5 , size = (number_of_
datapoints, 1))
# generate random y values in the range -5 to +5
y = np.random.uniform(-5 , 5 , size = (number_of_datapoints , 1))
# generate some random error in the range -1 to +1
noise = np.random.uniform(low =-1 , high =1, size = (number_of_
datapoints, 1))
z = 7 * x + 6 * y + 5 + noise

# Print x, y and z sample values for manual verification
x[:5,:].round(2)

y[:2,:].round(2)

z[:2,:].round(2)

# Stack x and y arrays for inputting to neural network
input = np.column_stack((x,y))

# Print few values of input array for demonstration purpose.
input[:2,:].round(2)

# Create a Keras sequential model consisting of single layer
with a single neuron.
model = tf.keras.Sequential([tf.keras.layers.Dense(units=1)])

# Compile the model with the spcified optimizier, loss function
and error metrics.
```

源码清单
链　接：https://pan.baidu.com/s/1NV0rimQ_8kRz22xfFHN-Cw
提取码：1218

```python
model.compile(optimizer = 'sgd' , loss = 'mean_squared_error' ,
metrics = ['mse'] )

# Import History module to record loss and accuracy on each
epoch during training
from tensorflow.keras.callbacks import History
history = History()

model.fit(input, z , epochs = 15 , verbose = 1, validation_
split=0.2, callbacks=[history])

# Print keys in the history just to know their names. These
will be used for plotting the metrics.
print(history.history.keys())

# Plot the loss metric on both training and validation
datasets.
plt.plot(history.history['loss'])
plt.plot(history.history['val_loss'])
plt.title('Accuracy')
plt.ylabel('loss')
plt.xlabel('epoch')
plt.legend(['train', 'validation'], loc='upper right')
plt.show()

#Plot the mean squared error on both training and validation
datasets.
plt.plot(history.history['mse'])
plt.plot(history.history['val_mse'])
plt.title('mean squared error')
plt.ylabel('mse')
plt.xlabel('epochs')
plt.legend(['train' , 'validation'] , loc = 'upper right')
plt.show()

plt.plot(np.squeeze(model.predict_on_batch(input)),
np.squeeze(z))
plt.xlabel('predicted output')
plt.ylabel('real output')
plt.show()

print("Predicted z for x=2, y=3 ---> ", model.predict([[2,3]]).
round(2))

# Checking from equation
# z = 7*x + 6*y + 5
print("Expected output: ", 7*2 + 6*3 + 5)
```

本节程序完成了基于 TensorFlow 2.x 进行深度学习算法开发的设置。下一节将深入了解实际的机器学习算法开发，了解机器学习算法开发的完整过程。下一节将使用真实的数据集，学习如何对输入数据进行预处理并输入神经网络，定义多级深度神经网络，训练模型，测试模型，并通过绘制正确率指标完成模型优化。不仅如此，读者还将学习如何在 Colab 环境中使用 TensorBoard 库可视化用于对模型训练进行分析的指标。

下面，让我们开始这个基于 TensorFlow 开发的真实项目。

2.2 使用 TensorFlow 解决二分类问题

在 2.1 节的示例中，我们使用内置数据集训练了一个简单的线性回归模型。现在，我们将使用真实的数据集完成一个机器学习实战示例，使用机器学习解决一个分类问题。本节数据集来自 Kaggle 竞赛，该数据集包含银行的客户数据。设想一下，银行准备开发机器学习预测模型，该模型可为银行提供有关客户离开银行的可能性分析。知道某个客户可能在不久的将来离开银行，银行可以采取一些预防措施来留住客户。

本节要解决的问题是开发一个二元分类模型，将使用 TensorFlow 的深度学习库和 Keras 库中高级 API 实现该模型。具体说来，读者将在本节示例中学习以下内容：

① 如何从本地或远程服务器加载 CSV 格式数据？
② 如何对数据进行预处理并使其适用于机器学习算法？
③ 如何使用 TensorFlow 的高级 Keras API 定义多层人工神经网络？
④ 如何训练模型？
⑤ 使用测试集数据评估模型性能。
⑥ 使用 TensorBoard 可视化预测结果。
⑦ 进行模型性能分析。
⑧ 对未知数据进行预测。

2.2.1 创建项目

创建一个新的 Colab 项目并将其命名为"Binary Classification"。本项目使用了本书源下载中提供的银行客户数据 (Churn_Modelling.csv)。将下载后的文件复制到选择的 Google Drive 文件夹中。

> 稍后在程序代码中需要适当地更改文件路径。

如果不想通过下载获取数据文件，仍然可以从本书的 GitHub 中获取数据，并运行本节程序。接下来，对本节项目进行调试。

2.2.2 导入

与 2.1 节的示例类似，在代码窗口使用如下语句导入附加依赖库。

```
%tensorflow_version 2.x
import tensorflow as tf
```

```
from tensorflow import keras
```

除此以外，使用 Pandas 的数据流（dataframes）加载外部数据库，使用 sklearn 库进行数据预处理并创建训练集与验证集数据，使用 Matplotlib 绘制图表。在项目代码中使用如下代码导入上述依赖库。

```
#loading data
import pandas as pd
#scaling feature values
from sklearn.preprocessing import StandardScaler
#encoding target values
from sklearn.preprocessing import LabelEncoder
#shuffling data
from sklearn.utils import shuffle
#splitting the dataset into training and validation

from sklearn.model_selection import train_test_split
#plotting curves
import matplotlib.pyplot as plt
```

接下来，在程序中挂载驱动器，以便程序可以访问存储在 Google Drive 中的文档。

2.2.3 挂载 Google 云盘

为了在程序中挂载 Google Drive，在新的代码片段中输入以下代码。

```
from google.colab import drive
drive.mount('/content/drive')
```

运行上述代码时，需要输入授权码以访问驱动器，显示如图 2-10 所示的界面。

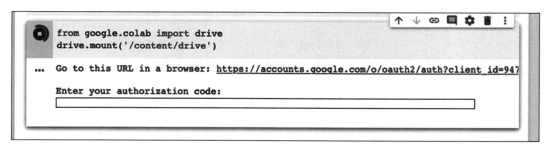

图 2-10　输入 Google Drive 授权码

单击图 2-10 中的链接，登录 Google 账户，将看到如图 2-11 所示的授权码。

单击授权码旁边的图标将其复制到剪贴板，将授权码粘贴到图 2-10 显示的授权窗口中。成功授权后，屏幕上将看到如下消息。

```
Mounted at /content/drive
```

现在，可以通过程序代码访问驱动器的内容。

![Google Sign in 图示]

图 2-11　登录授权

2.2.4　加载数据

加载数据前，在新的代码片段中输入以下语句并执行它。

```
data = pd.read_csv('/content/drive/
           <path to downloaded CSV>/Churn_Modelling.csv')
```

注意

需要为 csv 文件设置合适的路径。

如果使用本书在 GitHub 上的数据，使用以下代码而不是前面的代码片段。

```
data_url = 'https://raw.githubusercontent.com/Apress/artificialneural-
networks-with-tensorflow-2/main/ch02/Churn_Modelling.csv'
data=pd.read_csv(data_url)
```

其中，read_csv 函数从指定的文件加载数据并将其复制到 Pandas 数据流中。

1. 数据排序

根据数据收集者的个人习惯和操作便利性，原始数据可能按特定顺序排列。为了训练更优的网络模型，需要对原始数据进行随机化操作，以便模型不会受到原始数据中固定的排列模式所干扰。因此，使用以下方法进行数据随机化。

```
data=shuffle(data)
```

2. 数据检查

通过显示数据流中的内容来验证数据是否正确加载。本小节没有调用 data.head() 显示数据流最上层的数据，而显示了完整的数据集，以便显示全部数据细节，如图 2-12 所示。

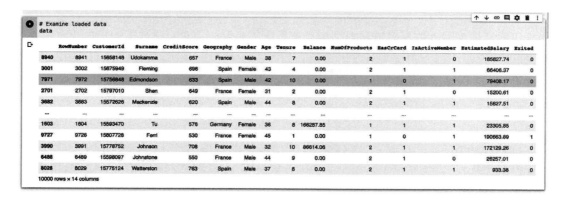

图 2-12　数据集

数据库由 10000 个数据组成，每个数据包括 14 个 Tag。下面给出各个 Tag 的简要说明。
☑ RowNumber：1～10000 之间的整数。
☑ CustomerId：客户的唯一标识。
☑ Surname：客户的姓氏。
☑ CreditScore：客户的信用评分。
☑ Geography：客户所在地。
☑ Gender：客户的性别。
☑ Age：客户的年龄。
☑ Tenure：客户与银行合作的时间。
☑ Balance：余额。
☑ NumOfProducts：使用银行产品数量。
☑ HasCrCard：是否持有信用卡。
☑ IsActiveMember：客户是否活跃。
☑ EstimatedSalary：客户当前的预估工资。
☑ Exited：整数 1 表示客户已离开银行。

现在，已将数据加载到内存中，在下一个任务输入网络之前需对已加载的原始数据进行预处理。

2.2.5　数据处理

原始的真实数据可能并不能满足人工神经网络 (ANN) 的训练要求。具体来说，通常需要对原始数据进行如下处理。
① 去除原始数据集可能包含的空数据。
② 筛选数据库中可能对模型学习有用的标签。
③ 某些数据标签的值存在较大差异，必须缩放到统一范围。
④ 某些数据标签可能包含类别信息，如男性和女性，将此类信息编码为 0 和 1。
⑤ 决定将哪些标签用作输入特征及确定相应的输出标签。

接下来，我们开始处理数据。

1. 检查空值数据

包含空值的输入数据将对网络训练产生严重的影响。检查空值数据的最简单方法是调用 isnull 函数，如下所示。

```
data.isnull().sum()
```

此语句的输出结果如下。

```
RowNumber          0
CustomerId         0
Surname            0
CreditScore        0
Geography          0
Gender             0
Age                0
Tenure             0
Balance            0
NumOfProducts      0
HasCrCard          0
IsActiveMember     0
EstimatedSalary    0
Exited             0
dtype: int64
```

计算空值数据的总和会产生错误。很明显，本节使用的数据集不包含任何空值，因此不存在从数据集中删除行（包含空字段的行）的问题。

2. 选择特征和标签建立

本节数据库中并非全部标签都可用于训练算法，如 CustomerId 和 Surname 这样的标签对机器学习没有任何意义，所以，需要删除这些列。使用以下语句完成。

```
X = data.drop(labels=['CustomerId', 'Surname',
            'RowNumber', 'Exited'], axis = 1)
```

> **注意**
>
> 预处理阶段从数据集中删除了四个标签，构成一个新数组 X，X 包含与本节模型构建相关的标签。客户是否退出银行代表了本节模型的输出结果。因此，Exited 成为本节模型的输出标签。使用如下语句将 Exited 提取到变量 y 中。

```
y = data['Exited']
```

此时，已经完成构建特征集 (X) 张量和标签集 (y) 张量。

3. 编码分类列

接下来，检查特征集 X 中选定的数据列是否具有与类别有关的标签。使用如下语句检查所有选定列的数据类型。

```
X.dtypes
```

上述语句的输出结果如下。

```
CreditScore         int64
Geography           object
Gender              object
Age                 int64
Tenure              int64
Balance             float64
NumOfProducts       int64
HasCrCard           int64
IsActiveMember      int64
EstimatedSalary     float64
Dtype:              object
```

> **注意** Geography 和 Gender 内的数据是对象类型，可以通过打印特征张量的前五行来检查它们包含的值，如图 2-13 所示。

图 2-13　特征向量的前五行

Gender 包含两个类别，即男性和女性，而 Geography 包含三个类别，即德国、西班牙和法国。在将其提供给网络之前，需要将它们转换为数值。使用 sklearn 预处理模块中的 LabelEncoder 完成编码如下所示。

```
from sklearn.preprocessing import LabelEncoder
label = LabelEncoder()
X['Geography'] = label.fit_transform(X['Geography'])
X['Gender'] = label.fit_transform(X['Gender'])
```

one-hot 编码为数据中的类别列创建了虚拟变量。例如，由于数据中心 Geography 列具有三个不同的值（Germany, France, Spain），因此 one-hot 编码将创建三个变量，即为上述每个国家创建一个变量。因此，训练集中将包含与国家相关的三个特征。特征数量过多将会增加训练时间，为了减少特征数量，可以去除部分与国家字段相关的虚拟变量，且能获得性能相近的结果。我们通过调用 get_dummies 方法删除第一个变量：

```
X = pd.get_dummies(X, drop_first=True, columns=['Geography'])
```

执行上述代码后，如果打印数据集前五行来检查此时的数据内容，可以注意到 Geography 只有两列数据，即 Geography_1 和 Geography_2。

在将数据输入网络进行模型训练前，还需要将特征值归一化到 [-1,1]。

4. 特征值归一化

由于真实数据中的特征值可以具有广泛的数值分布，如果将所有特征图标准化为相同的数据分布，网络模型将更加稳定。理想情况下，每类特征的平均值应为 0，标准差应为 1，此时最有利于模型训练。因此，使用如下方程进行特征值归一化。

```
z = (x - mu) / s
```

其中，mu 为特征平均值，s 为特征标准差。上述标准化过程可以使用 sklearn 库中的 StandardScaler 函数完成，如下所示。

```
from sklearn.preprocessing import StandardScaler
scaler = StandardScaler()
X = scaler.fit_transform(X)
```

至此，我们完成了全部数据的预处理，接下来，可以准备定义和训练模型。完成模型训练后，还需要验证训练模型的有效性。如果训练模型的输出结果不符合预期，需要对输入数据做进一步的预处理，如调整特征数量等。为了进行模型测试，训练模型前保留一部分数据作为测试集。因此，整个预处理后的数据被拆分为两部分，较大的部分为训练集，用于网络训练，较小的部分为测试集，用于测试训练后的模型。

5. 创建训练集与测试集

为了将数据分成两部分，使用 sklearn 库中的 train_test_split 方法，如下所示。

```
# Split dataset into training and testing
from sklearn.model_selection import train_test_split
X_train, X_test, y_train, y_test = train_test_split(X, y, 
test_size = 0.3)
```

其中，test_size 参数确定预留的测试集数据占全部数据的百分比。该函数返回一组用于训练和测试的向量。很多时候，验证集数据和测试集数据可以互换使用。为了避免混淆，下面对数据集术语给出广泛接受的明确定义。

- ☑ 训练集：用于模型拟合的数据集。
- ☑ 验证集：训练期间用于调整超参数的数据集。
- ☑ 测试集：训练后用于评估模型性能的数据集。

接下来，定义本节的网络结构。

2.2.6 定义 ANN

预处理后，本节使用的数据集共有 11 个特征。特征的数量可以通过计算训练数据集的形状确定，如下所示。

```
X_train.shape[1]
```

网络的预期输出为一个二进制值，表示客户离开银行的可能性，保存在向量 y_train 中。

接下来，创建一个网络深度为四层的深度学习网络模型。其中，第一层使用 128 个节点，第二层使用 64 个节点，第三层使用 32 个节点，第四层为 1 个输出节点。网络构建过程中，使用 TensorFlow 中的 tf.keras API 创建网络结构，使用 Sequential API 创建各层的线性堆栈，使用如下语句实例化模型。

```
model = keras.models.Sequential()
```

以第一层网络结构为例，使用如下语句将第一层添加到由 128 个节点组成的堆栈中。

```
model.add(keras.layers.Dense(128, activation = 'relu',
                             input_dim = X_train.shape[1]))
```

该层的输入维度在参数 input_dim 中设置，即由 X_train 向量的形状定义特征的数量。使用修正线性单元函数 (rectifier linear unit, ReLU) 作为激活函数。激活函数根据节点输入的加权和决定是否激活该节点。ReLU 是目前为止使用最广泛的激活函数，它将负值映射为 0，保持正值输出。类似地，可以使用如下语句将第二层添加到网络中。

```
model.add(keras.layers.Dense(64, activation = 'relu'))
```

该层的输入来自前一层，因此无须指定输入向量的维度。使用如下语句添加第三层结构。

```
model.add(keras.layers.Dense(32, activation = 'relu'))
```

最后，使用如下语句添加网络中的最后一层。

```
model.add(keras.layers.Dense(1, activation = 'sigmoid'))
```

本节使用 sigmoid 作为网络激活函数，将最后一层输出映射为一个二进制值。sigmoid 函数是一种激活函数，也称为挤压函数。sigmoid 函数将输出映射到 [0,1] 范围内，使网络输出表示当前输入的预测概率。网络构建后，通过调用 summary 函数显示网络主要结构，如下所示。

```
model.summary()
```

本节网络结构显示如下。

```
Model: "sequential"
_____
Layer (type)                 Output Shape              Param #
=================================================================
dense (Dense)                (None, 128)               1536
_____
dense_1 (Dense)              (None, 64)                8256
_____
```

dense_2 (Dense)	(None, 32)	2080
dense_3 (Dense)	(None, 1)	33

```
Total params: 11,905
Trainable params: 11,905
Non-trainable params: 0
```

编译模型

模型架构定义好后，需要进行模型编译，调用模型的 compile 方法实现：

```
model.compile(loss = 'binary_crossentropy',
              optimizer='adam', metrics=['accuracy'])
```

由于本节开发的模型是一个二元分类器，使用 binary_crossentropy 作为损失函数。在此情况下，适合使用 Adam 优化器进行模型优化。训练结束后，如果对模型的性能不满意，可以尝试使用其他优化器。得到预训练模型后，通过指定的评价指标参数进行模型有效性分析。

接下来展示如何使用 TensorBoard 分析网络性能。为此，需要定义一个回调函数，在训练期间的每个 epoch 中调用，并在训练过程中在 log 文件夹中保存训练过程。部分较早的 log 文件需要使用如下语句去除。

```
!rm -rf ./log/
```

使用如下语句定义上述回调函数。

```
#tensorboard visualization
import datetime, os
logdir = os.path.join("log",
                      datetime.datetime.now().
                      strftime("%Y%m%d-%H%M%S"))
tensorboard_callback = tf.keras.callbacks.TensorBoard(logdir,
                      histogram_freq = 1)
```

完成上述训练过程分析和模型编译设置后，可以开始网络训练了。

2.2.7 模型训练

为了进行模型训练，在预定义的模型实例上使用 fit 方法，如下所示。

```
r = model.fit(X_train, y_train, batch_size = 32, epochs = 50,
              validation_data = (X_test, y_test),
              callbacks = [tensorboard_callback])
```

其中，X_train 参数定义了特征向量，y_train 参数定义了标签，batch_size 参数定义了训练批量的大小，epochs 参数确定训练期间执行的迭代次数。数据预处理过程中生成的测试数据用于模型验证，并通过 validation_data 参数传递给模型拟合函数。最后，callbacks 参数指定每次迭代结束时将调用哪个回调函数。训练过程的部分输出如图 2-14 所示。

```
r = model.fit(X_train, y_train, batch_size = 32, epochs = 50, validation_data = (X_test, y_test), callbacks = [tensorboard_callback])
Train on 7000 samples, validate on 3000 samples
Epoch 1/50
7000/7000 [==============================] - 2s 218us/sample - loss: 0.4244 - accuracy: 0.8219 - val_loss: 0.3801 - val_accuracy: 0.8507
Epoch 2/50
7000/7000 [==============================] - 1s 99us/sample - loss: 0.3514 - accuracy: 0.8541 - val_loss: 0.3689 - val_accuracy: 0.8557
Epoch 3/50
7000/7000 [==============================] - 1s 97us/sample - loss: 0.3352 - accuracy: 0.8620 - val_loss: 0.3668 - val_accuracy: 0.8500
Epoch 4/50
7000/7000 [==============================] - 1s 95us/sample - loss: 0.3304 - accuracy: 0.8600 - val_loss: 0.3618 - val_accuracy: 0.8573
Epoch 5/50
7000/7000 [==============================] - 1s 98us/sample - loss: 0.3242 - accuracy: 0.8640 - val_loss: 0.3616 - val_accuracy: 0.8570
Epoch 6/50
7000/7000 [==============================] - 1s 98us/sample - loss: 0.3178 - accuracy: 0.8677 - val_loss: 0.3779 - val_accuracy: 0.8410
```

图 2-14　训练过程的部分输出

训练结束后,计算模型的评价指标来评估模型是否已训练到想要的精确度。

1. 性能评估

为了完成模型性能评估,在 Colab 环境中使用 %load_ext 语句加载 TensorBoard,然后使用 % tensorboard 语句启动 TensorBoard,如下所示。

```
%load_ext tensorboard
%tensorboard --logdir log #command to launch tensorboard on colab
```

通过上述语句运行 TensorBoard,将绘制预训练模型的正确率和损失指标,如图 2-15 所示。

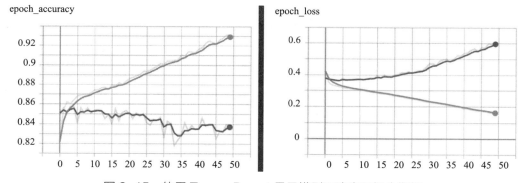

图 2-15　使用 TensorBoard 显示模型正确率和损失指标

图 2-15 显示的两条曲线分别通过训练集和验证集数据得到,展示了每个 epoch 训练后模型的正确率和损失指标。检查训练过程中模型的正确率和损失指标可辅助判断模型是否训练良好。如果每个 epoch 结束后模型的正确率在逐渐提高,那么说明模型的训练方向是正确的。类似地,模型的损失指标应该在每个 epoch 结束后逐渐减少。绘制训练过程中模型的性能评估图也可以检测模型是否出现过拟合等问题。如果出现对模型性能不满意的情况,可以调整模型参数并重新训练以提高正确率,尝试使用不同的优化器或引入正则化等方法均可提高模型的正确率。

除上述方法外,还可以调用 evaluation 方法。将测试集特征向量和标签向量作为参数传递到 evaluation 方法,通过如下语句实现。

```
test_scores = model.evaluate(X_test, y_test)
```

```
print('Test Loss: ', test_scores[0])
print('Test accuracy: ', test_scores[1] * 100)
Test Loss: 0.6143370634714762
Test accuracy: 83.96666646003723
```

可以看出,预训练模型在测试数据上的正确率约为 83%,这表明该模型能够正确分类 83% 的输入数据。

接下来,介绍如何使用 Matplotlib(一种传统的性能评估方法)在验证集数据上绘制模型性能评估图,使用如下代码段实现。

```
%matplotlib inline
import matplotlib.pyplot as plt #for plotting curves

plt.plot(r.history['val_accuracy'], label='val_acc')
plt.plot(r.history['val_loss'], label='val_loss')
plt.legend()
plt.show()
```

通过上述代码调用 Matplotlib 绘制的验证集数据正确率与损失函数变化曲线如图 2-16 所示。

除了使用正确率和损失指标外,还经常使用混淆矩阵评估模型性能,混淆矩阵将在后续内容介绍。

2. 预测测试集数据

输出混淆矩阵需要模型的预测值和数据的真实标签,因此,首先需要对测试集数据进行预测,使用 predict_classes 方法完成,如下所示。

```
y_pred = model.predict_classes(X_test)
```

predict_classes 方法将特征向量作为其参数并返回预测结果。得到预测结果后,可以在控制台上打印结果,使用如下语句实现。

```
y_pred
```

预测结果如下。

```
array([[1],
       [0],
       [0],
       ...,
       [1],
       [0],
       [0]], dtype=int32)
```

预测结果中值为 1 表示该客户将离开银行,值为 0 表示银行留住了该客户。可以使用这些预测结果创建和绘制混淆矩阵,以更好地可视化模型的性能。

3. 混淆矩阵

首先展示如何生成混淆矩阵,然后对混淆矩阵进行分析。使用 sklearn 库的内置函数生成混淆矩阵,如下所示。

```
from sklearn.metrics import confusion_matrix
cf = confusion_matrix(y_test, y_pred)
cf
```

显示如下混淆矩阵。

```
array([[2175, 209],
       [ 283, 333]])
```

使用以下代码绘制混淆矩阵,可以提供更佳的视觉效果。

```
from mlxtend.plotting import plot_confusion_matrix
plot_confusion_matrix(conf_mat = cf, cmap = plt.cm.cmapname)
```

混淆矩阵图像如图 2-17 所示。

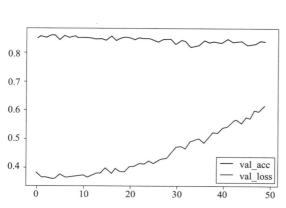

图 2-16　验证集数据的正确率和损失函数变化曲线

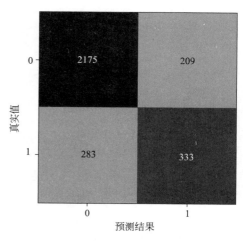

图 2-17　混淆矩阵

上图中,x 轴代表预测结果,y 轴代表真实标签。如图 2-17 所示,共有 2175 个真阳性和 333 个真阴性。真阳性表示该客户会离开银行,并且此类数据已经被模型正确分类。类似地,真阴性表示该客户不会离开银行,并且也被模型正确分类。真阳性和真阴性帮助我们确定模型的正确率。

sklearn 定义 accuracy_score 函数用了计算正确率。正确率通过将真阳性和真阴性的数量求和并与预测总数相除得到。以下代码段用于计算模型的正确率。

```
from sklearn.metrics import accuracy_score
accuracy_score(y_test, y_pred)
0.8396666666666667
```

执行上述代码得出了本节模型的正确率为 83.63%,此正确率在机器学习算法中是可接受的。

综上所述,本节模型的训练效果达到了我们的预期,接下来将使用此模型对未见过的数据进行模型测试。

4. 预测未知数据

首先创建一个未输入过模型的测试数据集。我们需要知道每个特征的数据类型，并为每个特征分配虚拟值。特征文件头显示了不同的类别名称和特征值范围，部分特征向量如图 2-18 所示。

	CreditScore	Geography	Gender	Age	Tenure	Balance	NumOfProducts	HasCrCard	IsActiveMember	EstimatedSalary
8940	657	France	Male	38	7	0.0	2	1	0	185827.74
3001	696	Spain	Female	43	4	0.0	2	1	1	66406.37

图 2-18 特征截图

因此，对于测试数据集，使用以下特征值。

```
CreditScore = 615
Gender = Male
Age = 22
Tenure = m5
Balance = 20000
NumOfProducts = 1
HasCrCard = 1
IsActiveMember = 1
EstimatedSalary = 60000
Geography = Spain
```

调用预定义模型的 predict 方法将上述数据集输入到训练好的模型中，通过传递数据列表索引完成参数传递，如下所示。

```
customer = model.predict([[615, 1, 22, 5, 20000, 5, 1, 1,
    60000, 0, 0]])
customer

if customer[0] == 1:
    print ("Customer is likely to leave")
else:
    print ("Customer will stay")
```

上述代码的输出结果如下。

```
Customer will stay
```

0 值表示该客户不太可能离开银行。使用此次数据集预测的模型正确率仍与之前的结果近似，约为 83%。

在模型完全训练到满意的效果后，可以将其保存到磁盘或将其部署在生产服务器上以供实际使用。模型保存与部署的实现方法将在后续介绍 tf.keras 时进行讨论。

2.2.8 完整源码

清单 2-2 中给出了本节项目的完整源码，仅供参考。

清单 2-2 二分类模型完整代码

源码清单
链　接：https://pan.baidu.com/s/1NV0rimQ_8kRz22xfFHN-Cw
提取码：1218

```python
%tensorflow_version 2.x
import tensorflow as tf
from tensorflow import keras
import pandas as pd

# Load data from Github
data_url = 'https://raw.githubusercontent.com/Apress/artificialneural-networks-with-tensorflow-2/main/ch02/Churn_Modelling.csv'
data=pd.read_csv(data_url)

# Shuffle data for taking care of patterns in data collection
from sklearn.utils import shuffle
data=shuffle(data) #shuffling the data

# Examine loaded data
data

# Check for null values
data.isnull().sum()

# Drop irrelevant columns to set up features vector
X = data.drop(labels=['CustomerId', 'Surname', 'RowNumber',
'Exited'], axis = 1)

# Set up labels vector
y = data['Exited']

# Check data types for finding categorical columns
X.dtypes

# Examine few records for finding values in categorical columns
X.head()

# Encode categorical columns
from sklearn.preprocessing import LabelEncoder
label = LabelEncoder()
X['Geography'] = label.fit_transform(X['Geography'])
X['Gender'] = label.fit_transform(X['Gender'])

# Drop the first column of Geography to reduce the number of features
X = pd.get_dummies(X, drop_first=True, columns=['Geography'])
X.head()

# Scale all data points to -1 to + 1
from sklearn.preprocessing import StandardScaler
scaler = StandardScaler()
X = scaler.fit_transform(X)
```

```python
# Split dataset into training and validation
from sklearn.model_selection import train_test_split
X_train, X_test, y_train, y_test = train_test_split(X, y, test_size = 0.3)

# Determine number of features
X_train.shape[1]

# Create a stacked layers sequential network
model = keras.models.Sequential() # Create linear stack of layers
model.add(keras.layers.Dense(128, activation = 'relu', input_dim = X_train.shape[1])) # Dense fully connected layer
model.add(keras.layers.Dense(64, activation = 'relu'))
model.add(keras.layers.Dense(32, activation = 'relu'))
model.add(keras.layers.Dense(1, activation = 'sigmoid')) # activation sigmoid for a single output

# Print model summary
model.summary()

# Compile model with desired loss function, optimizer and evaluation metrics
model.compile(loss = 'binary_crossentropy', optimizer='adam', metrics=['accuracy'])

#to clear any other logs if present so that graphs won't overlap with previous saved logs in tensorboard
!rm -rf ./log/

#tensorboard visualization
import datetime, os
logdir = os.path.join("log", datetime.datetime.now().strftime("%Y%m%d-%H%M%S"))
tensorboard_callback = tf.keras.callbacks.TensorBoard(logdir, histogram_freq = 1)

# Perform training
r = model.fit(X_train, y_train, batch_size = 32, epochs = 50, validation_data = (X_test, y_test), callbacks = [tensorboard_callback])

# Load tensorboard in Colab
%load_ext tensorboard
%tensorboard --logdir log #command to launch tensorboard on colab

# evaluate model performance on test data
test_scores = model.evaluate(X_test, y_test)
print('Test Loss: ', test_scores[0])
```

```python
print('Test accuracy: ', test_scores[1] * 100)

# Plot metrics in matplotlib
%matplotlib inline
import matplotlib.pyplot as plt #for plotting curves

plt.plot(r.history['val_accuracy'], label='val_acc')
plt.plot(r.history['val_loss'], label='val_loss')
plt.legend()
plt.show()

# Predict on test data
y_pred = model.predict_classes(X_test)
y_pred

# Create confusion matrix
from sklearn.metrics import confusion_matrix
cf = confusion_matrix(y_test, y_pred)
cf

# Plot confusion matrix
from mlxtend.plotting import plot_confusion_matrix
plot_confusion_matrix(conf_mat = cf, cmap = plt.cm.cmapname)

# Compute accuracy score
from sklearn.metrics import accuracy_score
accuracy_score(y_test, y_pred)

# Predict on unseen customer data
customer = model.predict([[615, 1, 22, 5, 20000, 5, 1, 1, 60000, 0, 0]])
customer

if customer[0] == 1:
    print ("Customer is likely to leave")
else:
    print ("Customer will stay")
```

总结

本章中，使用 TensorFlow 2.x 设置深度学习环境，使用 Colab 开发了 Python notebook。首先，通过一个简单的应用程序帮助读者了解开发环境。接下来，详细介绍了一个基于机器学习技术的二分类实战示例。在此示例中，通过加载外部数据库，学习了如何预处理数据以使其适合机器学习，学习了定义深度神经网络、编译和训练神经网络、使用 TensorBoard 和 Matplotlib 库进行图像绘制，最后学习使用训练好的模型对未输入过的数据集进行预测。本章实现了模型性能评估，但未讨论如何进行模型优化。在下一章中，将学习一些提高模型性能的技术，深入地了解 tf.keras，并讨论图像分类问题。

第 3 章
深入了解 tf.keras

Keras 是一个运行在 TensorFlow 之上的高级神经网络 API。过去许多年，开发者都在使用以 TensorFlow 为后端的 Keras API。在 TensorFlow 2.x 中，这种情况发生了改变。TensorFlow 现已经将 Keras 集成在了 tf.keras API 中。tf.keras 是 TensorFlow 对 Keras API 的规范实现，主要是为了在使用基于 Tensorflow 的 Keras 时保持一致。同时，这个改变还使得使用 Keras 时能够使用很多 TensorFLow 的功能，如 Eager 执行、分布式训练等。在撰写本书时，最新的 Keras 版本是 2.3.0，此版本增加了对 TensorFlow 2.x 的支持，也是多后端 Keras 的最后一个主要版本。今后，开发者将在所有深度学习应用程序中仅使用 tf.keras。读者已经在第 2 章使用了 tf.keras，本章将更深入地了解 tf.keras。

3.1 开始

在创建深度学习应用程序时，最重要的任务是定义一个神经网络模型。上一章构建了一个由单层和单个神经元组成的网络，使用了以下程序语句完成。

```
model = tf.keras.Sequential([tf.keras.layers.Dense(units=1)])
```

虽然没有明确指出，但上面的语句使用了 TensorFlow 中的 Keras 函数式 API。该语句构建了一个单层/单神经元网络模型。在早期的 Keras 实现方式中，可以使用如下两个语句执行相同的操作。

```
model = keras.Sequential()
model.add(Dense(1, input_dim = 1))
```

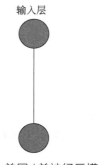

图 3-1 单层/单神经元模型

该模型如图 3-1 所示。

现在，建议在定义 Keras 深度学习模型时使用函数式 API。

3.2 用于模型构建的函数式 API

使用函数式 API，可以构建具有非线性拓扑结构的极其复杂的模型，可以在模型内共享特征层，还可以构建具有多个输入和输出的模型。因此，本书的所有应用程序中都将使用函数式 API 来定义 ANN 模型。接下来介绍如何使用函数式 API 定义神经网络架构。

3.2.1 序列化模型

假设想要构建如图 3-2 所示的模型。

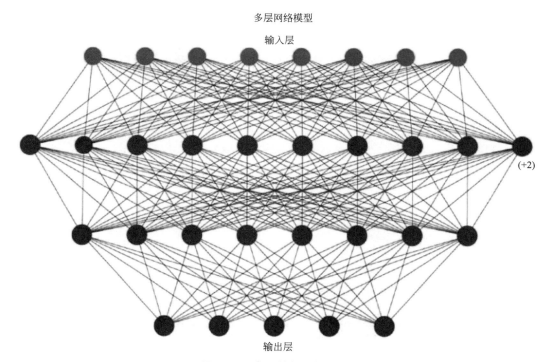

图 3-2　多层神经网络模型

如图 3-2 所示，所需的神经网络由多个层组成。函数式 API 提供了一系列工具用于构建具有多个层的图。

导入一些必要的模块，如下面的代码所示。

```
import tensorflow as tf
from tensorflow import keras
```

为了构建如图 3-2 所示的模型，首先使用以下语句创建一个输入层。

```
inputs = keras.Input(shape=(8,), name='image')
```

上述声明返回一个尺寸为 8 的输入张量（inputs）。接下来，使用如下声明定义一个具有 12 个节点的全连接层。

```
x = layers.Dense(12, activation='relu')(inputs)
```

> **注意**
>
> 该语句已将输入张量作为输入传递给新添加的层。新添加的层返回一个大小为 12 的张量,并可以输入到下一层。

使用以下语句再添加一个具有 8 个节点的层。

```
x = layers.Dense(8, activation='relu')(x)
```

来自前一层的张量(x)被输入到新层。同样,可以向网络添加任意数量的层,每个层都有自己的一组节点和激活函数。最后,使用以下语句将输出层添加到网络中。

```
outputs = layers.Dense(5)(x)
```

这一层的输出是一个尺寸为 5 的张量,因此,希望定义的神经网络在给定特定输入的情况下输出五个值之一。

可以使用上述输入(inputs)和输出(outputs)来定义网络模型,如下面的语句所示。

```
model = keras.Model(inputs=inputs, outputs=outputs,
        name='multilayer model')
```

使用如下声明生成神经网络的架构图。

```
keras.utils.plot_model(model, 'multilayer_model.png',
        how_shapes=True)
```

生成的网络架构图如图 3-3 所示。

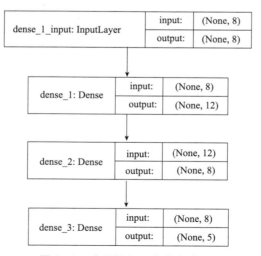

图 3-3 多层神经网络的架构图

函数式 API 能够构建复杂的网络架构，如具有多个输入和输出、允许共享网络层等。图 3-4 显示了一些上述复杂架构。

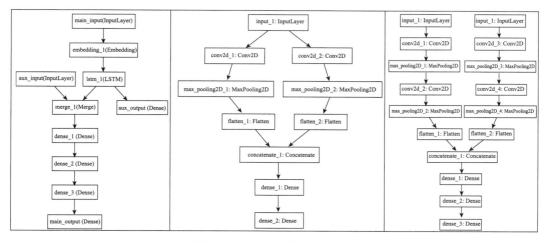

图 3-4 一些复杂的 ANN 架构

3.2.2 模型子类

有面向对象编程经验的读者（如具有 Java 或 C++ 编程经验）可能想知道 TensorFlow 是否有子类化的概念，使得可以重用已经构建好的模型。TensorFlow 支持模型子类化。要为模型创建自定义类，可以从 tf.keras.Model 继承。

```
class MyModel(tf.keras.Model):
```

继承 tf.keras.Model 时，需要在自定义类中提供两个重写的方法：__init__ 方法和 call 方法。__init__ 方法顾名思义，即在类别初始化时被调用。一个典型的 __init__ 定义方式如下面的代码片段所示。

```
def __init__(self, use_dp = False, num_output = 1):
    super(MyModel, self).__init__()
    self.use_dp = use_dp
    self.dense1 = tf.keras.layers.Dense(12, activation=tf.nn.relu)
    self.dense2 = tf.keras.layers.Dense(24, activation=tf.nn.relu)
    self.dense3 = tf.keras.layers.Dense(4, activation=tf.nn.relu)
    self.dense4 = tf.keras.layers.Dense(10, activation=tf.nn.sigmoid)

    if self.use_dp:
        use_dp = tf.keras.layers.Dropout(0.3)
```

在构造函数（__init__）中，为不同类型的层定义了变量，以构建神经网络。所有层都是 Dense 类型——全连接层。第一层由 12 个节点组成，第二层由 24 个节点组成，第三层由 4 个节点组成，最后一层由 10 个节点组成。前三层使用 ReLU 作为激活层，最后一层使用 sigmoid 作为激活层。构造函数中还定义了 dropout 操作，以备在上述任何一个全连接层增加 dropout 层。Dropout 函数传入的参数为 0.3，表示 dropout 的百分比为 30%。

接下来，定义 call 方法以创建对象。一个典型的 call 方法定义方式如下所示。

```python
def call(self, x):
    x = self.dense1(x)
    x = self.dense2(x)
    if self.use_dp:
        x = self.dp(x)
    x = self.dense3(x)
    if self.use_dp:
        x = self.dp(x)
    return self.dense4(x)
```

该神经网络包含四层，每一层的输出变量成为下一层的输入变量。代码的最后一行返回 dense4，其输出为 10 个类别信息。第二层和第三层使用了 dropout。

以上完成了对模型的子类化操作，接下来，将实例化该模型为一个对象。使用如下语句对模型进行实例化。

```python
model = MyModel()
```

创建模型对象之后，调用其 compile 方法来运行模型，并给 compile 方法传入需要传递的参数，如下面的代码所示。

```python
model.compile(loss = tf.losses.binary_crossentropy,
              optimizer = 'adam',
              metrics = ['accuracy'])
```

一般来说，如果用户喜欢面向对象编程，可以使用模型子类化，否则，不需要做子类化，函数式 API 能够满足用户构建复杂架构的所有需求。

3.2.3 预定义层

为了构建网络架构，tf.keras 提供了 Sequential API，以通过不断向 Sequential 中添加层来构建模型。TensorFlow 中提供了一些预定义层，每个层都定义为一个类，可以在代码中实例化该类并将其实例添加到模型中。以下是部分常用预定义层的列表。

- ☑ Dense：密集层（全连接层）。
- ☑ Conv2D：二维卷积层。
- ☑ InputLayer：网络入口点。
- ☑ LSTM：长短时记忆层（long short-term memory layer）。
- ☑ RNN：自定义循环层的基类。

预定义层非常详尽，除上述列出的层之外，还有 Dropout、Flatten、LayerNormalization、Multiply 等。此外，除了以上预定义的基类，用户还可以创建自定义层。

3.2.4 自定义层

自定义层继承自 tf.keras.layers.Layer，且需要重写四个方法：init、build、call 和 compute_output_shape 方法。典型的自定义类如下所示。

```python
class MyLayer(tf.keras.layers.Layer):
    def __init__(self, output_dim, ** kwargs):
        self.output_dim = output_dim
        super(MyLayer, self).__init__( ** kwargs)

    def build(self, input_shape):
        self.W = self.add_weight(name = 'kernel',
            shape = (input_shape[1], self.output_dim),
            initializer = 'uniform',
            trainable = True)
        self.built = True

    def call(self, x):
        return tf.matmul(x, self.W)

    def compute_output_shape(self, input_shape):
        return (input_shape[0], self.output_dim)
```

在 __init__ 方法中，调用了父类构造函数，并设置了网络的输出维度。在 buid 方法中，设置了初始权重矩阵，根据输入参数 input_shape 设置了矩阵的维度，并设置该矩阵为可训练的（trainable=True）。在 call 方法中，可以设置要对权重矩阵执行的任何操作。在当前代码中，将权重矩阵与输出向量 x 进行了矩阵乘法操作。最后，compute_output_shape 方法返回了输出维度。

接下来，将上述自定义层应用于任何神经网络模型中，如下面代码片段所示。

```python
model = tf.keras.Sequential([tf.keras.layers.Dense(256,
            input_shape=(784,)),
        tf.keras.layers.Dense(256, activation = 'relu'),
        MyLayer(10),
        tf.keras.layers.Dense(10, activation = 'softmax')])
```

自定义层是如何被添加到两个 Dense 层之间的？实际上，可以将自定层放在神经网络模型的任何位置。上述代码片段可由图 3-5 表示。

至此，调用模型的 compile 方法来编译模型，语句如下。

```python
model.compile(optimizer='adam',
            loss=tf.keras.losses.binary_crossentropy,
            metrics=['accuracy'])
```

最后，调用模型的 fit 方法来训练模型，语句如下。

```python
model.fit(x,y1, batch_size=32, epochs=30)
```

上述自定义层的完整项目代码可在本书的下载地址 (https://github.com/Apress/artificial-neural-networks-with-tensorflow-2) 中获取。

创建自定义层使得用户在设计复杂网络架构时拥有极大的灵活性，本书将使用函数式 API 构建各种各样的网络拓扑结构。

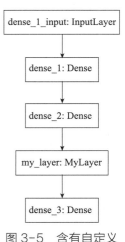

图 3-5　含有自定义层的神经网络

3.3 保存模型

通常，当模型训练完成后，需要将其保存到磁盘，以便后续将其部署到生产服务器上。不仅如此，tf.keras 中构建的模型可以在训练期间随时保存。使用这个保存的模型，可以从上次停止的位置继续训练，避免在一次会话中进行长时间训练。此外，还可以与其他人共享保存的模型，以便其他人可以在保存的模型基础上继续开发。tf.keras 提供了几种保存模型的方法，接下来将讨论这些方法。

1. 保存完整模型

当保存完整模型时，以下信息将被保存：模型的架构、模型的权重、传入模型 compile 方法中的训练参数、优化器及其状态。

使用如下代码保存完整模型。

```
model.save('filename.h5')
```

使用如下代码重新构建模型。

```
new_model = keras.models.load_model('filename.h5')
```

2. 导出为 SavedModel 格式

TensorFlow 提供了一种名为 SavedModel 的独立序列化格式，供 Python 以外的 TensorFlow 使用。TensorFlow Serving 框架也支持这种格式。使用如下代码将整个模型保存为 SavedModel 格式。

```
model.save('filename', save_format = 'tf')
```

使用如下代码加载保存的模型。

```
new_model = keras.models.load_model('filename')
```

3. 保存模型架构

某些情况下，用户只对保存模型的架构感兴趣，而不是其权重值或优化器状态。这种情况下，可调用模型的 get_config 方法获取模型的配置，然后使用该配置在保存的实例上重新构建模型，如下面代码所示。

```
# 获取和保存模型配置
config = model.get_config()

# 重新构建模型
model = keras.Model.from_config(config)
```

> 使用上述代码重新构建的模型会丢掉之前所有的训练信息。

4. 保存模型权重

某些情况下,用户希望只保存模型的训练状态,而不是模型的架构。这种情况下,可调用模型的 get_weights 和 set_weights 方法来保存和获取和恢复模型权重,如下面代码所示。

```
# 获取模型权重
weights = model.get_weights()
# 恢复模型权重
model.set_weights(weights)
```

> **注意**
>
> 默认情况下自定义层不支持 model.save 和 model.get_config 方法,因此,需要重写 get_config 方法以支持自定义层,但自定义层默认支持保存模型权重。

5. 保存为 JSON 格式

使用如下代码将模型保存为 JSON 格式。

```
from keras.models import model_from_json
# serialize to JSON
json_model = model_1.to_json()
with open("model_1.json", "w") as json_file:
    json_file.write(json_model)

# load json and re-create model
from keras.models import model_from_json
file = open('model_1.json', 'r')
buffer = file.read()file.close()
model = tf.keras.models.model_from_json(buffer)
```

例如,有以下模型定义。

```
model_1 = tf.keras.Sequential([
    Conv2D(32, (3, 3), activation = 'relu', padding = 'same',
           input_shape = (32, 32, 3)),
    Conv2D(32, (3, 3), activation = 'relu', padding = 'same'),
    MaxPooling2D((2, 2)),
    Dense(128, activation = 'relu'),
    Dense(10, activation = 'softmax')
])
```

将模型保存为 JSON 格式后,得到如下代码。

```
{
  "class_name":"Sequential",
```

```json
"config":{
  "name":"sequential_3",
  "layers":[
    {
      "class_name":"Conv2D",
      "config":{
        "name":"conv2d_2",
        "trainable":true,
        "batch_input_shape":[
          null,
          32,
          32,
          3
        ],
        "dtype":"float32",
        "filters":32,
        "kernel_size":[
          3,
          3
        ],
        "strides":[
          1,
          1
        ],
        "padding":"same",
        "data_format":"channels_last",
        "dilation_rate":[
          1,
          1
        ],
        "activation":"relu",
        "use_bias":true,
        "kernel_initializer":{
          "class_name":"GlorotUniform",
          "config":{
            "seed":null
          }
        },
        "bias_initializer":{
          "class_name":"Zeros",
          "config":{

          }
        },
        "kernel_regularizer":null,
        "bias_regularizer":null,
        "activity_regularizer":null,
        "kernel_constraint":null,
        "bias_constraint":null
      }
```

```
        },
    ]
},
...
"keras_version":"2.2.4-tf",
"backend":"tensorflow"
}
```

保存模型为 JSON 格式的项目代码可在本书的下载中获取。

对 tf.keras 进行简要介绍后,接下来将介绍 tf.keras 更实用的方法。下一节,将开发基于卷积神经网络 (convolutional neural network, CNN) 的图像分类模型。

3.4 卷积神经网络

本节在定义用于图像分类的神经网络之前,先介绍一些相关概念。

如图 3-6 所示,读者可以立刻辨认图中显示了一只鸟。该图像的大小为 275 像素 ×183 像素,每个像素代表一个颜色值,其中包含 RGB 分量。将上述彩色图像转换为黑白草图,如图 3-7 所示。

图 3-6　彩色图像 275 像素 × 183 像素

图 3-7　黑白草图 275 像素 × 183 像素

读者依然可以辨认出图中显示了一只鸟。接下来,把图像缩小到 32 像素 ×32 像素,如图 3-8 所示。

读者仍然可以看出图像显示的是一只鸟。试图理解当有一张鸟类的图片供人欣赏时为什么会发生这些变化?如果我们的目标是让机器学习解释(分类)给定图像中的对象,是否需要将如此高分辨率的图像提供给计算机?如果人类可以通过精简版的图像进行预测,为什么不能让计算机在这些精简的图像上学习图像识别的基本信息?上述做法的原因是高分辨率图像包含大量"比特"(bits) 信息,这些信息对于目标识别任务来说是多余的。将大量"比特"信息输入计算机需

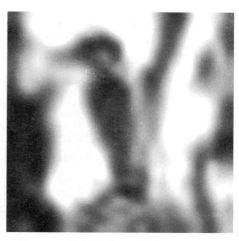

图 3-8　缩小图像 32 像素 × 32 像素

消耗大量内存和算力,而这些内存和算力都是非常稀缺的。因此,为了减少神经网络所需的训练时间并避免使用大量资源,本节将在数据集中使用精简图像。图 3-9 显示了一幅汽车图像,其原始尺寸为 200 像素×200 像素,但在 32 像素×32 像素的图像中仍然可以看出它是一辆汽车。

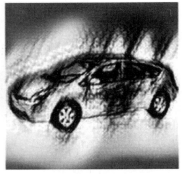

原始彩色图像200像素×200像素　　黑白草图200像素×200像素　　缩小图像32像素×32像素

图 3-9　汽车图像及其变换图

针对图像分类任务已有大量研究,且研究人员已成功设计了大量高精度 ANN 架构。上述图像变换的过程称为卷积。图像卷积的详细定义及其解释超出了本书的范围,本书主要关注使用 TensorFlow 2.x 构建高性能神经网络。研究人员设计了 CNN 来解决图像分类和目标识别问题。CNN 包含卷积层在内的多种不同类型的层。图 3-10 显示了具有两个卷积层的典型 CNN 架构。

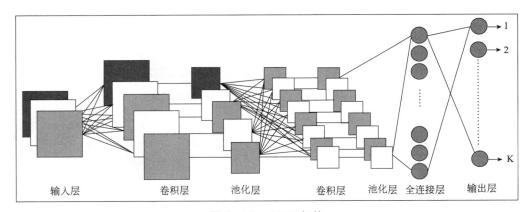

图 3-10　CNN 架构

如前所述,tf.keras 提供了几个现成的层,包括卷积、池化、展平(Flatten)等。接下来的任务就是按照正确的顺序和架构组装这些层。下一节,将创建 5 种不同的 CNN 架构并评价它们的性能。

3.5　使用 CNN 做图像分类

本节将介绍如何将给定的图像分类为已知类别。假设我们的任务是从给定的一幅特定的图像中识别显示的目标。对于人类来说，这通常是一项简单的任务，但是让机器解释图像并将其分类为已知的目标类别并不是一项容易的任务。幸运的是，该领域已有大量研究和开发项目，且已有大量预训练的即用型机器学习（ML）模型，可以精确地识别给定图像中的目标。本项目将介绍如何利用 tf.keras 中提供的库开发这样的模型。

3.5.1 创建项目

在 Colab 中打开一个新的 Python3 notebook，将其命名为 Ch3-imageClassifier。使用以下代码加载并导入 TensorFlow 2.x。

```
import tensorflow as tf
```

海量数据是机器学习项目的首要需求。幸运的是，对于图像分类任务，有人已在收集图像和数据清理方面付出很多努力，以便应用于图像分类算法。现在的任务则是开发一个模型并不断完善，以便该模型能够以可接受的正确率检测给定图像中的目标。在研究模型开发之前，先介绍哪些数据集可供使用。

3.5.2 图像数据

本项目使用的图像数据集由加拿大高等研究院 (Canadian Institute For Advanced Research, CIFAR) 创建。该数据集可供公众使用，以鼓励人们开发图像识别技术。CIFAR 的研究人员创建了两个数据集：CIFAR-10 和 CIFAR-100。这两个数据集都包含 60000 幅大小为 32 像素 ×32 像素的彩色图像。CIFAR-10 图像分为 10 类，包含猫、鸟、船、飞机等类别，每个类别各有 6000 幅图像，且被随机打乱以用于机器学习任务。CIFAR-10 数据集也在训练和测试数据集中进行了适当拆分——训练集由 50000 幅图像组成，其余 10000 幅图像作为测试数据集。图 3-11 显示了一组示例图像及其类别，以供读者快速了解 CIFAR-10。

所有图像的尺寸都是固定的——32 像素 ×32 像素，这是机器学习模型需要的。机器学习算法不支持变化尺寸的图像输入，神经网络的输入总是预先设定的固定大小。变化分辨率的图像包含可变数量的像素，因此无法输入到预设的网络架构中。CIFAR 数据集的创建者已将所有原始图像缩小到固定的微型尺寸。

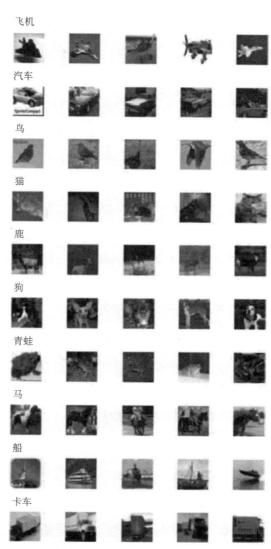

图 3-11　CIFAR-10 中的示例图像

> **注意**
>
> 如果在包含 512 像素 ×512 像素或更多像素的原始图像上训练模型,则 ANN 所需的输入像素点数量将非常大。这意味着需要训练海量"权重",且需要巨大的处理时间和算力。虽然将图像缩小到 32 像素 ×32 像素时会有明显的数据损失,但是,对这些小尺寸图像进行训练仍会产生很好的结果,且在现实世界中能够以极可接受的结果识别未知的物体。

CIFAR-100 与 CIFAR-10 类似,不同之处在于它拥有 100 个类别,而不是 10 个。每个类别进一步分为细类和粗类。例如,名为人(people)的超类包含婴儿、男孩、女孩、男人和女人等子类。

本项目使用 CIFAR-10 作为数据集,接下来将介绍如何在项目中加载数据集。

3.5.3 加载数据

tf.keras 模块提供了包含 CIFAR-10 在内的多个内置数据集,便于模型开发。这些数据集可以在 tf.keras.dataset 模块中调用。使用如下两行代码在控制台打印可用的数据集列表。

```
import tensorflow_datasets as tfds
print ("Number of datasets: ", len(tfds.list_builders()))
tfds.list_builders()
```

以下显示了输出的部分数据集列表。

```
Number of datasets: 141
['abstract_reasoning',
 'aeslc',
 'aflw2k3d',
 'amazon_us_reviews',
 'arc',
 'bair_robot_pushing_small',
 'big_patent',
 'bigearthnet',
 'billsum',
 'binarized_mnist',
 'binary_alpha_digits',
 'c4',
 'caltech101',
 'caltech_birds2010',
 ...
```

可以看到,目前有 141 个数据集可供使用。本节将使用 CIFAR10 开发当前应用程序。

3.5.4 创建训练、测试数据集

在上一章开发的二元分类器中,使用了 sklearn 的 train_test_split 方法来创建训练、

测试数据集。keras.datasets 提供了一个 load_data 方法，用于创建训练、测试数据集。使用如下语句从内置的 CIFAR-10 数据集创建训练、测试数据集。

```
(x_train, y_train), (x_test, y_test) = 
                tf.keras.datasets.cifar10.load_data()
```

加载数据后，将在 x_train、y_train、x_test、y_test 四个变量中获取到图像数据及其标签。使用以下代码检查可用于训练和测试的图像数量。

```
print("x_train dimensions : ",x_train.shape)
print("x_test dimensions  : ",x_test.shape)
print("y_train dimensions : ",y_train.shape)
print("y_test dimensions  : ",y_test.shape)
```

执行上述代码，得到如下结果。

```
x_train dimensions : (50000, 32, 32, 3)
x_test dimensions  : (10000, 32, 32, 3)
y_train dimensions : (50000, 1)
y_test dimensions  : (10000, 1)
```

根据 x_train 的维度，可知训练数据中包含 50000 幅图像，每幅图像有 32 像素 ×32 个像素，每个像素拥有与图像相关的 3 个值 RGB。x_test 的维度表明测试数据包含 10000 幅图像。对于 x_train 中的每幅图像，y_train 中都有一个范围在 0～9 之间的特定整数值与之对应。y_train 在机器学习中被称为标签。本小节将使用 x_train 来训练模型，使用 x_test 来评价模型的性能。

使用如下代码在控制台显示一幅图像，以便直观理解在计算机内存中加载了什么数据。

```
import matplotlib.pyplot as plt
plt.imshow(x_train[40])
```

执行上述代码后将显示一幅索引值为 40 的训练图像，如图 3-12 所示。

显然，这幅 32 像素 ×32 像素大小的图像几乎让人猜不出包含的是什么。然而，如前文所述，对于机器学习算法，这些尺寸的图像却足以用于训练。接下来，将为模型训练准备训练数据集。

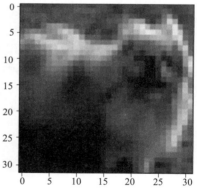

图 3-12　数据集中的一幅图像样本

3.5.5　准备模型训练数据

如前文所述，需要使用训练数据 x_train 来训练模型。同时，在开发机器学习算法时，还需要保留一部分数据以在训练阶段进行验证。本项目中，将保留 5% 的训练数据用于验证。

1. 创建验证集

使用 sklearn 工具包中的 split 方法将训练数据 x_train 拆分为训练集和验证集，如下面的代码所示。

```
from sklearn.model_selection import train_test_split
x_train, x_val, y_train, y_val = train_test_split(x_train,
                        y_train, test_size = 0.05,
                        random_state = 0)
```

这里,保留了 5% 的训练数据用于模型训练期间的验证。x_val 和 y_val 分别表示验证集中的图像和标签。接下来对数据进行扩增。

2. 数据扩增

在介绍如何进行数据扩增之前,首先阐述数据扩增是什么,以及数据扩增在机器学习中的必要性。数据扩增是数据科学家用来增加模型训练期间数据多样性的一种策略。通常情况下,收集的原始数据可能表现出部分一致性,从而导致不完全的模型训练。数据扩增避免了为求数据多样性而收集新样本的需要。一些常用的数据扩增技术有裁剪、填充和水平翻转。为了扩增图像数据,Keras 预处理模块提供了一个 ImageDataGenerator 类。使用以下代码实例化此类。

```
from tensorflow.keras.preprocessing.image import
ImageDataGenerator
datagen = ImageDataGenerator(
  rotation_range=15,
  width_shift_range=0.1,
  height_shift_range=0.1,
  horizontal_flip=True, )
```

上述代码中的各种参数决定了对图像进行扩增的不同类型。根据示例中的参数名可以看出,输入图像可以通过旋转、移动及翻转等操作获得所需的数据多样性。可以使用上述 datagen 实例来扩增训练数据。

每幅图像的 RGB 值范围在 0 ~ 255 之间,而在机器学习模型中需要将这些像素值缩放到 0 ~ 1 之间。因此,定义缩放函数如下。

```
def normalize(data):
    data = data.astype("float32")
    data = data/255.0
    return data
```

使用如下代码对训练、测试数据集进行扩充和缩放。

```
x_train = normalize(x_train)
datagen.fit(x_train)
x_val = normalize(x_val)
datagen.fit(x_val)
x_test = normalize(x_test)
```

上述代码中,normalize 函数用于缩放训练测试数据集。ImageDataGenerator 的 fit 方法扩增了数据。对于测试数据,只需进行归一化,而不对其进行扩增,因为需要获取真实图像的测试结果,而不是扩增图像的测试结果。

数据处理的最后一件事是调用 to_categorical 方法将标签的整数标量值转换为矩阵。生成的标签为 10 列,包含 10 个不同类别,如鸟类和汽车。调用 Keras.utils 中的 to_categorical 函数如下。

```
y_train = tf.keras.utils.to_categorical(y_train, 10)
y_test = tf.keras.utils.to_categorical(y_test, 10)
y_val = tf.keras.utils.to_categorical(y_val, 10)
```

完成对数据的所有变换后,可以使用如下所示的 imshow 函数在控制台上显示变换后的图像。

```
plt.imshow(x_train[40])
```

图 3-13 显示了变换后的结果图。

至此已完成了数据预处理。通过打印修改后的数据维度,可以了解数据预处理的最终结果,如下面的代码片段所示。

```
print("x_train dimensions : ",x_train.shape)
print("y_train dimensions : ",y_train.shape)
print("x_test dimensions  : ",x_test.shape)
print("y_test dimensions  : ",y_test.shape)
print("x_val dimensions   : ",x_val.shape)
print("y_val dimensions   : ",y_val.shape)
```

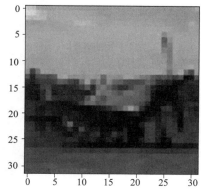

图 3-13 索引值为 40 的变换图像

上述代码的执行结果如下。

```
x_train dimensions : (47500, 32, 32, 3)
y_train dimensions : (47500, 10)
x_test dimensions  : (10000, 32, 32, 3)
y_test dimensions  : (10000, 10)
x_val dimensions   : (2500, 32, 32, 3)
y_val dimensions   : (2500, 10)
```

可以看到,特征向量(x_train, x_test, x_val)的维度保持不变,而标签向量(y_train, y_test, y_val)的维度已更改为 10,这是因为此处使用的数据集中有 10 个类别。

接下来是机器学习中最重要的部分:定义模型。在定义模型之前,将编写一个函数用来训练、评价和打印指定模型的评价指标。这样做的目的是展示不同模型架构对分类正确率的影响。然后,将从一个简单的模型架构开始,不断添加功能以期望提高分类正确率。如果对模型所做的改动导致分类正确率降低,将放弃这些更改并进行新的改动。

3.5.6 模型开发

在本节,首先定义几种具有不同架构的模型。其次,训练这些模型并对未知图像进行预测。再次,根据模型的分类正确率和预测结果对模型进行性能评价。最后,将最好的模型保存到磁盘,并在生产环境中使用该模型对未知图像执行分类操作。

为了对比不同模型的性能,将定义一个可在每个模型上调用的函数。该函数可以训练模型、评价模型性能并打印评价指标。

1. 训练、评价、显示函数

定义实现模型训练、模型评价和评价指标显示的函数如下。

```
def results(model):
```

该函数的输入参数类型为模型（model）类。接下来，将创建不同的模型并将其作为参数传递给此函数。

在函数体内部，调用模型的 fit 方法开始训练，如下。

```
epoch = 20
r = model.fit(x_train, y_train, batch_size = 32,
              epochs = epoch, validation_data =
              (x_val, y_val), verbose = 1)
```

在训练过程中，可以看到每个训练 epoch 的输出，如图 3-14 所示。

```
Train on 47500 samples, validate on 2500 samples
Epoch 1/20
47500/47500 [==============================] - 12s 262us/sample - loss: 1.8401 - accuracy: 0.3402 - val_loss: 1.5944 - val_accuracy: 0.4264
Epoch 2/20
47500/47500 [==============================] - 6s 120us/sample - loss: 1.4749 - accuracy: 0.4737 - val_loss: 1.3695 - val_accuracy: 0.5048
Epoch 3/20
47500/47500 [==============================] - 6s 122us/sample - loss: 1.3041 - accuracy: 0.5380 - val_loss: 1.2184 - val_accuracy: 0.5656
Epoch 4/20
47500/47500 [==============================] - 6s 124us/sample - loss: 1.1634 - accuracy: 0.5889 - val_loss: 1.1211 - val_accuracy: 0.5988
Epoch 5/20
47500/47500 [==============================] - 6s 123us/sample - loss: 1.0602 - accuracy: 0.6282 - val_loss: 1.1006 - val_accuracy: 0.6068
Epoch 6/20
47500/47500 [==============================] - 6s 123us/sample - loss: 0.9761 - accuracy: 0.6566 - val_loss: 1.0766 - val_accuracy: 0.6240
```

图 3-14 训练过程中每个训练 epoch 的输出

模型训练的迭代次数在名为 epoch 的 Python 变量中声明，该变量设置为 20。查看评价指标后可以更改此数字以重新训练网络。fit 方法的参数此处不再赘述，在第 2 章二元分类项目中已经讨论过这些参数。

训练结束后，通过调用模型的 evaluation 方法在测试数据上评价模型的正确率，如下所示。

```
acc = model.evaluate(x_test, y_test)
print("test set loss : ", acc[0])
print("test set accuracy :", acc[1]*100)
```

上述代码的一种典型输出如下。

```
test set loss : 1.4717637085914612
test set accuracy : 65.78999757766724
```

使用如下代码片段打印分类正确率。

```
# 打印训练和验证集正确率
# Plot training and validation accuracy
  epoch_range = range(1, epoch+1)
  plt.plot(epoch_range, r.history['accuracy'])
  plt.plot(epoch_range, r.history['val_accuracy'])
  plt.title('Classification Accuracy')
  plt.ylabel('Accuracy')
  plt.xlabel('Epoch')
  plt.legend(['Train', 'Val'], loc='lower right')
  plt.show()
```

训练集和验证集上的一个典型评价指标结果如图 3-15 所示。

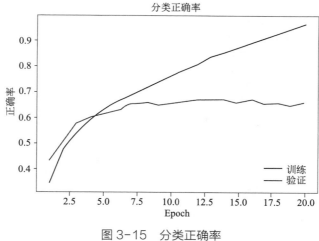

图 3-15　分类正确率

同理，使用如下代码片段打印损失函数信息。

```
# 打印训练集和验证集损失值
# Plot training & validation loss values
plt.plot(epoch_range,r.history['loss'])
plt.plot(epoch_range, r.history['val_loss'])
plt.title('Model loss')
plt.ylabel('Loss')
plt.xlabel('Epoch')
plt.legend(['Train', 'Val'], loc='lower right')
plt.show()
```

一个典型的损失值评价指标如图 3-16 所示。

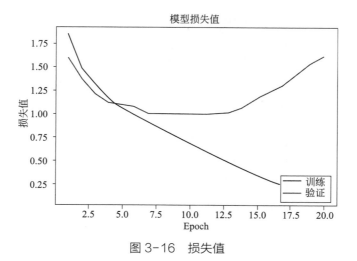

图 3-16　损失值

上述绘图代码已在第 2 章进行了解释，此处不再赘述。至此，完成了 results 函数的定义，以下列出了 results 函数的完整代码以供快速参考。

```
def results(model):
```

```python
epoch = 20
r = model.fit(x_train, y_train, batch_size = 32,
          epochs = epoch, validation_data =
          (x_val, y_val), verbose = 1)
acc = model.evaluate(x_test, y_test)
print("test set loss : ", acc[0])
print("test set accuracy :", acc[1]*100)

# Plot training and validation accuracy
epoch_range = range(1, epoch+1)
plt.plot(epoch_range, r.history['accuracy'])
plt.plot(epoch_range, r.history['val_accuracy'])
plt.title('Classification Accuracy')
plt.ylabel('Accuracy')
plt.xlabel('Epoch')

plt.legend(['Train', 'Val'], loc='lower right')
plt.show()

# Plot training & validation loss values
plt.plot(epoch_range,r.history['loss'])
plt.plot(epoch_range, r.history['val_loss'])
plt.title('Model loss')
plt.ylabel('Loss')
plt.xlabel('Epoch')
plt.legend(['Train', 'Val'], loc='lower right')
plt.show()
```

在定义各种模型之前,还需再定义一个函数,用于预测指定图像中目标的类别。

2. 预测函数

由于此处使用的数据集中包含 10 个类别,因此首先在代码中定义这些类的名称。

```python
classes = ['airplane','automobile', 'bird', 'cat',
           'deer','dog','frog', 'horse','ship','truck']
```

定义一个名为 predict_class 的函数如下。

```python
def predict_class(filename, model):
```

该函数有两个输入参数:第一个参数指定图像的文件名;第二个参数指定用于对给定图像分类的模型。因此,这是一个用于比较不同模型性能的通用函数。

在函数体内部,使用如下代码将图像加载入计算机内存。

```python
img = load_img(filename, target_size=(32, 32))
```

使用如下代码显示图像以供参考。

```python
plt.imshow(img)
```

需要将图像数据转换为可以输入到模型的 predict 方法的数组,img_to_array 方法结合 reshap 方法可以完成上述功能。

```
img = img_to_array(img)
img = img.reshape(1,32,32,3)
```

然后，使用如下代码将图像的像素值缩放至 0～1 之间。

```
img = img.astype('float32')
img = img/255.0
```

使用如下代码将图像数据输入模型的 predict 方法中。

```
result = model.predict(img)
```

模型的 predict 方法返回的输出结果是矩阵形式。使用 for 循环将模型输出复制到 Python 的字典（dict）中，如下面的语句所示。

```
dict2 = {}
for i in range(10):
    dict2[result[0][i]] = classes[i]
```

字典 dict2 的键（key）为模型的预测结果，字典的值（value）为其真实类别。

将预测结果 result[0] 复制到名为 res 的列表中，然后对 res 数组按升序排序如下。

```
res = result[0]
res.sort()
```

选取前三个预测结果进行展示，最佳预测结果存储在最后一个索引处：

```
res = res[::-1]
results = res[:3]
```

将结果打印在屏幕上，代码如下。

```
print("Top predictions of these images are")
for i in range(3):
    print("{} : {}".format(dict2[results[i]],
        (results[i]*100).round(2)))
```

使用如下代码展示实验中使用的图像以供参考。该图像已包含在 plt 对象中。

```
print('The image given as input is')
```

至此，我们完成了 predict_class 函数的定义，以下列出了整个函数的定义以供快速参考。

```
# Predict the class in a given image
from tensorflow.keras.preprocessing.image
import load_img, img_to_array

classes = ['airplane','automobile', 'bird', 'cat', 'deer',
           'dog','frog', 'horse','ship','truck']
def predict_class(filename, model):
  img = load_img(filename, target_size=(32, 32))

  plt.imshow(img)
```

```
    # convert to array
    # reshape into a single sample with 3 channels
    img = img_to_array(img)
    img = img.reshape(1,32,32,3)

# prepare pixel data
img = img.astype('float32')
img = img/255.0

#predicting the results
result = model.predict(img)

dict2 = {}
for i in range(10):
    dict2[result[0][i]] = classes[i]

res = result[0]
res.sort()
res = res[::-1]
results = res[:3]

print("Top predictions of these images are")
for i in range(3):
    print("{} : {}".format(dict2[results[i]],
        (results[i]*100).round(2)))

print('The image given as input is')
```

接下来将介绍如何定义模型。

3.5.7 定义模型

在本节，首先定义 5 个不同的模型，其复杂度逐渐递增。然后，对每个模型进行训练，评价其性能并对结果进行预测。最后，比较不同模型产生的结果。

模型 1：具有 2 个卷积层的模型

本节的第 1 个模型是一个仅包含 2 个卷积层的简单模型。为了定义模型，Keras 提供了一个第 2 章使用过的 Sequential API。首先使用以下语句导入 Sequential API。

```
from tensorflow.keras.models import Sequential
```

Sequential API 使得用户按照所需的顺序向其中添加不同层，以构建一个顺序的网络架构。Keras 不仅提供了多种类型的预定义层，同时它还可以在需要时创建自定义层。本节的应用程序使用了 Keras 提供的几类预定义层，包括 Conv2D、Dense、Dropout 和 Flatten。使用以下语句在代码中导入这些层。

```
from tensorflow.keras.layers
import Dense, Dropout, Conv2D, MaxPooling2D,
                Flatten, BatchNormalization
```

通过实例化 Sequential 类来定义网络模型，如下所示。

```
model_1 = Sequential([
```

```
        Conv2D(32, (3, 3), activation = 'relu', padding = 'same',
            input_shape = (32, 32, 3)),
        Conv2D(32, (3, 3), activation = 'relu', padding = 'same'),
        MaxPooling2D((2, 2)),
        Flatten(),
        Dense(128, activation = 'relu'),
        Dense(10, activation = 'softmax')
])
```

根据上述代码可以看出，网络模型由 6 层组成。第 1 层是卷积层，为 Conv2D 类型。该层由 32 个尺寸为 3×3 的滤波器组成，这些滤波器在 32×32×3 的图像上滑动计算。网络的输入是 32×32×3，为 x_train 的向量维度。第 2 层依然是卷积层，第 3 层是 MaxPooling2D 层，然后是 Flatten 层。最后两层是密集层——第一层由 128 个具有 ReLU 激活的节点组成，第二层包含 10 个具有 softmax 激活的节点。由于该神经网络模型应输出 10 个类别，因此最后一层包含 10 个节点。使用如下代码生成网络图。

```
from tensorflow.keras.utils import plot_model
plot_model(model_1, to_file='model1.png')
```

生成的网络架构图如图 3-17 所示。

调用 model_1 实例对象的 summary 方法可以获取到模型的摘要信息，如下所示。

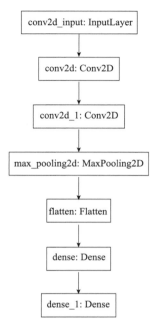

图 3-17　模型 1 的网络架构图

```
Model: "sequential"
_____
Layer (type)                 Output Shape              Param #
=================================================================
conv2d (Conv2D)              (None, 32, 32, 32)        896
_____
conv2d_1 (Conv2D)            (None, 32, 32, 32)        9248
_____
max_pooling2d (MaxPooling2D) (None, 16, 16, 32)        0
_____
flatten (Flatten)            (None, 8192)              0
_____
dense (Dense)                (None, 128)               1048704
_____
dense_1 (Dense)              (None, 10)                1290
=================================================================
Total params: 1,060,138
Trainable params: 1,060,138
Non-trainable params: 0
```

定义好网络模型后，将优化器设置为 SGD，并使用名为 categorical_crossentropy 的损失函数来编译模型，如下面的代码片段所示。

```
opt = tf.keras.optimizers.SGD(lr=0.001, momentum=0.9)
model_1.compile(optimizer=opt, loss = 'categorical_crossentropy',
```

```
                    metrics = ['accuracy'])
```

接下来,调用之前定义的 results 函数来训练、评价和显示评价指标,并将 model_1 作为它的参数:

```
results(model_1)
```

生成的输出结果如图 3-18 所示。

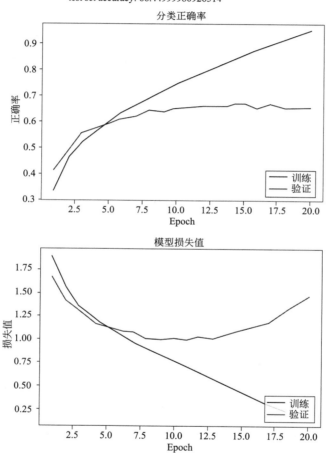

图 3-18　模型 1 的评价指标

由图 3-18 可以看出,分类正确率约为 66%。显然,在大多数情况下这个结果在不可接受的范围内。此外,模型的训练过程存在明显的过拟合。该过拟合现象可以通过参数优化和正则化来缓解。因此,接下来将使用这些技术继续提升模型的性能。

使用如下代码段对未知图像进行预测。

```
import urllib
resource = urllib.request.urlopen("https://raw.
githubusercontent.com/Apress/artificial-neural-networks-withtensorflow-
```

```
2/main/ch03/test01.png")
output = open("file01.jpg", "wb")
output.write(resource.read())
output.close()
predict_class("file01.jpg", model_1)
```

预测的输出结果如图 3-19 所示。

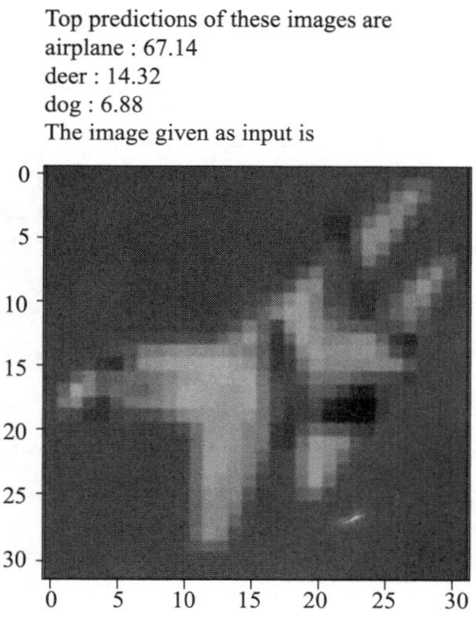

接下来,将在模型架构中额外添加 2 个卷积层以继续提升模型的分类性能。

模型 2:具有 4 个卷积层的模型

此处定义了一个具有 4 个卷积层的模型架构,如下面的代码片段所示。

```
model_2 = Sequential([
    Conv2D(32, (3, 3), activation = 'relu', padding = 'same',
            input_shape = (32, 32, 3)),
    Conv2D(32, (3, 3), activation = 'relu', padding = 'same'),
    MaxPooling2D((2, 2)),
    Conv2D(64, (3, 3), activation = 'relu', padding = 'same'),
    Conv2D(64, (3, 3), activation = 'relu', padding = 'same'),
    MaxPooling2D((2, 2)),
    Flatten(),
    Dense(128, activation = 'relu'),
    Dense(10, activation = 'softmax')
])

opt = tf.keras.optimizers.SGD(lr = 0.001, momentum = 0.9)
model_2.compile(optimizer = opt, loss =
                'categorical_crossentropy',
                metrics = ['accuracy'])
```

图 3-19 模型 1 分类性能

模型的摘要信息打印如下。

```
Model: "sequential_1"
_____
Layer (type)                 Output Shape              Param #
=================================================================
conv2d_2 (Conv2D)            (None, 32, 32, 32)        896

conv2d_3 (Conv2D)            (None, 32, 32, 32)        9248

max_pooling2d_1 (MaxPooling2 (None, 16, 16, 32)        0

conv2d_4 (Conv2D)            (None, 16, 16, 64)        18496

conv2d_5 (Conv2D)            (None, 16, 16, 64)        36928

max_pooling2d_2 (MaxPooling2 (None, 8, 8, 64)          0

flatten_1 (Flatten)          (None, 4096)              0
```

```
dense_2 (Dense)                (None, 128)                524416
_____
dense_3 (Dense)                (None, 10)                 1290
=================================================================
Total params: 591,274
Trainable params: 591,274
Non-trainable params: 0
```

生成的网络架构图如图 3-20 所示。

模型 2 的评价指标结果如图 3-21 所示。

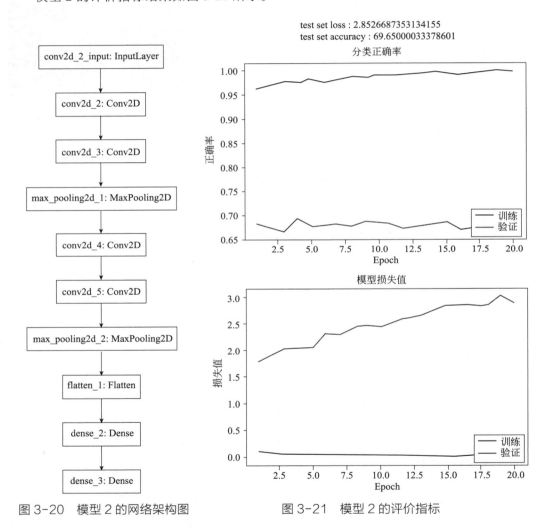

图 3-20　模型 2 的网络架构图　　　　图 3-21　模型 2 的评价指标

模型 2 在单幅图像（与模型 1 为同一幅图像）的预测结果显示如下。

```
Top predictions of these images are
airplane : 99.95
bird : 0.04
deer : 0.01
```

可以看到，预测结果的正确率从 66% 增加到 69.65%。对于给定图像的预测结果几乎保持不变。

接下来，将尝试添加更多卷积层，以查看模型的正确率是否会进一步提高。

模型 3：具有 6 个卷积层的模型（包含 32、64 和 128 个滤波器）

优化后的模型如下面代码片段所示。

```
model_3 = Sequential([
  Conv2D(32, (3, 3), activation = 'relu', padding = 'same',
      input_shape = (32, 32, 3)),
  Conv2D(32, (3, 3), activation = 'relu', padding = 'same'),
  MaxPooling2D((2, 2)),
  Conv2D(64, (3, 3), activation = 'relu', padding = 'same'),
  Conv2D(64, (3, 3), activation = 'relu', padding = 'same'),
  MaxPooling2D((2, 2)),
  Conv2D(128, (3, 3), activation = 'relu', padding = 'same'),
  Conv2D(128, (3, 3), activation = 'relu', padding = 'same'),
  MaxPooling2D((2, 2)),
  Flatten(),
  Dense(128, activation = 'relu'),
  Dense(10, activation = 'softmax')
])

opt = tf.keras.optimizers.SGD(lr = 0.001, momentum = 0.9)
model_3.compile(optimizer = opt, loss = 'categorical_
crossentropy',
      metrics = ['accuracy'])
```

模型 3 的摘要信息如下。

```
Model: "sequential_2"
```

Layer (type)	Output Shape	Param #
conv2d_6 (Conv2D)	(None, 32, 32, 32)	896
conv2d_7 (Conv2D)	(None, 32, 32, 32)	9248
max_pooling2d_3 (MaxPooling2	(None, 16, 16, 32)	0
conv2d_8 (Conv2D)	(None, 16, 16, 64)	18496
conv2d_9 (Conv2D)	(None, 16, 16, 64)	36928
max_pooling2d_4 (MaxPooling2	(None, 8, 8, 64)	0
conv2d_10 (Conv2D)	(None, 8, 8, 128)	73856
conv2d_11 (Conv2D)	(None, 8, 8, 128)	147584
max_pooling2d_5 (MaxPooling2	(None, 4, 4, 128)	0
flatten_2 (Flatten)	(None, 2048)	0

```
dense_4 (Dense)                  (None, 128)                 262272
_____
dense_5 (Dense)                  (None, 10)                  1290
=================================================================
Total params: 550,570
Trainable params: 550,570
Non-trainable params: 0
```

模型 3 的网络架构图如图 3-22 所示。

模型 3 的评价指标如图 3-23 所示。

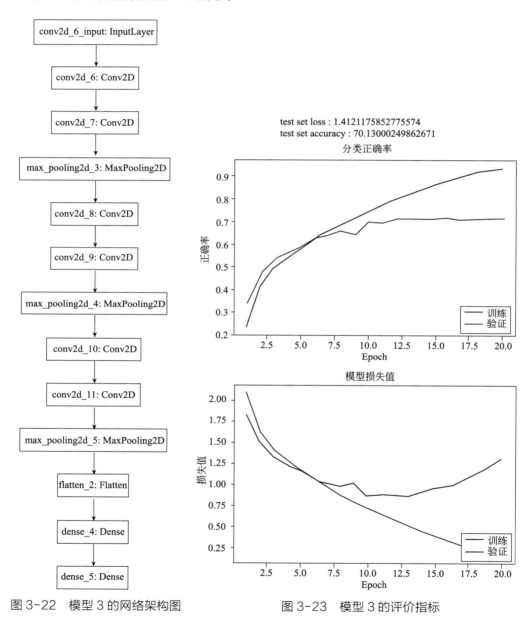

图 3-22　模型 3 的网络架构图　　图 3-23　模型 3 的评价指标

使用模型3对预测图像进行预测的结果如下（预测图像与模型1相同）。

```
Top predictions of these images are
deer : 87.18
bird : 8.07
airplane : 3.81
```

可以看到，该模型的正确率（70.13%）几乎与模型2的正确率（69.65%）持平。因此，仅仅添加更多的卷积层并不能帮助我们构建更好的模型。

接下来，尝试向模型中添加一个dropout层。

模型4：增加额外的dropout层

dropout的核心思想是在训练期间从网络模型中随机丢弃一些单元及其连接。训练过程中每一步参数数量的减少都起到了正则化的效果。修改后的模型定义如下。

```
model_4 = Sequential([
    Conv2D(32, (3, 3), activation = 'relu', kernel_initializer =
           'he_uniform', padding = 'same', input_shape =
           (32, 32, 3)),
    Conv2D(32, (3, 3), activation = 'relu', kernel_initializer =
           'he_uniform', padding = 'same'),
    MaxPooling2D((2, 2)),
    Dropout(0.2),
    Conv2D(64, (3, 3), activation = 'relu', kernel_initializer =
           'he_uniform', padding = 'same'),
    Conv2D(64, (3, 3), activation = 'relu', kernel_initializer =
           'he_uniform', padding = 'same'),
    MaxPooling2D((2, 2)),
    Dropout(0.2),
    Conv2D(128, (3, 3), activation = 'relu', kernel_initializer =
           'he_uniform', padding = 'same'),
    Conv2D(128, (3, 3), activation = 'relu', kernel_initializer =
           'he_uniform', padding = 'same'),
    MaxPooling2D((2, 2)),
    Dropout(0.3),
    Flatten(),
    Dense(128, activation = 'relu'),
    Dense(10, activation = 'softmax')
])

opt = tf.keras.optimizers.SGD(lr = 0.001, momentum = 0.9)
model_4.compile(optimizer = opt, loss = 'categorical_
crossentropy',
          metrics = ['accuracy'])
```

模型4的摘要信息如下。

```
Model: "sequential_3"
```

```
Layer (type)                    Output Shape              Param #
=================================================================
conv2d_12 (Conv2D)              (None, 32, 32, 32)        896
_____
conv2d_13 (Conv2D)              (None, 32, 32, 32)        9248
_____
max_pooling2d_6 (MaxPooling2    (None, 16, 16, 32)        0
_____
dropout (Dropout)               (None, 16, 16, 32)        0
_____
conv2d_14 (Conv2D)              (None, 16, 16, 64)        18496
_____
conv2d_15 (Conv2D)              (None, 16, 16, 64)        36928
_____
max_pooling2d_7 (MaxPooling2    (None, 8, 8, 64)          0
_____
dropout_1 (Dropout)             (None, 8, 8, 64)          0
_____
conv2d_16 (Conv2D)              (None, 8, 8, 128)         73856
_____
conv2d_17 (Conv2D)              (None, 8, 8, 128)         147584
_____
max_pooling2d_8 (MaxPooling2    (None, 4, 4, 128)         0
_____
dropout_2 (Dropout)             (None, 4, 4, 128)         0
_____
flatten_3 (Flatten)             (None, 2048)              0
_____
dense_6 (Dense)                 (None, 128)               262272
_____
dense_7 (Dense)                 (None, 10)                1290
=================================================================
Total params: 550,570
Trainable params: 550,570
Non-trainable params: 0
```

模型 4 的网络架构如图 3-24 所示。

模型 4 的评价指标如图 3-25 所示。

模型 4 对预测图像进行预测的结果如下（预测图像与模型 1 相同）。

```
Top predictions of these images are
cat : 26.33
dog : 25.52
airplane : 21.61
```

分类正确率现在提高到 78.24%，比之前的模型正确率高。接下来，将创建一个具有批量归一化和正则化的模型。

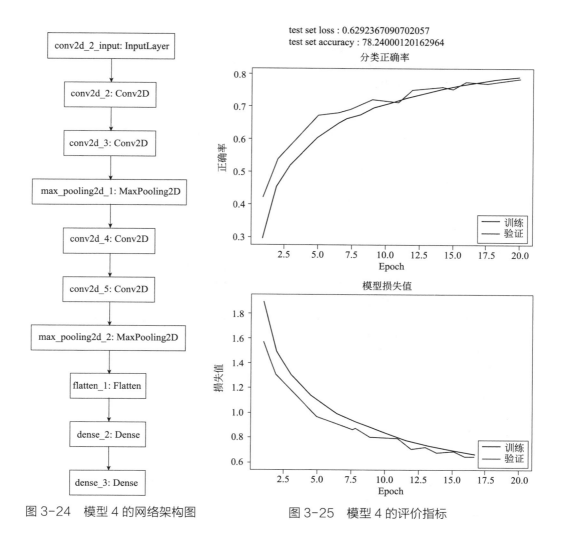

图 3-24　模型 4 的网络架构图　　图 3-25　模型 4 的评价指标

模型 5：增加正则化约束

BatchNormalization 的工作方式与常规的正则化方法相同，它适用于批量数据。BatchNormalization 处于卷积操作之后，以保持激活的平均值接近于 0，标准偏差接近于 1。因此，它减少了训练次数，进而加快了训练过程。在某些情况下，BatchNormalization 的使用可能允许我们无须使用之前模型中介绍的 dropout 操作。

正则化是一种通过在训练集上拟合适当的函数来减少误差以避免过度拟合的技术。本节将在新定义的模型中应用这项技术。模型定义如下面的代码所示。

```
weight_decay = 1e-4
model_5 = Sequential([
    Conv2D(32, (3, 3), activation = 'relu', padding = 'same',
        kernel_regularizer = tf.keras.regularizers.l2(weight_
        decay),
        input_shape = (32, 32, 3)),
```

```python
    BatchNormalization(),
    Conv2D(32, (3, 3), activation = 'relu', kernel_regularizer = 
        tf.keras.regularizers.l2(weight_decay), padding = 'same'),
    BatchNormalization(),
    MaxPooling2D((2, 2)),
    Dropout(0.2),
    Conv2D(64, (3, 3), activation = 'relu', kernel_regularizer = 
        tf.keras.regularizers.l2(weight_decay), padding = 'same'),
    BatchNormalization(),
    Conv2D(64, (3, 3), activation = 'relu', kernel_regularizer = 
        tf.keras.regularizers.l2(weight_decay), padding = 'same'),
    BatchNormalization(),
    MaxPooling2D((2, 2)),
    Dropout(0.3),
    Conv2D(128, (3, 3), activation = 'relu', kernel_regularizer = 
        tf.keras.regularizers.l2(weight_decay), padding = 'same'),
    BatchNormalization(),
    Conv2D(128, (3, 3), activation = 'relu', kernel_regularizer = 
        tf.keras.regularizers.l2(weight_decay), padding = 'same'),
    BatchNormalization(),
    MaxPooling2D((2, 2)),
    Dropout(0.3),
    Flatten(),
    Dense(128, activation = 'relu'),
    Dense(10, activation = 'softmax')
])

opt = tf.keras.optimizers.SGD(lr = 0.001, momentum = 0.9)
model_5.compile(optimizer = opt, loss = 'categorical_crossentropy',
        metrics = ['accuracy'])
```

模型 5 的摘要信息如下。

```
Model: "sequential_4"
```

Layer (type)	Output Shape	Param #
conv2d_18 (Conv2D)	(None, 32, 32, 32)	896
batch_normalization (BatchNo	(None, 32, 32, 32)	128
conv2d_19 (Conv2D)	(None, 32, 32, 32)	9248
batch_normalization_1 (Batch	(None, 32, 32, 32)	128
max_pooling2d_9 (MaxPooling2	(None, 16, 16, 32)	0

Layer (type)	Output Shape	Param #
dropout_3 (Dropout)	(None, 16, 16, 32)	0
conv2d_20 (Conv2D)	(None, 16, 16, 64)	18496
batch_normalization_2 (Batch	(None, 16, 16, 64)	256
conv2d_21 (Conv2D)	(None, 16, 16, 64)	36928
batch_normalization_3 (Batch	(None, 16, 16, 64)	256
max_pooling2d_10 (MaxPooling	(None, 8, 8, 64)	0
dropout_4 (Dropout)	(None, 8, 8, 64)	0
conv2d_22 (Conv2D)	(None, 8, 8, 128)	73856
batch_normalization_4 (Batch	(None, 8, 8, 128)	512
conv2d_23 (Conv2D)	(None, 8, 8, 128)	147584
batch_normalization_5 (Batch	(None, 8, 8, 128)	512
max_pooling2d_11 (MaxPooling	(None, 4, 4, 128)	0
dropout_5 (Dropout)	(None, 4, 4, 128)	0
flatten_4 (Flatten)	(None, 2048)	0
dense_8 (Dense)	(None, 128)	262272
dense_9 (Dense)	(None, 10)	1290

```
=================================================================
Total params: 552,362
Trainable params: 551,466
Non-trainable params: 896
```

模型 5 的网络架构如图 3-26 所示。

模型 5 的评价指标如图 3-27 所示。

模型 5 对预测图像进行预测的结果如下（预测图像与模型 1 相同）。

```
Top predictions of these images are
airplane : 83.3
cat : 5.25
ship : 4.34
```

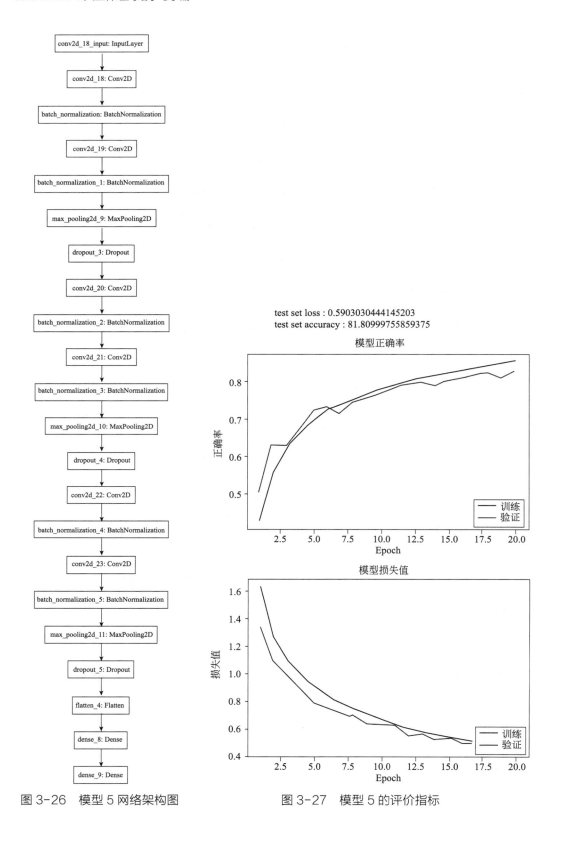

图 3-26 模型 5 网络架构图

图 3-27 模型 5 的评价指标

分类正确率现在提高到 81.80%，比之前所有的模型都有很大提升。表 3-1 列出了 5 种不同模型的分类性能。

表 3-1　5 种不同模型的性能对比

	模型 1	模型 2	模型 3	模型 4	模型 5
损失值	1.4560	2.8526	1.4121	0.6292	0.5903
分类正确率	66.4499	69.6500	70.1300	78.2400	81.8099

显然，在测试的所有模型中，模型 5 的分类正确率最高。此外，模型 3 和模型 2 的分类正确率非常接近，表明仅添加更多卷积层对于提高模型的分类正确率没有帮助。

3.5.8　保存模型

开发完成具有不同分类精度的模型之后，可以选择合适的模型应用于生产。为此，需要将模型保存到文件中。本章前面已经讨论了保存模型及其属性和状态的各种方法，这里不做赘述。通过调用模型实例上的 save 方法将上述模型保存至磁盘，如下面的代码所示。

```
model_5.save("model_5.h5")
```

该模型被保存为 HDF5 格式，使用以下语句重新加载。

```
m = load_model("model_5.h5")
```

然后可以使用新加载的模型来预测任何未知图像。

3.5.9　预测未知图像

使用保存的模型来预测三幅不同大小的未知图像，如图 3-28 所示。

unknown01.png 275×183

unknown02.png 334×151

unknown03.png 200×200

图 3-28　用于预测的未知图像

> **注意**
>
> 上述所有图像都不是方形的，由于模型是在大小为 32 像素 ×32 像素的方形图像上训练的，因此需要相同大小的输入图像。接下来，查看模型是否准确地预测了这些图像中的物体，使用以下代码片段从项目的 GitHub 地址加载图像并调用 predict_class 函数进行预测。

```
# unseen image 1
resource = urllib.request.urlopen("https://raw.
githubusercontent.com/Apress/artificial-neural-networks-withtensorflow-
2/main/ch03/unknown01.png")
output.write(resource.read())
output.close()
predict_class("/content/unknown01.jpg", m)

# unseen image 2
resource = urllib.request.urlopen("https://raw.
githubusercontent.com/Apress/artificial-neural-networks-withtensorflow-
2/main/ch03/unknown02.png")
output = open("/content/unknown02.jpg","wb")
output.write(resource.read())
output.close()
predict_class("/content/unknown02.jpg", m)

# unseen image 3
resource = urllib.request.urlopen("https://raw.
githubusercontent.com/Apress/artificial-neural-networks-withtensorflow-
2/main/ch03/unknown03.png")
output = open("/content/unknown03.jpg","wb")
output.write(resource.read())
output.close()
predict_class("/content/unknown03.jpg", m)
```

The predictions made by the model on the three images are given below:

```
Top predictions of these images are
airplane : 98.56
bird : 0.66
deer : 0.64

Top predictions of these images are
automobile : 99.94
airplane : 0.06
truck : 0.0

Top predictions of these images are
bird : 99.9
cat : 0.08
airplane : 0.01
```

可以看到，在 3 种情况下，模型都以近 99% 的正确率正确预测了目标。该项目的完整源码可在本书的下载中获取。

到目前为止，本章已创建了自定义模型，这是一个耗时的过程，需要付出很多努力。那么，为什么不根据需要重用他人在相同数据集上训练的模型呢？这就是下一章的内容。

总结

  Keras API 现已完全在 TensorFlow 中实现，并可通过 tf.keras 模块访问。使用 tf.keras 中提供的函数式 API，可以创建具有多个输入和输出、非线性拓扑结构、具有共享层的模型等复杂网络架构。本章介绍了如何使用预定义层和自定义层来定义神经网络架构；介绍了如何为面向对象编程的模型子类化；介绍了如何保存模型及其状态和权重。为了应用上述知识，本章介绍了一个使用 CNN 进行图像分类的完整示例。借助 CNN 的简要理论，该示例开发了一个成熟的基于 CIFAR-10 数据集的图像分类应用程序。从一个简单的模型开始，通过应用不同的技术不断提高模型性能，如增加 Conv2D 层的数量、添加 dropout、BatchNormalization 和正则化等操作。此外，本章还介绍了如何保存模型并利用保存的模型对未知图像进行预测。

第 4 章
迁移学习

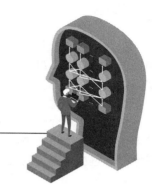

第 3 章开发了一个图像分类器，在 60000 幅训练图像上训练一段时间后，得到了一个分类模型，其分类准确率为 80%～90%。如果想要更高的准确度，则需要对更多的图像进行训练。事实上，在数据量越多的情况下，深度学习模型的学习效果越好。ImageNet（https://devopedia.org/imagenet）是同类任务中规模最大的一个数据集，由 14197122 幅图像组成，分为 21841 个子类，这些子类进一步分为 27 个子树。为了对 ImageNet 数据集中的图像进行分类，至今已开发了大量机器学习模型，这些模型主要以研究和竞赛的方式开发。2017 年，上述其中一个机器学习模型实现了低至 2.3% 的错误率，其底层网络非常复杂。由于训练这种复杂模型所需的数据十分庞大，可想而知训练模型所需的资源和时间是巨大的。

现在的问题是，能否重用上述神经网络所学到的知识，以用于我们自己的模型呢？这就是本章将介绍的全部内容，其背后的技术被称为迁移学习（transfer learning），类似于人类将所学的知识传授给年轻一代。本章将介绍如何将其他神经网络学习到的知识为我们所用，并开发自己的模型。

简而言之，本章涵盖以下内容。
- ☑ 什么是知识迁移？
- ☑ 什么是 TensorFlow Hub？
- ☑ 有哪些可用的预训练模型？
- ☑ 如何使用预训练模型？
- ☑ 如何使用迁移学习技术构建犬种分类器？

让我们从什么是知识迁移开始吧！

4.1　知识迁移

过去几十年间，开发人员一直在重用自己的代码或他人共享的代码。在机器学习中，能否采用相同的概念,即通过重用他人开发的模型来提升我们自己的模型呢?答案是可以,但这并不像听起来那么简单。在软件库中,只有代码被共享,而在机器学习中,需要共享的不仅仅是代码。以下列出了训练模型所需的四个重要组成部分:算法(algorithm)、数据(data)、训练(training)、专业知识(expertise)。

首先是算法,由开发者开发,用于训练神经网络,是知识迁移的代码部分。其次是数据,通常情况下,神经网络模型需在海量数据的支持下进行学习。再次是训练,训练一个模型需要巨大的处理能力和时间,且会消耗大量资源。最后,领域专家的专业知识也会融入训练模型的过程中。那么,如何将上述四个方面迁移到一个新模型中呢?为此,TensorFlow 的研究人员开发了一个 TensorFlow Hub 平台。

4.2 TensorFlow Hub

TensorFlow Hub 是 TensorFlow 提供的一个用于发布预训练模型的平台,可以从这些模型中重用部分机器学习模块(module)。那么,什么是模块呢?图 4-1 展示了一个模块的示意图。

模块及其训练权重本质上是 TensorFlow 图(graph)的一个独立部分,该图可以在其他地方重用以执行类似的任务。用户可以向该图中添加更多层以创建自己的模型,然后使用较小的数据集训练新模型,从而加快训练速度。在某种程度上,这促进了模块的泛化,使得泛化模块可以在多个模型中重复使用。可以将模型类比为二进制文件,即软件工程中最终的可执行文件,而模块则可类比为用于创建可执行文件的通用库。

模块是可以组合的,如图 4-2 所示。

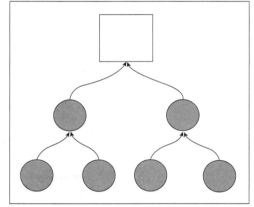

图 4-1 模块示意图

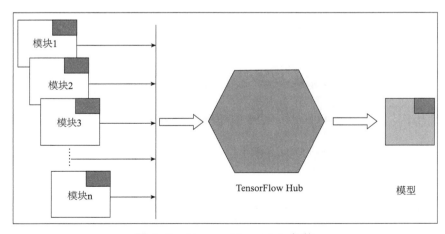

图 4-2 TensorFlow Hub 架构

可以在预训练模块上添加自己的层创建一个自定义的神经网络,然后即可训练包含预训练模块在内的整个神经网络。

通过给模块的相应函数传递一个参数即可重用和重新训练这些模块。这里,可重新训练的含义是可以像正常的神经网络一样对模块进行反向传播。

注意

如果对预训练模块进行重新训练,需确保使用一个较低的学习率,否则,已学到的权重可能会失控,导致完全出乎意料的结果。

当创建自定义模型后,如果认为该模型能够被其他人使用,则可以将其提交给Google,使其在TensorFlow Hub上发布。目前TensorFlow Hub中包含大量可用的第三方模型。但Google提示用户谨慎使用第三方模型,尤其是在不信任来源的情况下。

那么,我们现在可以使用哪些模型呢?

4.2.1 预训练模型

Google和TensorFlow Hub中的渠道合作伙伴提供了多个预训练模型。这些模型分为三类——图像、文本和视频。图4-3展示了其中的图像类。

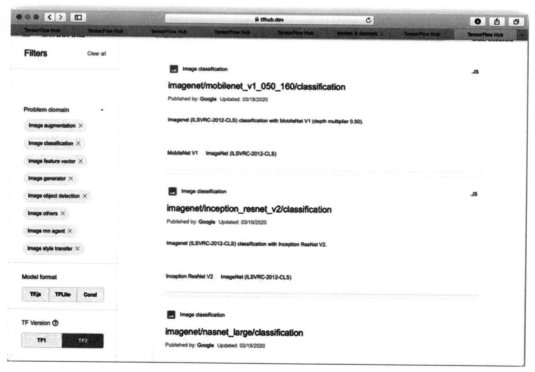

图4-3 TensorFlow中的预训练模型(图像类)

图像类中的一些模块如下。

① 使用 MobileNet、Inception ResNet V2、NASNet-A 的 ImageNet 分类模块。

② 使用 MobileNet、Inception、ResNet、PNASnet 的图像特征提取模块。

上述模块支持 TensorFlow 2.0。目前有一个由 Google 及其他合作伙伴开发的、在 TensorFlow 1.0 下运行的模块列表，正在等待迁移至 TensorFlow 2.0，此类别下的部分模块如下。

① 渐进式生成对抗网络（Progressive GAN）。

② 任意图像的快速风格迁移（style transfer）。

③ 图像扩增模块——随机裁剪、小幅度旋转和颜色失真等。

④ 预测任意照片粗略地理位置的 PlaNet。

⑤ Google 标志点（Google Landmarks）——深度局部特征(deep local features, DELF)。

在文本类中，有如下模块。

① Embeddings from Language Models, ELMO。

② Bidirectional Encoder Representation from Trasformers, BERT。

③ 通用语句编码器（universal sentence encoder）。

④ 基于 token 的文本嵌入——在维基百科英语语料库上训练。

在视频类中，截至撰写本文，尚没有用于 TensorFlow 2.0 的模型。在 TensorFlow 1.0 下，有少量由 Google 和 DeepMind 创建的模型。

TensorFlow 1.0 下具有大量可用的模块，可通过访问 TensorFlow Hub (https://tfhub.dev/) 查看可用模块的完整列表。

至此，我们已经知道有哪些模块可供使用，下一个问题是如何使用这样的预训练模块。

4.2.2 模型的使用

要在自己的程序中使用预训练模型，可从 tfdev 网站（https://tfhub.dev/）选择所需的模型。首先寻找模型的格式，然后在 Saved Model 下，找到模型的网址链接，如图 4-4 所示。

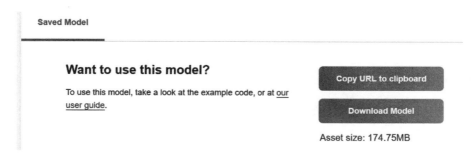

图 4-4　模型的网址链接

复制上述网址并在程序代码中使用它，如下所示。

```
module_url = "https://tfhub.dev/google/imagenet/mobilenet_
v2_100_160/feature_vector/4"
```

```
my_model = hub.KerasLayer(module_url)
```

可以向 my_model 添加更多层，然后，像训练其他模型一样，使用 fit 方法训练新模型。训练完成后，即可将其用于预测未知数据。

了解了预训练模型的功能之后，是时候尝试一下了。

4.3 ImageNet 分类器

本项目将使用由 Google 提供的名为 mobilenet_v2 的 ImageNet 分类器，该分类器提供 1001 种不同的分类类别。该模型在超过一百万张图像上进行了训练，初始学习率为 0.045，学习衰减率为每个 epoch 0.98，批量大小 (batch size) 为 96，使用了 16 个 GPU 异步线程训练。可以想象创建此模型所用的资源量是巨大的。

在此模型的基础上，我们将使用迁移学习将这些知识迁移到自己构建的图像分类器中。本章接下来的两个项目中都将执行此操作。第一个项目介绍如何使用 MobileNet 模型对自己的图像进行分类。第二个项目介绍如何扩展 MobileNet，以添加更多自定义的分类。

首先展示如何使用 ImageNet 分类器对自己的图像进行分类。在这个项目中，我们按原样使用 MobileNet 分类器，而不向其添加任何层。我们将学习加载预训练模型并将其应用于自己的图像。本章第二个项目（犬种分类器）将详细介绍如何将分类层加入 ImageNet 模型，然后进行大量实验以了解迁移学习带来的好处。

首先在 Colab 中创建一个名为 ImageNetClassifier 的项目。

4.3.1 创建项目

使用如下代码片段设置 TensorFlow 并导入所需的包。

```
import tensorflow as tf

# other imports
import tensorflow_hub as hub
import numpy as np
import matplotlib.pyplot as plt
```

我们需要导入 tensorflow_hub。另外，导入在之前项目中使用过的 numpy 和 matplotlib 包，用于矩阵运算和图像显示。

4.3.2 分类器 URL

本项目使用预训练的 MobileNet 分类器。在 TensorFlow Hub 上找到这个分类器，tfhub.dev 站点上搜索结果的屏幕截图如图 4-5 所示。

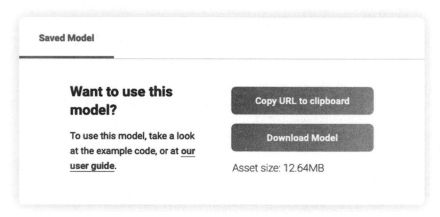

图 4-5 TensorFlow Hub 上的 MobilNet 模型

单击"Copy URL to clipboard"（复制 URL 到剪贴板）按钮复制模型的 URL，获得网址为"https://tfhub.dev/google/tf2-preview/mobilenet_v2/classification/4"。

> **注意**
>
> 读者获得模型的版本可能会比此处显示的更高。

在项目代码中声明如下两个变量。

```
classifier_url = "https://tfhub.dev/google/tf2-preview/mobilenet_v2/classification/4"
IMAGE_SHAPE = (224, 224)
```

该模型的输入需要尺寸为 224 像素 ×224 像素的图像，如模型描述中所述。模型描述可在用户指南中获取，通过单击如图 4-5 中所示的"our user guide"（用户指南）阅读。

4.3.3 创建模型

接下来，使用 Sequential API 创建模型，如下所示。

```
# create a model
classifier = tf.keras.Sequential([
    hub.KerasLayer(classifier_url, input_shape = IMAGE_SHAPE+(3,))
])
```

根据上面的代码，hub.KerasLayer 返回一个模块，并添加到 Sequential 模型中。上述模块中包含的所有层现在都可应用于自定义的网络模型。本节不向此模型中添加更多自己的层，因此，模型定义现已完成。由于这是一个经过充分训练的模型，因此不需要在分类器上调用 fit 方法来进一步训练，只需调用模型的 predict 方法即可预测任何给定的输入图像。在调用 predict 方法之前，先准备一些图像以输入到模型中。

4.3.4 准备图像

本书的下载网站已上传了一些图像，以下列出了这些图像的 URL。将此代码片段添加到项目中。

```
# set up URLs for image downloads
image_url1 = "https://raw.githubusercontent.com/Apress/artificialneural-
networks-with-tensorflow-2/main/Ch04/bulck_cart.jpg"
image_url2 = "https://raw.githubusercontent.com/Apress/artificialneural-
networks-with-tensorflow-2/main/Ch04/flower.jpg"
image_url3 = "https://raw.githubusercontent.com/Apress/artificialneural-
networks-with-tensorflow-2/main/Ch04/swordweapon.jpg"
image_url4 = "https://raw.githubusercontent.com/Apress/artificialneural-
networks-with-tensorflow-2/main/Ch04/tiger.jpg"
image_url5 = "https://raw.githubusercontent.com/Apress/artificialneural-
networks-with-tensorflow-2/main/Ch04/tree.jpg"
```

使用以下代码将图像下载到磁盘驱动器中。

```
# download images
!pip install wget
import wget
wget.download(image_url1,'image1.jpg')
wget.download(image_url2,'image2.jpg')
wget.download(image_url3,'image3.jpg')
wget.download(image_url4,'image4.jpg')
wget.download(image_url5,'image5.jpg')
```

使用 wget 下载上述五个图像文件，分别以 image1，image2，…，image5 为文件名存储在磁盘驱动器的 content 文件夹中。

由于这些图像大小不一，将它们加载到内存后需要调整图像尺寸为 224 像素 ×224 像素。还需要将图像数据的数值范围归一化为 0～1，以便更好地进行机器学习。以下代码完成了上述所有操作。

```
# load images and reshape to 224x224 required by the model
import PIL.Image as Image

image1 = tf.keras.utils.get_file("/content/image1.jpg",
image_url1)
image1 = Image.open(image1).resize(IMAGE_SHAPE)
# load images and reshape to 224x224 required by the model
import PIL.Image as Image

image1 = tf.keras.utils.get_file("/content/image1.jpg",
```

```
image_url1)
image1 = Image.open(image1).resize(IMAGE_SHAPE)
# scale the array
image1 = np.array(image1)/255.0

image2 = tf.keras.utils.get_file("/content/image2.jpg",
image_url2)
image2 = Image.open(image2).resize(IMAGE_SHAPE)
image2 = np.array(image2)/255.0

image3 = tf.keras.utils.get_file("/content/image3.jpg",
image_url3)
image3 = Image.open(image3).resize(IMAGE_SHAPE)
image3 = np.array(image3)/255.0

image4 = tf.keras.utils.get_file("/content/image4.jpg",
image_url4)
image4 = Image.open(image4).resize(IMAGE_SHAPE)
image4 = np.array(image4)/255.0

image5 = tf.keras.utils.get_file("/content/image5.jpg",
image_url5)
image5 = Image.open(image5).resize(IMAGE_SHAPE)
image5 = np.array(image5)/255.0
```

至此，已准备好预测图像。从第一幅图像开始，调用模型的 predict 方法来预测图像：

```
result = classifier.predict(image1[np.newaxis, ...])
```

现在，可以从结果张量（Tensor）中获得概率预测结果。使用 result 变量的 shape 方法打印预测结果的维度如下。

```
result.shape
(1, 1001)
```

可以看到，结果数组（result）中有 1001 个值，表明输出结果中有 1001 个类别。概率预测结果按升序排列，所以我们选择最后一个值作为模型的最大预测值。

```
predicted_class = np.argmax(result[0], axis=-1)
predicted_class
293
```

可以看到，预测的类别是 293。这个整数对我们来说没有任何意义，但幸运的是，Google 提供了一个从整数到图像类别标签的映射，以将预测结果对应的数值映射为图像标签。

4.3.5 加载标签映射

图像类别标签的名称可在 Google 站点上的 ImageNetLabels.txt 文件中找到。使用以下代码将图像类别标签加载到程序中。

```
labels_path = tf.keras.utils.get_file('ImageNetLabels.
txt','https://storage.googleapis.com/download.tensorflow.org/
data/ImageNetLabels.txt')
imagenet_labels = np.array(open(labels_path).read().
splitlines())
```

使用以下两个语句打印图像标签的名称和标签数组的长度。

```
print (imagenet_labels)
print ("Number of labels: " , len(imagenet_labels))
```

```
['background' 'tench' 'goldfish' ... 'bolete' 'ear' 'toilet
tissue']
Number of labels: 1001
```

输出信息的第一行显示了标签数组开头和结尾的部分名称，第二行信息说明共有 1001 个标签名称，每个名称都代表了模型输出的类别。

4.3.6 显示预测结果

使用以下代码绘制图像和预测类别名称，以显示预测结果。

```
plt.imshow(image1)
plt.axis('off')
predicted_class_name = imagenet_labels[predicted_class]
_ = plt.title("Prediction: " + predicted_class_name.title())
```

预测结果图如图 4-6 所示。

可以看到，分类模型正确地预测了图像，显示图像中为一只老虎。接下来，编写类似于预测 image1 的代码对其他四张图像进行预测。

为此，使用以下预测和显示图像的函数。

```
def predict_display_image(imagex):
    result = classifier.predict(imagex[np.newaxis, ...])
    predicted_class = np.argmax(result[0], axis=-1)
    plt.imshow(imagex)
    plt.axis('off')
    predicted_class_name = imagenet_labels[predicted_class]
    _ = plt.title("Prediction: " + predicted_class_name.title())
```

图 4-6　测试图像及其预测结果

该函数接收预处理后的图像作为参数，对其进行预测，并同时显示预测结果与图像。在剩余的四幅图像上调用此函数，如下所示。

```
# predict and print results for the specified image
predict_display_image(image2)
predict_display_image(image3)
```

```
predict_display_image(image4)
predict_display_image(image5)
```

四幅图像的预测结果如图 4-7 所示。

图 4-7　剩余四幅图像的预测结果

在讨论图 4-7 中的预测结果之前，首先查看一下 ImageNet 分类器的类别列表。

4.3.7　列出所有类别

使用如下代码查看 ImageNet 预测的所有类别列表。

```
for i in range (len(imagenet_labels)):
 print (imagenet_labels[i])
```

输出类别的前几项如下：background、tench、goldfish、great white shark、tiger shark、hammerhead、electric ray、stingray、cock、Hen。

接下来，将讨论如图 4-7 所示的预测结果。

4.3.8　结果讨论

如图 4-7 所示的第一幅图像被预测为牛车，这个结果是正确的。第二幅图像被预测为一个开信刀，但它的外观更接近一把剑（该图像的实际类别是剑）。由于开信刀的外观类似于一把剑，所以这个预测结果是可接受的。第三幅图像是一棵树，但被预测为花盆。最后一幅图像是一束玫瑰花，但被预测为粉饼，完全错误。那么，从以上讨论中可以得到什么结论呢？为了正确区分一个粉饼和一束玫瑰花，需要使用更多图像对 ImageNet 进行进一步训练，这显然超出了我们的能力范围。但是，可以扩展 ImageNet，以添加我们自己的分类。这就是下一个话题中即将展示的内容。

4.4　犬种分类器

假设有一个以品种为标签的犬类图片数据集，但可能无法对每一类品种收集足够的图像。如果想要开发一个模型来根据品种对犬类进行区分，可以考虑使用前面程序中讨论的 ImageNet 分类器来提取犬类图像的特征。然后，在此基础上添加一个分类层，根据提取的特征来区分犬类的品种。简言之，我们将迁移 ImageNet 中学习到的知识对犬的品种进行分类，并省去训练模型过程中的特征提取部分。

图像分类模型通常有数百万个参数，因此从头开始训练模型是一项极具挑战的任务。

这需要海量的训练数据，且计算成本十分高。迁移学习通过获取一个已经预训练的模型来缩短训练任务，只需在预训练模型之上添加分类层即可。

本项目将演示如何基于 TensorFlow Hub 上的预训练模型 MobileNet V2 来构建自定义的 Keras 模型以对犬种进行分类。MobileNet V2 用于图像的特征提取，其输入图像尺寸为 (224,224,3)。最后，MobileNet V2 的输出将作为自定义模型密集层的输入用于图像分类。

4.4.1 项目简介

本项目将使用迁移学习来构建对不同犬种进行分类的新模型。该模型是一个多类别分类器，将给定的犬类图像分为多个预定义的类别。我们可以通过具体示例区分二元分类和多元类，犬与猫的分类或者人与马的分类是二元分类，而特斯拉在其自动驾驶汽车中使用的是多元图像分类。

本项目将使用来自 Kaggle 犬种识别竞赛（https://www.kaggle.com/c/dog-breedidentification/overview）的数据集。该数据集包含 10000 多张标记犬类图像，共 120 个不同品种。在数据预处理过程中，需要将图像数据转换为张量，因为机器学习模型需要在输入张量中提取特征。作为一个竞赛，Kaggle 分别提供了训练和测试数据，但 Kaggle 没有为测试数据提供类别标签。

4.4.2 创建项目

创建一个新的 Colab 项目，将其重命名为 DogBreedClassifier。使用以下两个语句导入 TensorFlow 和 TensorFlow Hub。

```
import tensorflow as tf
import tensorflow_hub as hub
```

接下来，将数据加载到项目中。

4.4.3 加载数据

本书已将整个数据集（来源：www.kaggle.com/c/dog-breedidentification/data）保存在了项目站点上，可以运行以下代码片段将所需数据下载到项目中（需在 Mac 或 Linux 下执行）。

```
! wget --no-check-certificate -r 'https://drive.google.com/uc?export=download&id=11t-eBwdXU9EWriDhyhFBuMqHYiQ4gdae' -O dogbreed
```

注意

> 在前面的下载中，假定下载站点是受信任的。读者若有任何疑虑，可以使用自己的工具下载文件，并存入项目的相应路径下以输入到模型中。

下载的 ZIP 文件包含用于训练和测试的标签及图像。ZIP 文件的结构如图 4-8 所示。

图 4-8　下载数据的文件夹结构

使用以下命令解压缩下载的文件（需在 Mac 或 Linux 下执行）。

!unzip dogbreed

ZIP 文件还包含训练图像的标签。可以通过加载 labels.csv 文件并打印前五个记录来查看标签。

```
# Checkout the labels
import pandas as pd
labels_csv = pd.read_csv("/content/labels.csv")
labels_csv.head()
```

图 4-9 显示了上述代码的输出结果。

	id	breed
0	000bec180eb18c7604dcecc8fe0dba07	boston_bull
1	001513dfcb2ffafc82cccf4d8bbaba97	dingo
2	001cdf01b096e06d78e9e5112d419397	pekinese
3	00214f311d5d2247d5dfe4fe24b2303d	bluetick
4	0021f9ceb3235effd7fcde7f7538ed62	golden_retriever

图 4-9　部分犬种标签

要描述表格的内容，可在数据流（dataframe）上调用 describe 方法。执行的命令及其对应的输出如图 4-10 所示。

从输出可以看出，有 10222 幅图像和 120 个类别。要打印数据的分布信息，可以使用如下语句。

```
# How many images are there of each breed?
labels_csv["breed"].value_counts().plot.bar(figsize=(20, 10));
```

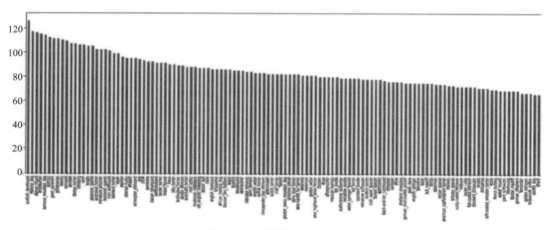

图 4-10 标签的表结构

上述命令产生如图 4-11 所示的图表。

图 4-11 图像数据分布

如图 4-11 所示，每个类别都有超过 60 幅图像。Google 建议在图像分类任务中每个类别至少使用 10 幅图像。图像的数量越多，找出图像之间模式的概率就越大。

使用如下命令从数据集中打印示例图像。

```
from IPython.display import
display, Image
Image("/content/train/000bec180
eb18c7604dcecc8fe0dba07.jpg")
```

输出的图像如图 4-12 所示。

至此，我们获取到了用于训练和测试模型的所有数据，这些数据需要转换为特定格式后才能输入到神经网络模型中。

图 4-12 下载数据集的示例图像

4.4.4 设置图像和标签

现在，设置图像的存储路径以准备训练数据，使用以下两个语句完成。

```
# Define our training file path for ease of use
train_path = "/content/train"

# Create path names from image ID's
filenames = [train_path + '/'+fname + ".jpg" for fname in labels_csv["id"]]
```

通过显示一幅图像来验证文件路径设置是否正确，语句如下。

```
# Check an image directly from a file path
Image(filenames[9000])
```

可以看到如图 4-13 所示的图像。

我们已经设置了要处理的图像数据，接下来设置图像标签。从 labels_csv 读取标签并将其转换为 numpy 数组代码如下。

```
import numpy as np
labels = labels_csv["breed"].to_numpy()
```

我们知道，标签的数量是 10222，需要从这些标签中提取唯一值，使用以下代码。

图 4-13　新路径下的示例图像

```
unique_breeds = np.unique(labels)
len(unique_breeds)
```

上述代码输出为 120，表明该数据集中有 120 个类别，即 120 个犬类品种。通过调用 list 方法打印此列表代码如下。

```
# print class names
list(unique_breeds)
```

部分输出如图 4-14 所示。

接下来，使用以下代码对数值介于 0 到 n_classes-1 之间的标签进行编码。

```
# Encode target labels with values between 0 and 120
from sklearn.preprocessing import LabelEncoder
labels = LabelEncoder().fit_transform(labels).reshape(-1,1)
labels
```

输出如图 4-15 所示。

由图 4-15 可以看出，第一个犬种被归类为 #19，第二个犬种被归类为 #37，以此类推。接下来使用 OneHotEncoder 将这些类别数值转换为用于模型训练的列（columns），代码如下。

```
# Use one-hot encoding to convert categorical values
from sklearn.preprocessing import OneHotEncoder
```

```
boolean_labels = OneHotEncoder().fit_transform(labels).
toarray()
```

```
['affenpinscher',
 'afghan_hound',
 'african_hunting_dog',
 'airedale',
 'american_staffordshire_terrier',
 'appenzeller',
 'australian_terrier',
 'basenji',
 'basset',
 'beagle',
 'bedlington_terrier',
 'bernese_mountain_dog',
 'black-and-tan_coonhound',
 'blenheim_spaniel',
 'bloodhound',
 'bluetick',
```

```
array([[19],
       [37],
       [85],
       ...,
       [ 3],
       [75],
       [28]])
```

图 4-14　部分类别列表　　　　　　　　图 4-15　已编码的标签

通过打印数组中的一组数值来查看独热编码（one-hot）的效果，代码如下。

```
Boolean_labels[5]
```

上述代码的输出如图 4-16 所示。

```
array([0., 0., 0., 0., 0., 0., 0., 0., 0., 0., 1., 0., 0., 0., 0., 0.,
       0., 0., 0., 0., 0., 0., 0., 0., 0., 0., 0., 0., 0., 0., 0., 0.,
       0., 0., 0., 0., 0., 0., 0., 0., 0., 0., 0., 0., 0., 0., 0., 0.,
       0., 0., 0., 0., 0., 0., 0., 0., 0., 0., 0., 0., 0., 0., 0., 0.,
       0., 0., 0., 0., 0., 0., 0., 0., 0., 0., 0., 0., 0., 0., 0., 0.,
       0., 0., 0., 0., 0., 0., 0., 0., 0., 0., 0., 0., 0., 0., 0., 0.,
       0., 0., 0., 0., 0., 0., 0., 0., 0., 0., 0., 0., 0., 0., 0., 0.,
       0.])
```

图 4-16　one-hot 编码后的一个标签数据示例

注意

编码完成后，图像标签现在有 120 个字段（列）。在标签数组中，给定图像的所属类别对应列的数值为 1，其余列的数值为 0。

对标签进行预处理并为机器学习做好准备后，接下来将对图像进行预处理并为模型的学习做好准备。

4.4.5 图像预处理

到目前为止，已将标签转换为数字格式。但是，图像仍然只是文件名的形式，因此需要读取图像数据并将其转换为适合模型输入的格式。我们以张量的形式表示图像数据，以便使用 GPU 进行更快的处理。为此，需编写一个函数来执行以下操作。

① 以图像文件名作为函数输入。
② 加载二进制格式的图像（JPEG 文件）。
③ 将图像数据转换为张量。
④ 将图像尺寸调整为 (224, 224)。
⑤ 返回图像张量。

由于 MobileNet 模型需要尺寸为 (224, 224, 3) 的图像输入，其中 3 表示 RGB 维度，因此，需要将图像大小调整为 224 像素 ×224 像素。函数的分步分析如下。

首先，为特征（features）和标签设置变量，代码如下。

```
# Setup variables
X = filenames
y = boolean_labels
```

其次，将训练数据拆分为训练集和验证集，代码如下。

```
from sklearn.model_selection import train_test_split
# Split them into training and validation
X_train, X_val, y_train, y_val = train_test_split(
    X, y, test_size=0.2, random_state=42)
```

在训练文件夹中的 10222 幅图像中，使用 2045 幅图像用于验证。

接下来，将编写一个对图像进行预处理的函数，即从给定路径加载图像、读取其数据并将其转换为张量。

4.4.6 处理图像

首先，为图像的尺寸定义一个变量。

```
IMG_SIZE = 224
```

定义一个名为 "process_image" 的函数，代码如下。

```
def process_image(image_path):
```

该函数以一个参数作为函数输入，即图像文件的路径。要读取图像，使用 tf.io 的 read_file 函数，代码如下。

```
image = tf.io.read_file(image_path)
```

read_file 函数将 JPEG 图像中的数据读取到二进制数组中。读取到的数据需使用 tf.image 中的 decode_jpeg 方法进行解码，代码如下。

```
image = tf.image.decode_jpeg(image, channels=3)
```

调用 tf.image 的 convert_image_dtype 方法，将图像数据转换为浮点型数值，代码如下。

```
image = tf.image.convert_image_dtype(image, tf.float32)
```

调用 tf.image 的 resize 方法来调整图像的尺寸,代码如下。

```
image = tf.image.resize(image, size=[IMG_SIZE, IMG_SIZE])
```

最后,将处理后的图像数据返回给调用者,代码如下。

```
return image
```

process_image 函数的完整代码如清单 4-1 所示。

清单 4-1 process_image 函数

```
# Define image size
IMG_SIZE = 224

def process_image(image_path):
    """
    Takes an image file path and turns it into a Tensor.
    """
    # Read image file
    image = tf.io.read_file(image_path)
    # Turn the jpeg image into numerical Tensor
    image = tf.image.decode_jpeg(image, channels=3)
    # Convert the color channel values from 0-225 values to 0-1 
    values
    image = tf.image.convert_image_dtype(image, tf.float32)
    # Resize the image to our desired size (224, 244)
    image = tf.image.resize(image, size=[IMG_SIZE, IMG_SIZE])
    return image
```

源码清单
链　接:https://pan.baidu.com/s/1NV0rimQ_8kRz22xfFHN-Cw
提取码:1218

接下来,将编写一个函数将图像与其对应标签关联起来。

4.4.7　关联图像与标签

本节编写一个函数,将图像路径作为参数,通过调用 process_image 函数处理图像,并将标签与图像关联。函数定义如下。

```
# Create a simple function to return a tuple (image, label)
def get_image_label(image_path, label):
    """
    Takes an image file path name and the associated label,
    processes the image and returns a tuple of (image, label).
    """
    image = process_image(image_path)
    return image, label
```

该函数返回一个 Python 元组(tuple),包含张量形式的图像及其对应的标签。

下一步,编写一个函数将数据转换为输入管道。

4.4.8 创建数据批次

本节首先介绍什么是批次（batch）。批次即整个数据（图像及其标签）的一小部分。通常情况下，一个批次的大小为 32，这意味着一个批次中有 32 幅图像和 32 个对应的标签。在深度学习中，通常不是在整个数据集中同时查找模式（pattern），而是分批次地查找。一次性加载和处理多于 10000 幅的图像需要大量内存和处理能力，因此，需要以小批量的方式处理数据。本节将编写一个函数，用于将数据分成批次并创建由图像数据及其对应标签组成的张量。创建数据批次的函数如清单 4-2 所示。

清单 4-2 用于创建数据批次的函数

源码清单
链　接：https://pan.baidu.com/s/1NV0rimQ_8kRz22xfFHN-Cw
提取码：1218

```
# Define the batch size, 32 is a good default
BATCH_SIZE = 32

# Create a function to turn data into batches
def create_data_batches(x, y = None, batch_size = BATCH_SIZE, data_type = 1):
  """
  Creates batches of data out of image (x) and label (y) pairs.
  Shuffles the data if it's training data but doesn't shuffle it
  if it's validation data.
  Also accepts test data as input (no labels).
  """
  # If the data is a test dataset, we don't have labels
  if data_type == 3:
    print("Creating test data batches...")
    data = tf.data.Dataset.from_tensor_slices((tf.constant(x)))
        # only filepaths
    data_batch = data.map(process_image).batch(BATCH_SIZE)
    return data_batch
  # If the data if a valid dataset, we don't need to shuffle it
  elif data_type == 2:
    print("Creating validation data batches...")
    data = tf.data.Dataset.from_tensor_slices((tf.constant(x),
                                              # filepaths
                                              tf.constant(y)))
                                              # labels
    data_batch = data.map(get_image_label).batch(BATCH_SIZE)
    return data_batch

  else:
    # If the data is a training dataset, we shuffle it
    print("Creating training data batches...")
    # Turn filepaths and labels into Tensors
    data = tf.data.Dataset.from_tensor_slices((tf.constant(x),
                                              # filepaths
                                              tf.constant(y)))
                                              # labels
```

```
# Shuffling pathnames and labels before mapping image
  processor function
# is faster than shuffling images
data = data.shuffle(buffer_size = len(x))

# Create (image, label) tuples
# (this also turns the image path into a preprocessed image)
data = data.map(get_image_label)

# Turn the data into batches
data_batch = data.batch(BATCH_SIZE)

return data_batch
```

根据输入数值,该函数在训练集、验证集和测试集上创建数据批次。在训练集中,通过打乱数据顺序以实现数据的一些随机性。from_tensor_slices 函数将数据转换为输入管道;map 函数将 (image, label) 元组转换为数据输入管道;batch 函数将数据集拆分为数据批次。

下一小节,将编写一个函数来显示集合中的一些图像,这个函数将在测试期间起到作用。

4.4.9 显示图像函数

本节编写一个函数用来显示数据集中的 25 幅图像。清单 4-3 显示了函数 show_25_images 的完整代码。假设读者熟悉 matplotlib 绘图工具包,因此代码没有给出过多注释。

清单 4-3 图像显示函数

```
# Function for viewing images in a data batch
import matplotlib.pyplot as plt
def show_25_images(images, labels):
    """
    Displays 25 images from a data batch.
    """
    # Setup the figure
    plt.figure (figsize = (10, 10))
    # Loop through 25 (for displaying 25 images)
    for i in range(25):
      # Create subplots (5 rows, 5 columns)
      ax = plt.subplot(5, 5, i+1)
      # Display an image
      plt.imshow(images[i])
      # Add the image label as the title
      plt.title(unique_breeds[labels[i].argmax()])
      # Turn grid lines off
      plt.axis("off")
```

源码清单

链　接:https://pan.baidu.com/s/1NV0rimQ_8kRz22xfFHN-Cw

提取码:1218

至此,我们已准备好用于数据预处理和显示的各种功能。下一小节,将构建用于犬种分类的机器学习模型。

4.4.10　选择预训练模型

如前文所述，我们将在模型定义过程中使用迁移学习。为此，需要在 TensorFlow Hub 中寻找合适的模型。TensorFlow Hub 将模型分为不同的类别，在图像分类类别下可以发现有一些模型供我们选择。其中一些模型只支持 TensorFlow，由于本书使用的是 TensorFlow 2.0，因此需选择支持 TensorFlow 2.0 的模型。在撰写本书时，mobilenet_v2_130_224 是一个满足需求的模型。模型选择的屏幕截图如图 4-17 所示。

图 4-17　从 TensorFlow Hub 中选择 MobileNet

查看模型文档，可以发现模型的输出图像尺寸为 (224, 224, 3)。

4.4.11　定义模型

在定义模型时，需要以下三个重要信息：输入数据的维度（张量形式的图像）、所需的类别数量（犬种数量）、要使用的预训练模型 URL。

使用以下代码片段声明这些变量。

```
# Setup input shape to the model
INPUT_SHAPE = [None, IMG_SIZE, IMG_SIZE, 3] # batch, height, width, colour channels

# Setup output shape of the model
OUTPUT_SHAPE = len(unique_breeds) # number of unique labels

# Setup model URL from TensorFlow Hub
MODEL_URL = "https://tfhub.dev/google/imagenet/mobilenet_v2_130_224/classification/4"
```

To define the model, we will write a function called create_model:

```
def create_model(input_shape=INPUT_SHAPE,
                 output_shape=OUTPUT_SHAPE,
                 model_url=MODEL_URL):
  print("Building model with:", MODEL_URL)
```

可以看到，create_model 函数以输入维度、输出维度和预训练模型的 URL 作为参数。create_model 是一个构建模型的函数，以便稍后试验不同的预训练模型，这些模型可能需要不同的输入和输出维度。

接下来，使用 Sequential API 在模型中定义两个层，代码如下。

```
# Setup the model layers
model = tf.keras.Sequential([
  hub.KerasLayer(MODEL_URL), # TensorFlow Hub layer
  tf.keras.layers.Dense(units=OUTPUT_SHAPE,
                        activation="softmax") # output layer
])
```

第一层是源自 TensorFlow Hub 的整个预训练模型。第二层为 softmax 分类层，将犬种分为 120 个类别。

使用以下语句编译模型。

```
# Compile the model
model.compile(
    loss=tf.keras.losses.CategoricalCrossentropy(),
    optimizer=tf.keras.optimizers.Adam(),
    metrics=["accuracy"]
)
```

可以看出，模型以 CategoricalCrossentropy 为损失函数，以 Adam 为优化器，以准确率为评价指标评估模型性能。

调用 build 方法来构建模型，代码如下。

```
model.build(INPUT_SHAPE)
```

将编译后的模型返回给调用者，代码如下。

```
return model
```

清单 4-4 显示了 create_model 函数的完整定义。

清单 4-4 create_model 函数

```
# Create a function which builds a Keras model
def create_model(input_shape=INPUT_SHAPE,
                 output_shape=OUTPUT_SHAPE,
                 model_url = MODEL_URL):
  print("Building model with:", MODEL_URL)
F
  # Setup the model layers
```

源码清单

链　接：https://pan.baidu.com/s/1NV0rimQ_8kRz22xfFHN-Cw

提取码：1218

```python
model = tf.keras.Sequential([
    hub.KerasLayer(MODEL_URL), # TensorFlow Hub layer
    tf.keras.layers.Dense(units=OUTPUT_SHAPE,
                          activation="softmax") # output layer
])

# Compile the model
model.compile(
    loss=tf.keras.losses.CategoricalCrossentropy(),
    optimizer=tf.keras.optimizers.Adam(),
    metrics=["accuracy"]
)

# Build the model
model.build(INPUT_SHAPE)

return model
```

创建模型变量并按如下方式打印模型摘要来测试此函数。

```python
model = create_model()
model.summary()
```

上述代码生成的模型摘要如图 4-18 所示。

```
Building model with: https://tfhub.dev/google/imagenet/mobilenet_v2_130_224/classification/4
Model: "sequential"
_____
Layer (type)                 Output Shape              Param #
=================================================================
keras_layer (KerasLayer)     multiple                  5432713
_____
dense (Dense)                multiple                  120240
=================================================================
Total params: 5,552,953
Trainable params: 120,240
Non-trainable params: 5,432,713
_____
```

图 4-18　模型摘要

至此，我们已配置好分类模型准备训练。下一节，将创建数据集来训练此模型。

4.4.12　创建数据集

我们已将训练数据拆分为训练集和验证集，现在只需要使用之前定义的 create_data_batches 函数对这些数据进行预处理，代码如下。

```python
train_data = create_data_batches(X_train, y_train)
Val_data = create_data_batches(X_val, y_val)
```

如果想直观查看数据集中的一些图像，可以使用之前开发的 show_25_images 函数，代码如下。

```
train_images, train_labels = next(train_data.as_numpy_
iterator())
show_25_images(train_images, train_labels)
```

显示结果如图 4-19 所示。

图 4-19 数据集中的示例图像

读者运行代码时可能会显示不同的图像，因为程序每次运行都会打乱训练数据的顺序。

至此，用于模型训练的数据已准备就绪。在训练模型之前，还需要做最后一件事，即设置 TensorBoard 用于训练过程的分析。

4.4.13 设置 TensorBoard

首先，加载 TensorBoard。

```
%load_ext tensorboard
```

然后，清除 TensorBorad 上次运行的日志（如果有）：

```
!rm -rf ./logs/  # cleaning the previous log
```

创建一个回调函数，在每个 epoch 之后调用：

```python
import datetime
import os

# Create a function to build a TensorBoard callback
def create_tensorboard_callback():
    # Create a log directory for storing TensorBoard logs
    logdir = os.path.join("logs",
                          # Timestamp the log
    datetime.datetime.now().strftime("%Y%m%d-%H%M%S"))
    return tf.keras.callbacks.TensorBoard(logdir)
```

> **注意**
>
> 该函数创建了一个日志路径并将当前时间存储在了每条日志上。

接下来，使用 EarlyStopping 函数监控验证集的分类准确率。以下代码创建了两个变量，用于日志的记录和验证集准确率的监控。

```python
# TensorBoard callback
model_tensorboard = create_tensorboard_callback()

# Early stopping callback
model_early_stopping = tf.keras.callbacks.EarlyStopping(monitor
="accuracy",
# stops after 3 rounds
# of no improvements
                                    patience=3)
```

将这些值传递给 fit 方法中的 callbacks 参数。如果训练期间验证集的准确率在最近的 3 次循环中没有太大提升（由 patience 参数定义），则停止训练。

接下来，将编写训练模型的代码。

4.4.14 训练模型

首先，调用之前定义的 create_model 函数构建模型：

```
model = create_model()
```

其次，为模型训练期间使用的迭代轮数（epochs）定义一个变量：

```
NUM_EPOCHS = 100
```

之前提到，训练 epoch 的数值非常大。本节将向读者展示 TensorFlow 2.0 的一个重要特性，即当模型准确率达到饱和时，模型会自动停止训练，表明进一步的训练无助于获得更高的准确率。 实际进行模型训练时，我们会注意到训练周期会在设置的 100 个 epoch 之前停止。

接下来，调用 fit 方法开始训练模型：

```
model.fit(x = train_data,
          epochs = NUMBER_OF_EPOCHS,
          validation_data = val_data,
          callbacks = [model_tensorboard,
                       model_early_stopping])
```

训练过程中，屏幕上会详细地显示训练输出。本书的模型训练在 14 次迭代后停止，停止训练前的进度截图如图 4-20 所示。

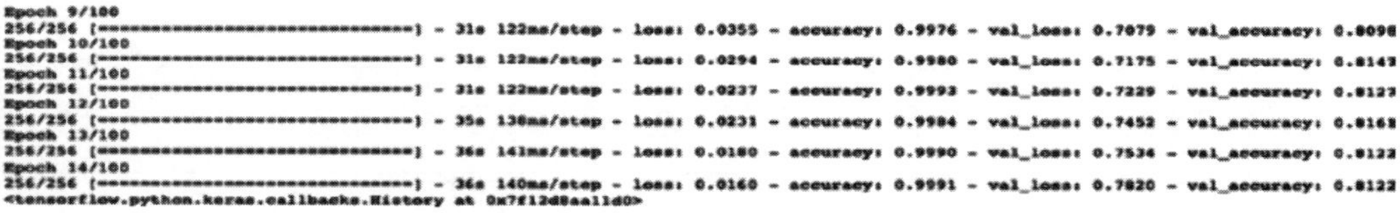

```
Epoch 9/100
256/256 [==============================] - 31s 122ms/step - loss: 0.0355 - accuracy: 0.9976 - val_loss: 0.7079 - val_accuracy: 0.8098
Epoch 10/100
256/256 [==============================] - 31s 122ms/step - loss: 0.0294 - accuracy: 0.9980 - val_loss: 0.7175 - val_accuracy: 0.8143
Epoch 11/100
256/256 [==============================] - 31s 122ms/step - loss: 0.0237 - accuracy: 0.9993 - val_loss: 0.7229 - val_accuracy: 0.8127
Epoch 12/100
256/256 [==============================] - 35s 138ms/step - loss: 0.0231 - accuracy: 0.9984 - val_loss: 0.7452 - val_accuracy: 0.8161
Epoch 13/100
256/256 [==============================] - 36s 141ms/step - loss: 0.0180 - accuracy: 0.9990 - val_loss: 0.7534 - val_accuracy: 0.8122
Epoch 14/100
256/256 [==============================] - 36s 140ms/step - loss: 0.0160 - accuracy: 0.9991 - val_loss: 0.7820 - val_accuracy: 0.8122
<tensorflow.python.keras.callbacks.History at 0x7f12d8aa11d0>
```

图 4-20 早停（EarlyStopping）之前的训练输出

注意

在 12、13、14 次迭代中，验证集的准确率分别为 0.8161、0.8122、0.8122，表明训练达到饱和。验证数据的准确率会显示模型对于训练集之外样本的分类性能。

4.4.15 训练日志

使用以下命令在 Colab 环境中打开 TensorBoard。

```
%tensorboard --logdir logs
```

如图 4-21 所示的屏幕截图显示了在训练集和验证集上的准确率和损失值。

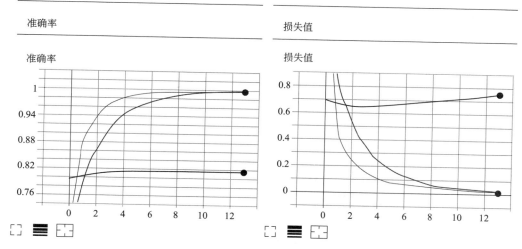

图 4-21　准确率和损失值

4.4.16　验证模型性能

通过调用模型的 evaluate 方法来评估模型的性能：

```
model.evaluate(val_data)
```

运行结果显示了 81.22% 的准确率，如图 4-22 所示。

```
64/64 [==============================] - 7s 108ms/step - loss: 0.7830 - accuracy: 0.8122
[0.7830359155777842, 0.8122249]
```

图 4-22　模型评价结果

下一步，我们将对测试图像进行预测，以查看模型的性能是否令我们满意。

4.4.17　预测测试图像

下载的数据集中有一个单独的 test 文件夹，包含用于测试的所有图像。在程序中设置测试图像的存储路径并调用之前的 create_data_batches 函数来准备数据集。

```
# set up path to test images
test_path = "/content/test"
test_filenames = [test_path +'/'+ fname for fname in os.listdir(test_path)]
# prepare test dataset
test_data = create_data_batches(test_filenames, data_type = 3)
```

调用已训练模型的 predict 方法对图像进行预测：

```
# Make predictions
test_predictions = model.predict(test_data,
                                 verbose=1)
```

运行上述代码后将得到一组预测结果，通过打印其维度来查看此数组的大小：

```
test_predictions.shape
```

输出结果为 (10357,120)，表明共分析了 10357 幅图像。数组中的每一行都包含一个 120 列的数组，其对应的索引值表示犬类概率。例如，以下代码行打印了测试数据中第一幅图像的预测结果。

```
test_predictions[0]
```

输出结果如图 4-23 所示。

```
array([[1.71625504e-06, 2.09769784e-08, 4.42356409e-08, 3.60783137e-09,
        6.83360646e-10, 3.30916805e-08, 2.28535613e-09, 5.69529739e-08,
        1.12093312e-09, 1.28260762e-08, 8.87338480e-10, 1.12767259e-06,
        2.72967480e-08, 1.75928699e-06, 7.62196764e-11, 2.46849332e-08,
        2.15978602e-07, 2.86165952e-10, 3.18566293e-08, 5.58064350e-09,
        3.46805145e-08, 2.05301305e-08, 1.93678787e-07, 1.10779412e-08,
        4.09313365e-08, 5.94809269e-09, 2.17959713e-08, 5.79143240e-08,
        4.13444889e-10, 1.95744946e-07, 1.79607351e-09, 1.95080951e-09,
        1.01097652e-09, 6.40962469e-07, 4.43499848e-10, 2.93363597e-11,
        1.50097392e-08, 3.50538730e-11, 1.70451464e-09, 8.80133584e-08,
        1.56203242e-07, 7.27263028e-10, 9.05519479e-11, 8.22270516e-11,
        2.50760905e-07, 3.76196425e-08, 4.52130638e-10, 3.47058382e-09,
        1.23620314e-09, 3.41306472e-09, 7.42496198e-09, 4.11943439e-08,
        7.67591057e-09, 1.03463926e-09, 5.51114709e-09, 3.05814609e-08,
        7.21103248e-08, 5.35525457e-10, 3.54033114e-09, 5.31774624e-10,
        8.41837249e-08, 5.17785429e-06, 1.08501794e-07, 1.11155103e-08,
        4.11728540e-10, 1.00259268e-09, 2.02230996e-08, 7.91160970e-10,
        5.65018032e-11, 8.55700288e-10, 1.55159352e-09, 6.42869891e-09,
        4.19546797e-09, 3.46384006e-08, 5.19582727e-07, 2.08382822e-08,
        1.77174229e-07, 3.15960480e-09, 3.40442496e-08, 9.31709287e-09,
        1.71757530e-09, 1.13689913e-09, 8.17984847e-10, 1.01354489e-08,
        9.99965549e-01, 1.57419588e-08, 2.75883593e-07, 1.29112177e-05,
        3.02005532e-09, 9.60315050e-10, 4.79819462e-10, 2.03560968e-09,
        1.59662164e-07, 3.72057656e-08, 1.54300839e-09, 1.36166378e-07,
        4.08241085e-09, 2.83839636e-08, 1.39534592e-07, 2.19978847e-06,
        1.49835174e-07, 1.88222664e-08, 3.91109688e-06, 6.00249006e-09,
        5.31374695e-08, 2.38341569e-10, 1.84611926e-09, 1.26122968e-09,
        5.42101697e-10, 2.70874096e-07, 1.22152613e-07, 1.76050214e-08,
        2.92270439e-08, 4.73288964e-09, 2.40588860e-09, 6.44375362e-08,
        3.84022556e-08, 1.86857574e-09, 2.89643296e-08, 6.90489230e-07],
       dtype=float32)
```

图 4-23　第一幅图像的 120 个预测概率值

使用 numpy 的 argmax 函数找出最大概率及其对应的索引值。使用以下代码片段打印第一幅图像的预测结果。

```
# the max probability value predicted by the model
print(f"Max value: {np.max(predictions[0])}")
# the index where the max value in predictions[0] occurs
print(f"Max index: {np.argmax(predictions[0])}")
# the predicted label
print(f"Predicted label: {unique_breeds[np.argmax(predictions[0])]}") # the predicted label
```

输出结果如下。

```
Max value: 0.9999655485153198
Max index: 84
Predicted label: papillon
```

下一节，我们将编写代码，以打印犬类的图像，输出预测的类别，并显示前 10 个预测结果的分布图。这样，我们可以更好地可视化与解释测试结果。

4.4.18 可视化测试结果

本节将在结果图中的每一行创建两个图。第一个图显示输入犬类的图像及其预测的类别名称，第二个图显示类别概率分布的条形图。

第一个结果图的绘图函数如清单 4-5 所示。

清单 4-5 用于显示输入犬类的图像及其预测的类别名称的函数

```
def plot_pred(prediction_probabilities, images):
    image = process_image(images)
    pred_label = unique_breeds[np.argmax(prediction_
probabilities)]
    plt.imshow(image)
    plt.axis('off')
    plt.title(pred_label)
```

源码清单
链　接：https://pan.baidu.com/s/1NV0rimQ_8kRz22xfFHN-Cw
提取码：1218

函数 plot_pred 的输入参数为预测的概率（prediction_probabilities）和预测图像（images）。在函数体内，通过调用 process_image 函数检索和绘制图像；通过在 unique_breeds 数组中检索图像的标签，然后将检索到的标签作为绘图标题打印出来。

第二个结果图的绘图函数如清单 4-6 所示。

清单 4-6 用于显示类别概率分布条形图的函数

```
def plot_pred_conf(prediction_probabilities):
    top_10_pred_indexes = prediction_
probabilities.argsort()
[-10:][::-1]
    top_10_pred_values = prediction_probabilities[top_10_pred_
indexes]
    top_10_pred_labels = unique_breeds[top_10_pred_indexes]
    top_plot = plt.bar(np.arange(len(top_10_pred_labels)),
                       top_10_pred_values,
                       color="grey")
    plt.xticks(np.arange(len(top_10_pred_labels)),
               labels=top_10_pred_labels,
               rotation="vertical")
    top_plot[0].set_color("green")
```

根据清单 4-6，首先对预测结果数组进行排序，从排序数组中选取最后 10 个结果即可获得前 10 个预测类别的索引。然后绘制条形图，并将对应标签作为垂直文本显示于 x 轴。

使用如下代码片段打印前三幅图像及其预测结果。

```
num_rows = 3
plt.figure(figsize = (5 * 2, 5 * num_rows))
for i in range(num_rows):
    plt.subplot(num_rows, 2, 2*i+1)
```

```
    plot_pred(prediction_probabilities=predictions[i],
              images=test_filenames[i])
    plt.subplot(num_rows, 2, 2*i+2)
    plot_pred_conf(prediction_probabilities=predictions[i])
plt.tight_layout(h_pad=1.0)
plt.show()
```

结果如图 4-24 所示。

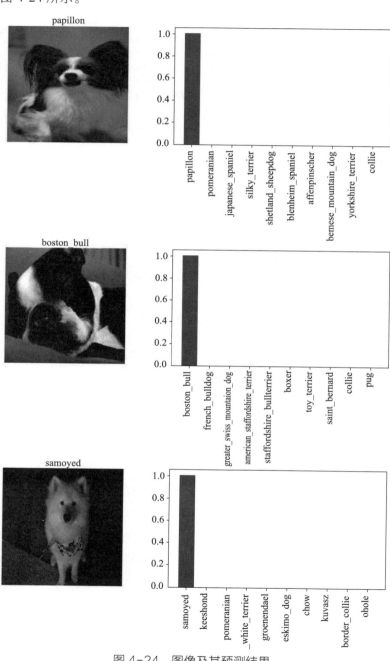

图 4-24　图像及其预测结果

4.4.19 预测未知图像

给训练完的模型输入一幅虎的图像，如图 4-25 所示，神经网络会预测出什么结果呢？

为此，首先使用以下代码片段从本书的下载网站加载图像。

```
!pip install wget
url='https://raw.githubusercontent.com/Apress/artificialneural-networks-with-tensorflow-2/main/Ch04/tiger.jpg'
import wget
wget.download(url,'tiger.jpg')
```

图 4-25 测试图像

其次，调用 create_data_batches 函数为模型准备图像：

```
data=create_data_batches(['/content/tiger.jpg'],batch_size=1,data_type=3)
```

然后，调用模型的 predict 方法进行预测：

```
result = model.predict(data)
```

获取预测结果的类别及其名称：

```
predict_class_index = np.argmax(result[0],axis=-1)
predict_class_name = unique_breeds[(predict_class_index)]
```

调用模型的 predict_proba 方法来获得预测的概率结果：

```
result_proba = model.predict_proba(data,batch_size=None)
```

然后，查看预测结果的最大值。如果该值小于某个阈值，则可以得出给定图像不是犬类的结论。

```
if result_proba.max() > 0.7:
 print(pred_label)
else:
 print('Not a dog breed as the predicted probability is {}'.format(result_proba.max()))
```

上述代码的运行结果如图 4-26 所示。

```
1  # Check the threshold for prediction value
2  if result_proba.max() > 0.7:
3    print(pred_label)
4  else:
5    print('Not a dog breed as the predicted probability is {}'.format(result_proba.max()))

Not a dog breed as the predicted probability is 0.24044956266880035
```

图 4-26 未知图像的预测结果

现在，我们能够体会到迁移学习的作用。由于 MobileNet 模型是针对各种类别进行训练的，因此可以利用其学习到的知识推断给定的输入图像不是犬。如果输入的是犬类图像，模型将使用我们自定义的分类层将其品种分为已知的 120 个类别之一。

下一节，将展示迁移学习的另一个重要用途，即使用较小的数据集训练模型。

4.4.20 使用小数据集训练

由于在很多情况下难以收集海量的数据，因此本节将验证是否能够通过迁移学习来辅助我们在小数据集上开发可应用的模型。

首先，编写一个名为 train_model 的函数，使其可以被多个模型调用。函数定义如下。

```
model_performances = []

# function for training the given model on specified
# number of images
def train_model (model, NUM_IMAGES):
 model.fit(x=train_data,
           epochs=NUM_EPOCHS,
           validation_data=val_data,
           callbacks=[model_tensorboard,
                      model_early_stopping])
 # append the results
 model_performances.append(model.evaluate(val_data))
```

函数 train_model 接收两个参数：第一个参数 (model) 指定要训练的模型；第二个参数（NUM_IMAGES）指定训练模型使用的图像数量。该函数通过调用模型的 fit 方法进行训练，然后调用模型的 evaluate 方法来评价分类性能，最后将评价结果添加到一个数组中以供比较。

使用以下代码片段调用该函数。

```
# Training
NUM_EPOCHS = 100
# Create models and test for 1000,2000, 3000, 4000 images
for NUM_IMAGES in range(1000, 5000, 1000):
 model = create_model()
 x_train,x_val,y_train,y_val=train_test_split(X[:NUM_
IMAGES],y[:NUM_IMAGES],test_size=0.2,random_state=10)
 train_data=create_data_batches(x_train,y_train,batch_size=10)
 val_data=create_data_batches(x_val,y_val,batch_size=10,data_
type=2)
 train_model(model,NUM_IMAGES)
 model_performances.append(model.evaluate(val_data))
```

根据上述代码，在调用 train_model 函数之前，首先通过调用 create_model 函数创建了一个模型，然后将训练数据拆分为指定数量的图像作为训练集和验证集。我们像前文一样预处理图像数据，然后在处理后的图像数据上训练创建的模型。在训练完成之后，评价结果存储在一个数组中。最后，创建一个 pandas 包下的数据流（dataframe），用于存储损失值和准确率这两个评价指标，代码如下。

```
import pandas as pd
comp = pd.DataFrame(model_performances,index =
[1000,2000,3000,4000], columns = ['val_loss', 'val_acc'])
```

使用如下代码打印和显示预测结果。

```
# plot the table
import matplotlib.pyplot as plt
plt.style.use('ggplot')
comp.plot.bar()
plt.xlabel('number of images')
plt.ylabel('performance values')
plt.show()
```

输出结果如图 4-27 所示。

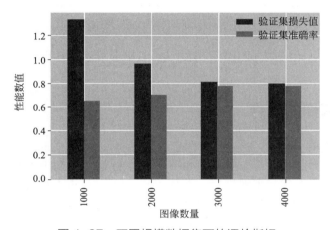

图 4-27　不同规模数据集下的评价指标

从图 4-27 可以看出，分类准确率在大约 2000 幅图像后饱和。因此，与在 10000 幅图像的完整数据集上训练模型相比，我们可以在更小的数据集上训练模型，且能获得可接受的预测结果。这是迁移学习的一大优势，特别是当没有充足的数据用于训练模型时。

至此，我们已经完成了构建犬种分类的新模型，并希望将其投入实际应用。下一节，将展示如何保存经过训练的模型，再利用该模型来预测未知的图像。

4.4.21　保存、加载模型

调用模型的 save 方法将训练好的模型保存为 h5 格式：

```
model.save('model.h5') #saving the model
```

调用 load_model 方法加载已保存的模型：

```
from tensorflow.keras.models import load_model
model=load_model('model.h5',custom_objects={"KerasLayer":hub.
KerasLayer})
```

通过查看模型的摘要信息确保模型已正确加载：

```
model.summary()
```

输出结果如图 4-28 所示。

```
Model: "sequential_5"
_____
Layer (type)                 Output Shape              Param #
=================================================================
keras_layer_5 (KerasLayer)   multiple                  5432713
_____
dense_5 (Dense)              multiple                  120240
=================================================================
Total params: 5,552,953
Trainable params: 120,240
Non-trainable params: 5,432,713
_____
```

图 4-28　加载模型的摘要信息

至此，我们可以利用前文的方法和已加载的模型对未知图像进行预测，请读者自行尝试。

4.5　提交你的工作

TensorFlow 允许用户提交自己的作品（网络模型）到 TensorFlow Hub 存储库中。tensorflow_hub 库用于从 TensorFlow Hub 存储库加载模型。基于 HTTP 的协议允许用户检索模型文档并提供获取模型的链接。要从 TensorFlow Hub 存储库中加载模型，可使用本章示例中介绍的 load_model 方法。我们可以创建自己的模型存储库，这些模型可使用 tensorflow_hub 库加载。为此，我们的 HTTP 分发服务需要遵循某种协议。

当网络模型训练完成且达到满意的程度后，使用如下代码保存计算图（graph）和模型参数值。

```
saver = tf.train.saver()
saver.save (sess, 'my_model')
```

然后，可以将保存的模型放在生产服务器上以供公共服务使用。经过 Beta 测试后，可以通过 TensorFlow Hub 将模型添加到存储库中以获取社区利益。规范的更多细节信息超出了本书的范围，感兴趣的读者可通过 TensorFlow 站点上的 "Hosting Your Own Models"（托管您自己的模型）了解更多相关信息。

4.6　进一步工作

本章，我们演示了一个基于 TensorFlow Hub 预训练模型的图像分类示例。如前文所述，TensorFlow Hub 也提供了很多其他领域的模型。例如，我们可以轻松开发自己的目标检测分类器、文本分类器、通用语句编码器等。建议读者访问 TensorFlow Hub 站点以探索更多预训练模型。

TensorFlow 站点上列出了一些商业部署的案例。例如，Airbnb 通过重新训练 ResNet50 模型对其网站上的图像进行分类，以改善用户体验。有人使用 TensorFlow.js 使得 Amazon Echo 可以响应手语以造福残疾人。可口可乐在其移动购买凭证应用程序中使用 SqueezeNet CNN 进行光学字符识别（OCR）。谷歌像素（Google Pixel）手机使用 TensorFlow Hub 的 MobileNet 模块进行相机图像识别以增强其相机功能。在我们自己的应用程序中，使用迁移学习将有无限可能。因此，请读者继续试验 TensorFlow Hub 中提供的预训练模型，并将想法用到自己的应用程序中。

总结

迁移学习是机器学习中重要的知识迁移技术。在软件工程中，人们使用二进制库来重用代码。在机器学习中，经过训练的模型包含算法、数据、处理能力和专家的专业知识，所有这些都需要迁移到新模型中，这就是迁移学习所提供的内容。在 TensorFlow Hub 中，有众多可用的预训练模型，供不同领域和开发人员使用。本章，我们首先学习了重用一个预训练模型来对自己的图像进行分类，然后我们开发了一个多元图像分类器，该分类器运行在由 Google 预训练的模型之上，以利用其专业知识、处理时间和处理能力。

下一章，我们将通过一个回归模型的开发示例继续探索机器学习方法。

第 5 章
使用神经网络处理回归问题

到目前为止，我们研究了深度学习中的分类模型，那么我们能否将迄今为止在深度学习中学到的技术应用于回归问题呢？由于深度学习需要较大开销，那么是否值得在回归领域尝试深度学习技术呢？与传统统计技术相比，特别是在回归建模的问题上，深度学习是否具有优势呢？本章中将对上述问题及类似问题做出解答。

朴素的线性回归是机器学习中最简单的问题，且通常情况下回归问题是机器学习中的第一个话题。统计回归模型已经存在多年，并在许多实际应用中提供了有效的解决方案。这些模型同样适用于工业、商业及科学领域。同时，我们迄今为止在本书中所学的深度学习网络被认为已较成功地解决了很多复杂的问题。这些神经网络模型提供了非常准确的预测结果，其训练过程类似于我们大脑的学习过程。现在的问题是我们是否可以使用深度学习网络来解决回归问题？除了可以区别于传统统计编码技术之外，使用深度学习解决回归问题是否还能获得优势？对上述问题的简要回答是肯定的，其原因和方式是本章将要讨论的内容。

首先，让我们定义什么是回归。

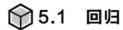

5.1 回归

在本节，我们首先介绍回归的定义，然后介绍其在现实中的应用，接下来解释回归的统计建模，最后介绍各种回归类型。

5.1.1 定义

在统计建模中，回归分析是一组统计过程。这些统计过程试图在因变量（目标）和自变量（特征）之间建立关系。因变量也称为结果变量，自变量有时称为预测变量或协变量。在简单的情况下，只有一个自变量；在复杂的数据分布中，会有多个自变量。最简单的回归模型是线性回归,即自变量和因变量之间为线性关系。但是,在很多实际情况下,自变量和因变量之间的关系为非线性,可以由一组直线的多项式组合表示。确定复杂数据集中的超平面是数据分析的一个巨大挑战。存在多种类型的回归,如线性回归、逻辑回归、逐步回归等。对于开发人员来说,估计变量之间的关系类型并对其进行编码并不是一项简单的任务。

那么,回归模型在哪些地方应用呢?

5.1.2 应用

回归分析主要应用于预测和推断因果关系。

使用线性回归模型估计房屋的价格,使用多元模型预测一只股票的未来价格,这些都是回归分析在预测领域的应用。而对于以下问题——什么事件导致了另一个事件或什么因素导致了某种变化?导致网站流量激增的原因、装配线故障的原因、药物是否导致某些医疗条件的改善——以上都是建立因果关系的例子。为了解决这些问题,数据分析师必须首先调查一种关系是否具有预测关系,以及为什么两个变量之间存在或者不存在因果关系?对于现代统计学家或数据分析师来说,回答上述问题并不是一件容易的事。神经网络将帮助我们回答这些问题,这就是本章即将介绍的内容。

5.1.3 回归问题

在回归问题中,我们的目标是根据给定的一组连续值,预测因变量的值或概率。与之前研究过的分类问题进行对比,分类问题的目标是在预定义的类别列表中选择一个类别。例如,在第 4 章中,我们构建了一个犬种分类器,其中输入图像被预测为一组 120 个预定义品种中的一个类别。在回归分析中,我们对因变量与一个或多个自变量之间的关系进行建模,这种关系可以用一个简单的数学方程表示:

$$\gamma = \beta_1 X_1 + \beta_2 X_2 + \beta_3 X_3 + ... + \beta_k X_k + \varepsilon$$

其中,γ 是要预测的因变量;X_1, X_2, …, X_k 是自变量;β_1, β_2, …, β_k 是神经网络中的系数或权重 ε 是机器学习术语中的偏置。接下来,让我们看一下回归问题的类型。

5.1.4 回归问题的类型

回归分析可分为线性回归、多项式回归、逻辑回归、逐步回归、岭回归、Lasso 回归、ElasticNet 回归。

线性回归适用于因变量和自变量之间存在线性关系(一条直线)的情况。在多项式回归中,因变量最好由多项式拟合,即一条曲线或一系列曲线。在此类模型中,异常值会扭曲预测,因此容易出现机器学习中所谓的过拟合。在逻辑回归中,预测值是二元的并且严格遵循二项分布。当有大量自变量也即数据维度较高时,可以使用逐步回归来检测哪些变量更重要,并删除不重要的变量以最大化模型的预测能力。当自变量高度相关时(也称为多重共线性),这些自变量会使方差大到足以导致预测值出现较大偏差。为解决此问题,岭回归技术在回归估计中添加了一个偏置项,以对模型的系数或权重值进行惩罚。岭回归使用最小二乘法来缩小系数以确保它们不会达到零,这是一种正则化形式——L2 正则化。Lasso 回归类似于岭回归,即通过缩小回归系数以解决多重共线性问题。但是,Lasso 缩小的是权重的绝对值而不是最小二乘。这也意味着一些模型权重可以缩小为零，以完全消除该特定节点的输出。这在特征选择中很有用，因为其本质是从一组因变量中挑选出一个因变量。这也是一种正则化——L1 正则化。最后，ElasticNet 回归是岭回归和 Lasso 回归的组合，该模型依次使用 L1 和 L2 正则化进行训练，从而在两种技术之间进行权衡。因此，ElasticNet 回归可能会选择多个相关变量。

5.2 神经网络中的回归问题

为了展示确实可以使用神经网络来解决最简单的回归问题，我们用一个简单的例子来验证这一点。使用来自 Kaggle 竞赛的数据集进行简单的线性回归，从以下链接下载数据集 (www.kaggle.com/luddarell/101-simple-linear-regressioncsv)。该数据集只包含两列内容——GPA 和 SAT。GPA 表示学生的平均绩点，SAT 表示学生的学术能力测验分数。我们将开发一个线性回归模型来建立学生 GPA 与 SAT 分数之间的关系。一旦模型训练完成，我们使用它来预测学生在给定 GPA 的情况下的 SAT 分数。

5.2.1 创建项目

创建一个新的 Colab 项目并将其命名为 LinearRegression。在项目代码中导入以下模块。

```
import tensorflow as tf
from tensorflow import keras
import pandas as pd
```

数据文件可在本书的下载站点上找到。由于需要使用 wget 下载项目中的文件，因此添加以下代码安装 wget。

```
!pip install wget
import wget
```

使用如下代码下载文件。

```
url = 'https://raw.githubusercontent.com/Apress/artificialneural-
networks-with-tensorflow-2/main/Ch05/student.csv'
wget.download(url,'data.csv')
```

下载的文件存储在 content 文件夹中，文件名为"data.csv"。首先将文件加载到 Pandas 模块的 dataframe 中，然后使用以下代码打印前几条内容以查看数据。

```
import pandas as pd
df=pd.read_csv('/content/data.csv')
df.head(10)
```

输出结果如图 5-1 所示。

接下来，我们将从该数据集中提取样本的特征和标签。

```
import pandas as pd
df=pd.read_csv('/content/data.csv')
df.head(10)
```

	SAT	GPA
0	1714	2.40
1	1664	2.52
2	1760	2.54
3	1685	2.74
4	1693	2.83
5	1670	2.91
6	1764	3.00
7	1764	3.00
8	1792	3.01
9	1850	3.01

图 5-1　已加载数据集中的样本

5.2.2　提取特征和标签

该数据集仅包含两列数据，即 GPA 和 SAT。我们使用 GPA 作为特征，使用 SAT 作为标签。注意：我们的任务是根据学生的 GPA 来预测学生的 SAT 分数。使用以下代码获取特征和标签。

```
# Extract features and label
dataset = df.values
x = dataset[:,1]
y = dataset[:,0]
```

为了简化程序，我们不创建训练集和验证集，并且不保留任何数据用于测试。接下来，我们将定义回归模型。

5.2.3　定义、训练模型

使用前文使用过的 Sequential API 来定义模型，代码如下。

```
model = tf.keras.Sequential([tf.keras.layers.Dense
           (units=1, input_shape=[1])])
```

该模型仅由一个单层神经元组成，模型输入为一维张量。使用如下代码编译模型。

```
model.compile(optimizer = 'sgd',
              loss = 'mean_squared_error')
```

该模型以随机梯度下降为优化器，以均方误差为损失函数。使用 fit 方法训练模型，代码如下。

```
model.fit(X, y, epochs = 15)
```

注意

在这个简单的示例中，我们没有捕获用于评价模型性能的指标。

5.2.4 预测

模型训练完成后，接下来进行预测。假设我们想了解 GPA 为 5.0 的学生在 SAT 上的分数，可以使用模型的 predict 方法进行预测，并按如下方式打印预测结果。

```
result = model.predict([5.0])
print("Expected SAT score for GPA 5.0: {:.0f}"
                .format(result[0][0]))
```

假设要了解 GPA 为 3.2 的学生在 SAT 上的分数，可以运行如下命令。

```
result = model.predict([3.2])
print("Expected SAT score for GPA 3.2: {:.0f}"
                .format(result[0][0]))
```

这里我们没有验证预测结果的准确性，但它向我们证明了一点：即使是最简单的线性回归问题，也可以使用神经网络来创建机器学习模型。

接下来，我们将处理回归分析中一个更实际的多重共线性的问题。

5.3 分析葡萄酒质量

在本项目中，我们使用回归分析根据某些特征来确定葡萄酒的质量。我们将使用 UCI 机器学习存储库（UCI machine learning repository）中提供的白葡萄酒质量数据集。该回归模型的目标是根据给定的基于物理、化学测试的输入特征预测葡萄酒的质量，这些特征包括非挥发性酸（fixed acidity）、挥发性酸（volatile acidity）、柠檬酸（citric acid）、剩余糖分（residual sugar）、氯化物（Chlorides）、游离二氧化硫（free sulfur dioxide）、总二氧化硫（total sulfur dioxide）、密度（density）、pH 值（pH）、硫酸盐（sulfates）、酒精（alcohol）。

数据集中的质量（Quality）字段作为标签，其数值介于 0 ～ 10 之间。

```
Output Label:
    Quality
```

5.3.1 创建项目

创建一个新的 Colab 项目并将其重命名为 WineQuality。加载 TensorFlow 2.x 并使用如下代码导入所需的模块。

```
import tensorflow as tf

import pandas as pd
import requests
import io
import matplotlib.pyplot as plt
```

5.3.2 数据准备

白葡萄酒质量数据集可在 UCI 网站上获得。

5.3.3 下载数据

我们将在代码中声明一个变量来引用 UCI 机器学习数据集，该数据集链接如下。

https://raw.githubusercontent.com/Apress/artificial-neuralnetworks-with-tensorflow-2/main/Ch05/winequality-white.csv

5.3.4 准备数据集

使用 pandas 模块将 CSV 文件读入数据流：

dataset = pd.read_csv(url , sep = ';')

可以调用 head 或 tail 方法显示数据样本，调用 tail 方法的结果如图 5-2 所示。

	fixed acidity	volatile acidity	citric acid	residual sugar	chlorides	free sulfur dioxide	total sulfur dioxide	density	pH	sulphates	alcohol	quality
4893	6.2	0.21	0.29	1.6	0.039	24.0	92.0	0.99114	3.27	0.50	11.2	6
4894	6.6	0.32	0.36	8.0	0.047	57.0	168.0	0.99490	3.15	0.46	9.6	5
4895	6.5	0.24	0.19	1.2	0.041	30.0	111.0	0.99254	2.99	0.46	9.4	6
4896	5.5	0.29	0.30	1.1	0.022	20.0	110.0	0.98869	3.34	0.38	12.8	7
4897	6.0	0.21	0.38	0.8	0.020	22.0	98.0	0.98941	3.26	0.32	11.8	6

图 5-2 数据集中的最后几行

从图 5-2 可以看出，数据库集有 4897 条记录（行），每条记录包含 12 个字段（列）。最后一列将作为标签使用，为葡萄酒的质量信息。使用以下两个语句提取特征和标签。

```
x = dataset.drop('quality', axis = 1)
y = dataset['quality']
```

接下来，我们将创建数据集。

5.3.5 创建数据集

首先，我们需要构建训练数据集和测试数据集，而训练数据集需进一步拆分为训练集和验证集。为了创建这些数据集，我们使用 sklearn 的 train_test_split 方法，代码如下。

```
# creating training, validation and testing datasets
from sklearn.model_selection import train_test_split
x_train_1 , x_test , y_train_1 , y_test =
    train_test_split
    (x , y , test_size = 0.15 , random_state = 0)
x_train , x_val , y_train , y_val =
    train_test_split(x_train_1 , y_train_1 ,
    test_size = 0.05 , random_state = 0)
```

可以看到，我们保留了 15% 的数据用于测试，并以 95∶5 的比例拆分训练数据——95% 用于训练，5% 用于验证。

一般来说，不同领域的数据在数值上会表现出很大的差异，如果将这些数据项缩小到固定比例，那么神经网络的学习效果会更好。因此，我们需要对整个数据进行预处理，主要包括以标准化或 Z-score 的方式对特征进行归一化操作。在本项目，我们重新调整特征值，以得到均值为 0、标准差为 1 的标准正态分布，如图 5-3 所示。

接下来，我们将通过缩放数据，以获得如图 5-3 所示的数据分布。

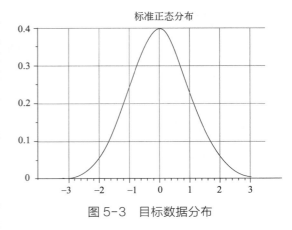

图 5-3　目标数据分布

5.3.6　数据归一化

为了缩放输入数据，更精确地说是对数据进行中心化，我们可以减去所有数据的均值，然后将结果再除以标准差：

$$x' = \frac{x - \mu}{\sigma}$$

其中，μ 为平均值，σ 是标准差。

首先对训练数据进行转换。使用 sklearn 的 StandardScaler 类并调用其 fit_transform 方法，如下面的代码所示。

```
from sklearn.preprocessing import StandardScaler
sc_x = StandardScaler()
x_train_new = sc_x.fit_transform(x_train)
```

通过绘制转换前后的原始数据来检查数据转换的效果。使用以下代码为非挥发性酸特征生成两个图。

```
fig, (ax1, ax2) = plt.subplots(
                  ncols = 2, figsize = (20, 10))

ax1.scatter(x_train.index,
            x_train['fixed acidity'],
            color = c,
            label = 'raw',
            alpha = 0.4,
            marker = m
            )

ax2.scatter(x_train.index,
            x_train_new[: , 1],
            color = c,
            label = 'adjusted',
            alpha = 0.4,
            marker = m
            )

ax1.set_title('Training dataset')
```

```
ax2.set_title('Standardized training dataset')

for ax in (ax1, ax2):
    ax.set_xlabel('index')
    ax.set_ylabel('fixed acidity')
    ax.legend(loc ='upper right')
    ax.grid()

plt.tight_layout()

plt.show()
```

上述代码的输出结果如图 5-4 所示。

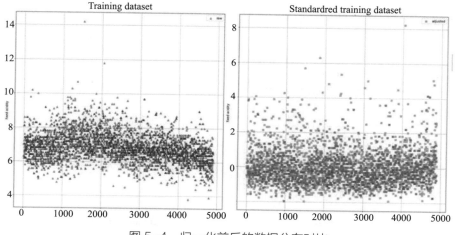

图 5-4　归一化前后的数据分布对比

图 5-4 中的左图显示了原始数据，可以看到非挥发性酸的数值范围为 0 ～ 14 之间，平均值为 7 左右。右图显示了转换后的相同数据，可以看到平均值约为 0，数值变化范围介于 –2 ～ +2 之间，表明标准偏差为 ±1。如果查看其他特征的数据分布图，也可以观察到平均值为 0，并且在 0 的两侧都等于标准差的均匀分布。

我们再尝试绘制一幅图——这次将 y 轴表示为总二氧化硫的数值，将 x 轴表示为剩余糖分的数值。使用以下代码生成图像。

```
fig, (ax1, ax2) = 
    plt.subplots(ncols = 2, figsize = (20, 10))

for l, c, m in zip(range(0, 2), 
                   ('blue', 'red'), ('^', 's')):

    ax1.scatter(x_train['residual sugar'],
                x_train['total sulfur dioxide'],
                color = c,
                label = 'class %s' % l,
                alpha = 0.4,
```

```
                marker = m
                )
for l, c, m in zip(range(0, 2),
                   ('blue', 'green'), ('^', 's')):
    ax2.scatter(x_train_new[: , 3],
                x_train_new[: , 6],
                color = c,
                label = 'class %s' % l,
                alpha = 0.4,
                marker = m
                )

ax1.set_title('Training dataset')
ax2.set_title('Standardized training dataset')

for ax in (ax1, ax2):
    ax.set_xlabel('residual sugar')
    ax.set_ylabel('total sulfur dioxide')
    ax.legend(loc ='upper right')
    ax.grid()

plt.tight_layout()

plt.show()
```

输出结果如图 5-5 所示。

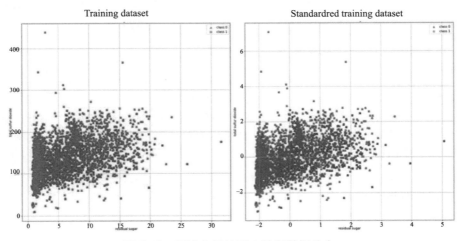

图 5-5　标准化后的两个特征数据分布

从图 5-5 中左图可以看出，剩余糖分数值在 0～30 之间变化，而总二氧化硫数值在 0～400 之间变化。图 5-5 中右图显示两个特征的平均值均为 0，其余数据点的平均分布在 −2～+2 之间。

这种数据转换有助于神经网络更好地学习，因此，我们对所有特征都进行了归一化转换。fit_transform 方法不仅转换了数据，还将 μ 和 σ 的数值存储在了内部变量中。显然，我们还需将训练集上的转换同样应用于测试集和验证集，因此，我们保留了 μ 和 σ 的数值以使用 fit_transform 方法。fit_transform 方法首先拟合数值，然后记住这些数值，最后进行数据转换。接下来，要转换测试集和验证集，只需调用 transform 方法而非 fit_transform 方法即可。使用以下两行代码完成上述操作。

```
x_test_new = sc_x.transform(x_test)
x_val_new = sc_x.transform(x_val)
```

至此，我们已将所有特征都归一化为平均值为 0，且由训练数据标准差决定的均匀分布。接下来，将介绍如何构建模型。

5.3.7 创建模型

构建模型之前，我们将编写一个小函数来可视化即将训练的几个模型的训练结果。

5.3.8 可视化评价函数

我们使用 Matplotlib 绘制分析指标，而没有使用 TensorBoard。读者如果想要换为 TensorBoard 显示，请注意可能需要在每次绘图和模型运行之后重置其状态，或使用单独的日志文件夹记录。对比发现，使用 Matplotlib 的绘图功能更方便。

清单 5-1 列出了完整的绘图函数。

清单 5-1 绘图函数

源码清单
链　接：https://pan.baidu.com/s/1NV0rimQ_8kRz22xfFHN-Cw
提取码：1218

```
import matplotlib.pyplot as plt
epoch = 30
def plot_learningCurve(history):
  # Plot training & validation accuracy values
  epoch_range = range(1, epoch+1)
  #plotting the mae vs epoch of training set
  plt.plot(epoch_range, history.history['mae'])
  #plotting the val_mae vs epoch of the validation dataset.
  plt.plot(epoch_range, history.history['val_mae'])
  
  plt.ylim([0, 2])
  plt.title('Model mae')
  plt.ylabel('mae')
  plt.xlabel('Epoch')
  plt.legend(['Train', 'Val'], loc = 'upper right')
  plt.show()
  
  print("--------------------------------
              -----------------------")
  
  # Plot training & validation loss values
  plt.plot(epoch_range, history.history['loss'])
  plt.plot(epoch_range, history.history['val_loss'])
  plt.ylim([0, 4])
  plt.title('Model loss')
  plt.ylabel('Loss')
  plt.xlabel('Epoch')
```

```
plt.legend(['Train', 'Val'], loc = 'upper right')
plt.show()
```

接下来,我们将开始练习构建回归模型。我们将定义三个复杂度逐渐增高的模型(小模型、中模型、大模型),然后训练不同模型并评估它们在测试数据上的性能。

> 需要使用相同的测试数据来评估不同模型的性能。

在本案例研究中,我们将为大模型尝试使用不同的优化器,同时介绍什么时候发生过拟合,以及如何检测过拟合,最后介绍一些关于如何减轻过拟合的策略。

5.3.9 小模型

在小模型中,我们构建一个只含有一个隐藏层的模型。

(1) 定义模型

该模型的输入为一个具有 11 维输入特征的张量,输出是一个一维数据,表示葡萄酒的质量。我们在隐藏层中设置神经元数量为 16,并在每个神经元之后使用 ReLU 函数激活。该模型使用如下语句定义。

```
small_model = tf.keras.Sequential([
                tf.keras.layers.Dense(16 ,
                    activation = 'relu' ,
                    input_shape = (11 , )),
                tf.keras.layers.Dense(1)
])
```

我们设置优化器为 Adam,设置损失函数为均方误差,并使用平均绝对误差进行分析。使用如下语句编译模型。

```
small_model.compile(optimizer = 'adam' ,
                    loss = 'mse' ,
                    metrics = ['mae'])
```

(2) 训练模型

我们调用模型的 fit 方法来训练模型,训练期间设置批量大小为 32。由于数据集中有多于 4000 条的样本,因此在训练期间可以创建足够多数量的批次。我们使用以下语句执行训练,并在 history_small 变量中捕获结果。

```
history_small = small_model.fit
                (x_train_new, y_train ,
                 batch_size = 32,
                 epochs = 30,verbose = 1 ,
                 validation_data =
                    (x_val_new , y_val))
```

(3) 评估模型

训练结束后,调用之前定义的绘图函数来绘制评价指标:

plot_learningCurve(history_small)

输出结果如图 5-6 所示。

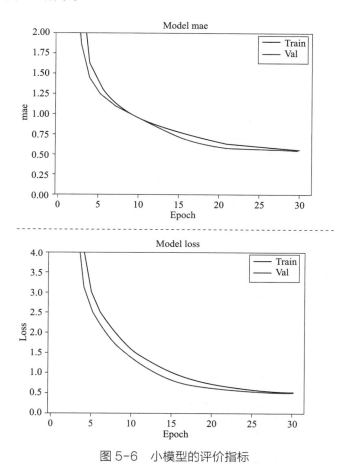

图 5-6 小模型的评价指标

调用模型的 evaluate 方法来评价模型在测试数据上的性能。调用方法及其输出如下所示。

```
s_test_loss , s_test_mae = small_model.evaluate
              (x_test_new , y_test ,
              batch_size = 32 , verbose = 1)
print("small model test_loss : {}"
              .format(s_test_loss))
print("small model test_mae : {} "
              .format(s_test_mae))

small model test_loss : 0.6353380084037781
small model test_mae : 0.6149751543998718
```

(4) 预测未知数据

为了评估模型在未知数据上的性能,我们需要创建一个未知的数据。为此,我们提取了 id 等于 2125 的测试数据,删除了其标签(葡萄酒质量),并使用以下代码创建了一个未知数据。

```
unseen_data = np.array([[6.0 , 0.28 , 0.22 , 12.15 ,
                         0.048 , 42.0 , 163.0 ,
                         0.99570 , 3.20 , 0.46 ,
                         10.1]])
```

对测试数据进行预测并打印结果,代码如下。

```
y_small = small_model.predict
                (sc_x.transform(unseen_data))
print ("Wine quality on unseen data
                (small model): ", y_small[0][0])

Wine quality on unseen data (small model): 5.618517
```

可以看到,葡萄酒质量的预测值为 5.618517,与数据集中 id 为 2125 的实际标签值 5.0 相近。

接下来,我们将继续构建一个更复杂的模型。

5.3.10 中模型

在这个模型中,我们将隐藏层的数量从 1 增加到 3,并且将每一层的神经元数量增加至 64。我们继续使用之前的 Adam 优化器,将 MSE 作为损失函数,将 MAE 作为模型评价指标。

(1) 定义、训练模型

以下代码用于定义模型、编译模型并对其进行训练。

```
medium_model = tf.keras.Sequential([
                tf.keras.layers.Dense
                  (64 , activation = 'relu' ,
                        input_shape = (11, )),
                tf.keras.layers.Dense
                  (64 , activation = 'relu'),
                tf.keras.layers.Dense
                  (64 , activation = 'relu'),
                tf.keras.layers.Dense(1)
])

medium_model.compile(loss = 'mse' ,
                optimizer = 'adam' ,
                metrics = ['mae'])

history_medium = medium_model.fit
                (x_train_new , y_train ,
                  batch_size = 32,
                  epochs = 30, verbose = 1 ,
                  validation_data =
```

```
                   (x_val_new , y_val))
```

在 medium_model 的定义中,我们简单地添加了两个 Dense 层,使其总共有三个隐藏层,每个隐藏层由 64 个神经元组成。和以前一样,每个神经元都以 ReLU 函数激活。添加上述这些层的原因是验证向神经网络中添加更多层是否能够获得更高的正确率。接下来,我们通过查看评价结果来测试一下。

(2) 评估模型

与前文一样,我们通过调用绘图函数来绘制评价指标:

```
plot_learningCurve(history_medium)
```

输出结果如图 5-7 所示。

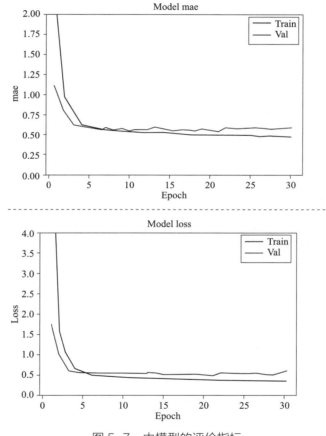

图 5-7 中模型的评价指标

由图 5-7 可以看出,训练在大约 5 个 epoch 之后饱和。回顾小模型的情况,我们直到大约 20 个 epoch 才看到饱和现象。在中模型中,我们也观察到一些过拟合的现象。如果图中显示的两条曲线出现分歧(训练集指标一直下降,验证集指标先下降后上升),则可能存在过拟合。

使用如下代码评估模型在测试数据上的性能。

```
m_test_loss , m_test_mae = medium_model.evaluate
                (x_test_new , y_test ,
                batch_size = 32 , verbose = 1 )
print("medium model test_loss : {}".format
                        (m_test_loss))
print("medium model test_mae : {}".format
                        (m_test_mae))
```

输出结果如下。

```
medium model test_loss : 0.6351445317268372
medium model test_mae : 0.6231803894042969
```

与小模型相比，中模型的 test_loss 和 test_mae 都降低了。最后，查看对未知数据的评价结果。

```
y_medium = medium_model.predict
                (sc_x.transform(unseen_data))
Print ("Wine quality on unseen data
                (medium model): ", y_medium[0][0])

Wine quality on unseen data (medium model): 5.246436
```

预测的葡萄酒质量数值 5.246436 与实际值 5.0 更加接近，且优于小模型的预测数值 5.618517。

接下来，我们继续定义一个更复杂的模型。

5.3.11 大模型

我们在中模型的基础上再添加两个隐藏层，使其总共有四个隐藏层，并且将每层的神经元数量增加到 128 个。

> 添加层数和神经元的数量会增加要调整的参数数量。

我们继续像以前一样使用 Adam、MSE 和 MAE。
（1）定义、训练模型

在定义和训练模型的代码中，除了增加两个隐藏层和神经元的数量之外，没有太大变化。以下列出了定义、训练模型的代码。

```
large_model = tf.keras.Sequential([
            tf.keras.layers.Dense
                (128 , activation = 'relu' ,
                input_shape = (11, )),
            tf.keras.layers.Dense
                (128 , activation = 'relu'),
```

```
                tf.keras.layers.Dense
                    (128 , activation = 'relu'),
                tf.keras.layers.Dense
                    (128 , activation = 'relu'),
                tf.keras.layers.Dense(1)
])

large_model.compile
            (loss = 'mse' , optimizer = 'adam' ,
            metrics = ['mae'])

history_large = large_model.fit
                (x_train_new , y_train ,
                batch_size = 32, epochs = 30,
                verbose = 1 , validation_data =
                (x_val_new , y_val))
```

接下来,我们观察一下模型的评估性能。

(2) 评估模型

调用 plot_learningCurve 方法,得到如图 5-8 所示的结果。

```
plot_learningCurve(history_large)
```

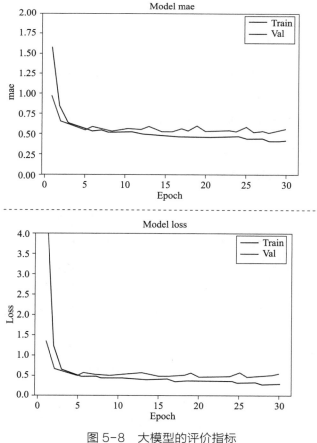

图 5-8　大模型的评价指标

从图 5-8 中，可以看到在大约 3 个 epoch 之后模型达到饱和，比前两种模型都早。我们还注意到，该模型的过拟合现象更加明显，下一节将介绍克服这种过拟合问题的技术。让我们首先观察测试数据上的评价结果：

```
l_test_loss , l_test_mae = large_model.evaluate
              (x_test_new , y_test ,
              batch_size = 32 , verbose = 1)
print("large model test_loss : {}"
              .format(l_test_loss))
print("large model test_mae : {}"
              .format(l_test_mae))

large model test_loss : 0.5520739555358887
large model test_mae : 0.5552783012390137
```

可以看到，损失值为 0.57，MAE 也为 0.57。稍后，我们将列出一个表格来比较三种模型的结果。让我们先查看模型在未见数据上的预测结果。

```
y_large = large_model.predict(sc_x.transform
              (np.array([[6.0 , 0.28 , 0.22 , 12.15 ,
                    0.048 , 42.0 , 163.0 ,
                    0.99570 , 3.20 , s0.46 ,
                    10.1]])))
Print ("Wine quality on unseen data (large model): ",
          y_large[0][0])

Wine quality on unseen data (large model): 5.389405
```

该模型预测的葡萄酒质量为 5.389405，略高于中模型预测的质量。本章最后将对不同结果进行全面讨论。接下来，让我们查看如何解决过拟合问题。

5.3.12 解决过拟合

在介绍解决过拟合问题的技术之前，我们首先讨论什么是过拟合、适拟合及欠拟合。
（1）什么是过拟合
图 5-9 直观地展示了欠拟合、适拟合和过拟合之间的区别。

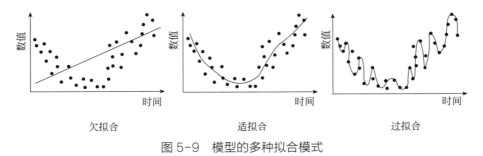

图 5-9 模型的多种拟合模式

通常情况下，当模型在训练期间正确率很高，但是在新数据上的正确率显著下降时，就说明模型发生了过拟合。换言之，该模型过分地拟合了训练数据，导致无法在新数据上泛化其预测结果。从图 5-9 可以看出，如果曲线不平衡，则有可能出现过拟合。减轻过拟合问题的技术之一是添加 dropout 层。

接下来，让我们尝试使用这个技术。

(2) 添加 dropout 层

在大模型中可以非常明显地观察到过拟合。我们尝试向该模型添加 dropout 层，以验证其是否可以减轻过拟合。以下代码用于建立添加了 dropout 层的新模型。

```
large_model_overfit = tf.keras.Sequential([
            tf.keras.layers.Dense
            (128 , activation = 'relu' ,
                input_shape = (11, )),
            tf.keras.layers.Dropout(0.4),
            tf.keras.layers.Dense
            (128 , activation = 'relu'),
            tf.keras.layers.Dropout(0.3),
            tf.keras.layers.Dense
            (128 , activation = 'relu'),
            tf.keras.layers.Dropout(0.2),
            tf.keras.layers.Dense
            (128 , activation = 'relu'),
            tf.keras.layers.Dense(1)
])
large_model_overfit.compile(loss = 'mse' ,
        optimizer = 'adam' , metrics = ['mae'])
history_large_overfit = large_model_overfit.fit
    (x_train_new , y_train , batch_size = 32,
     epochs = 30,verbose = 0 , validation_data =
    (x_val_new , y_val))
plot_learningCurve(history_large_overfit)
```

可以看到，我们在前三个 Dense 层之后分别添加了 40%、30% 和 20% 的 dropout 层，其余代码与大模型相同。然后即可查看这个模型产生的评价指标。图 5-10 显示了两个网络的评价指标图——没有 dropout 层的模型结果在左侧显示，带有 dropout 层的模型结果在右侧显示。

从图 5-10 中可以清楚地看到，添加 dropout 层消除或至少减轻了过拟合。

dropout 层通常被添加到大型网络模型中，因为它们在每一层都有充足数量的可用神经元，使得我们可以在训练模型过程中删除一些神经元以防止过拟合。

(3) 使用 RMSprop 优化

前文构建的小模型没有表现出任何过拟合，但仍然可以使用 RMSprop 优化器来优化其训练过程。以下代码片段设置了 RMSprop 优化器。

```
model_small = tf.keras.Sequential([
            tf.keras.layers.Dense(16 ,
                activation = 'relu' ,
                input_shape = (11 , )),
```

```
            tf.keras.layers.Dense(1)
])
optimizer = tf.keras.optimizers.RMSprop(0.001)
model_small.compile(loss = 'mse' , optimizer =
            optimizer , metrics = ['mae'])
history_small_overfit = model_small.fit
        (x_train_new , y_train , batch_size = 32,
            epochs = 30, verbose = 0 ,
            validation_data =
            (x_val_new , y_val))
plot_learningCurve(history_small_overfit)
```

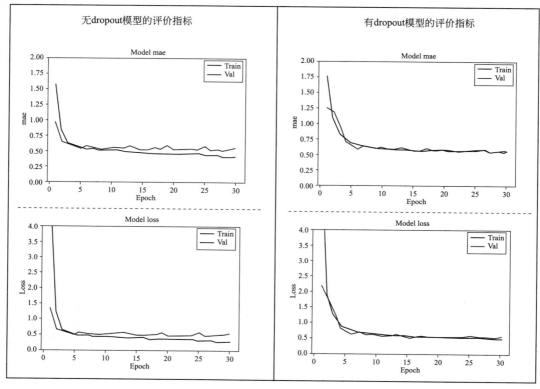

图 5-10　评价指标对比图

我们对小模型案例研究所做的唯一更改是将优化器由 Adam 更改为 RMSprop，并使用一个非常慢的学习率 0.001。图 5-11 显示了使用 RMSprop 优化器的结果。

图 5-11 的左图显示了使用 Adam 优化器的模型指标，右侧显示了使用 RMSprop 优化器的模型指标。与左图中的曲线相比，右图中的曲线更平滑，表明过拟合较少。此外，图 5-11 清晰地表明，RMSprop 显著减少了训练 epoch 的数量，这将减少整体训练时间，在大型数据集的情况下至关重要。此外，减少训练时间还可以使用第 4 章介绍的早停法（EarlyStopping）。

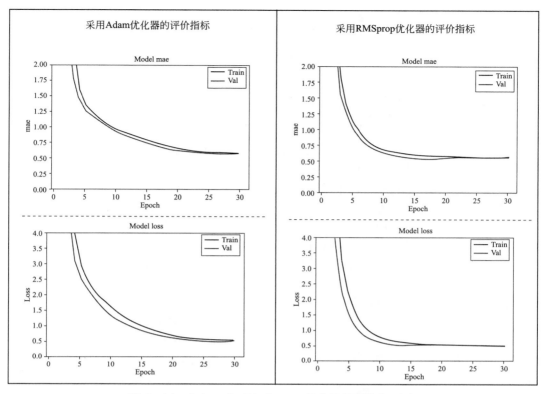

图 5-11　Adam 与 RMSprop 优化器误差指标对比

5.3.13　结果讨论

以下，我们展示了测试的三个模型的运行结果。注意，每次运行的结果会有所不同。

```
small model test_loss : 0.6353380084037781
small model test_mae : 0.6149751543998718
Wine quality on unseen data (small model): 5.618517
Trainable params: 209

medium model test_loss : 0.6351445317268372
medium model test_mae : 0.6231803894042969
Wine quality on unseen data (medium model): 5.246436
Trainable params: 9,153

large model test_loss : 0.5520739555358887
large model test_mae : 0.5552783012390137
Wine quality on unseen data (large model): 5.389405
Trainable params: 51,201
```

表 5-1 列出了不同模型的预测结果，以便进行快速比较。

表 5-1 不同模型的评价指标对比

	MSE	MAE	未知数据的葡萄酒质量预测结果	可训练的网络参数数量
小模型	0.6353	0.6149	5.6185	209
中模型	0.6351	0.6231	5.2464	9153
大模型	0.5520	0.5552	5.3894	51201

从表 5-1 中可以看出，随着模型复杂度的增加，MAE 的值先升高，后降低。但是，这种差异很小。同样地，不同模型对未知数据的预测结果几乎在同样的范围内。同时，增加模型的复杂度会导致过拟合。最后，如果查看模型的可训练参数，可以发现模型从 209 增加到惊人的 51201。那么，在这样一个简单的回归案例中，是否有必要使用复杂的模型呢？可以得出结论，在这种小数据集的情况下使用小模型即可满足要求。

总结本章至此的研究，我们可以得出结论，神经网络和深度学习技术可以用于解决回归问题。当存在大量具有不同范围的数值特征时，作为数据预处理的一部分，需要确保每个特征都独立地缩放到相同范围内。此外，如果没有充足的训练数据，则建议使用具有少量隐藏层的小型网络以避免过拟合。

在本章的回归项目中，我们使用 MSE 作为损失函数，使用 MAE 作为评价指标，这些指标在回归模型中非常常见。然而，tf.keras 库为回归模型提供了更多损失函数。接下来，我们将讨论其中的一些损失函数，使得在实验中可以尝试使用它们创建性能更好的模型。

5.4 损失函数

损失函数的主要功能是用来衡量机器学习模型预测结果的效果。没有一种损失函数可以应用于所有类型的数据，因此，根据数据的分布及异常值，我们需要使用不同的损失函数。用于回归问题的损失函数可能与用于分类问题的损失函数有所区别。本节将描述一些专用于回归问题的损失函数，以及它们适用于什么样的数据分布。主要包括以下五个损失函数。

① 均方误差 (mean squared error, MSE)、二次损失（quadratic loss）、L2 损失（L2 loss）。

② 平均绝对误差 (mean absolute error, MAE)、L1 损失（L1 loss）。

③ Huber 损失 (Huber loss)。

④ Log cosh 损失（Log cosh loss）。

⑤ 分位数损失（quantile loss）。

5.4.1 均方误差

MSE 可能是最常用的损失函数。在数学上，MSE 可表示为如下公式。

$$\text{MSE} = \frac{1}{N}\sum_{i=1}^{N}(y_i - \hat{y}_i)^2$$

可以看到，该损失函数的计算方式为首先将预测结果与真实标签之间的差值取平方，然后在整个数据集上取平均值。均方误差对于消除异常值非常有效,因为误差会因平方而被放大。MSE 在 tf.keras 库中以 tf.keras.losses.MSE 函数提供。

5.4.2 平均绝对误差

数学上,MAE 的表达方式如下。

$$\text{MAE} = \frac{1}{n}\sum_{j=1}^{n}\left|y_j - \hat{y}_j\right|$$

可以看到,MAE 首先计算预测结果和真实标签之间差的绝对值,然后在整个数据集上取平均值。该函数不会对异常值给予过多的关注,而是给所有训练数据点一个均匀度量。MAE 在 tf.keras 库中以 tf.keras.losses.MAE 函数提供。

5.4.3 Huber 损失

MSE 检测了异常值,而 MAE 忽略了异常值,在某些情况下,这两种损失都不能训练出理想的预测结果。假设有这样一个数据分布,其中 80% 的数据以 y_1 为真实标签,其余 20% 的数据以 y_2 为真实标签。由于 MAE 将所有数据点看作同等重要,因此会将 20% 的数据视为异常值;由于 MSE 放大了误差,因此在许多情况下可能会将 y_2 作为预测结果。Huber 损失提供一种介于 MAE 和 MSE 之间的解决方案。在数学上,Huber 损失表示为:

$$L_\delta\left(y, f(x)\right) = \begin{cases} \frac{1}{2}\left(y - f(x)\right)^2 & y - f(x) \leqslant \delta \\ \delta\left|y - f(x)\right| - \frac{1}{2}\delta^2 & \text{其他} \end{cases}$$

可以看到,当 δ 趋近于 0 时,Huber 损失趋近于 MAE;当 δ 趋近于无穷大时,Huber 损失趋近于 MSE。Huber 损失对异常值不太敏感,并提供了一种介于 MAE 和 MSE 之间的解决方案。Huber 损失函数在 tf.keras 库中以 tf.keras.losses.Huber 提供。

5.4.4 Log Cosh 损失

在数学上,Log Cosh 损失表示为:

$$L\left(y, y^p\right) = \sum_{i=1}^{n}\log\left(\cosh\left(y_i^p - y_i\right)\right)$$

可以看到,Log Cosh 损失计算了预测误差的双曲余弦的对数。对于小误差,log(cosh(x)) 趋近于 (error)^2/2;对于大误差,损失函数趋近于 abs(error) – log(error)。因此,在大多数情况下,Log Cosh 损失近似于 MSE。此外,该损失函数在任何位置都是二次可微的,且具有 Huber 损失的所有优点。Log Cosh 损失在 tf.keras 库中以 tf.keras.losses.LogCosh 函数提供。

5.4.5 分位数损失

我们可能希望使用多条回归线对如图 5-2 中左图所示的数据进行建模,而不是使用单条回归线建模,如图 5-12 中右图所示。在这种情况下,可以使用分位数损失。在数学上,分位数损失表示为:

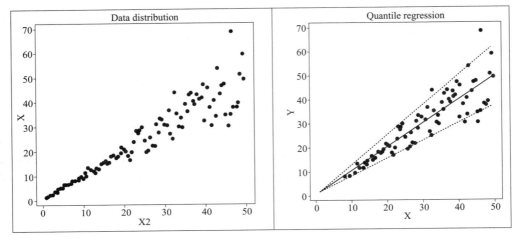

图 5-12　分位数回归图

$$L(y, y^p) = \sum_{i = y_i < y_i^p} (\gamma - 1)|y_i - y_i^p| + \sum_{i = y_i \geqslant y_i^p} \gamma |y_i - y_i^p|$$

其中，γ 为 0～1 之间的值。分位数损失实际上是 MAE 的扩展，当 γ 取 0.5 时，分位数损失即变为 MAE。分位数损失会对高估和低估的值都提供惩罚。当我们想要的预测结果为一个区间而不是一个点时，可以使用分位数损失函数，这可以显著改善许多业务问题的决策。

至此，我们介绍了回归任务中使用的各种损失函数。接下来，将介绍 tf.keras 中有哪些可用的优化器，这些优化器可以与上述损失函数一起使用。

5.5　优化器

优化器本质上是用于减小模型训练过程中损失的算法。优化器不断更新网络权重参数以最小化损失函数，使得当前的网络权重逐渐朝着正确的方向移动以达到全局最小值。与损失函数不同，优化器并不特定应用于回归，但所有的优化器都可以应用于回归问题。tf.keras 中可用的优化器包括 Adagrad、RMSprop、Adam、SGD、Adadelta、Adamax、Nadam。

我们已经在之前的项目中使用了其中一些优化器。每种优化器的设计都有其特殊目的，本书强烈建议读者阅读 tf.keras 文档以了解它们的重要性。根据需要，可以尝试使用不同的优化器来提高模型的性能。再次注意，这些优化器并非特定应用于回归问题，而是适用于所有深度学习项目。

总结

可以说，如果现有统计回归模型已经符合我们的需求，那么就无须使用神经网络。如果需要对一个复杂的数据集进行建模，且该数据集具有大量特征、特征与真实标签之间具有复杂的非线性关系，那么可以使用深度神经网络以获得更好的预测性能。传统的回归函数可以在 R 语言、scikit-learn 工具包及其他类似的库中获取。例如，除了线性回归之外，scikit-learn 工具包还提供了一些其他的回归器，如 KNeighboursRegressor、DecisionTreeRegressor 和 RandomForestRegressor。当然，相比于传统的回归函数，深度神经网络会有一些额外开销，但其提供的预测能力是任何传统回归函数都无法比拟的。通常情况下，我们很难找到一个完全适合给定数据集的回归方程，但是，深度神经网络会尝试自行找到最合适的回归方程，而无须我们付出额外努力。总而言之，即使是很小的回归问题，今后我们也可以考虑使用神经网络来解决，因为这个小问题（小数据集）在一段时间之后可能会变得很大。在这种情况下，如果坚持使用传统统计回归技术将会成为一场噩梦。

下一章，我们将介绍一些更快的机器学习技术。

第 6 章
Estimators
（估算器）

任何机器学习项目都包含多个阶段，包括训练、评价、预测，最后将其导出在生产服务器上以提供服务。我们在前面讨论分类与回归模型的章节中已经学习了上述阶段。为了开发最佳性能的模型，我们使用了不同的人工神经网络架构，尝试了多种不同的模型原型以获得所需的结果。在 TensorFlow 2.0 之前，实验的每一次改动并不像本书介绍的那么容易，而是需要先构建一个计算图然后在会话中运行该计算图。本章中，我们即将介绍的 Estimators（估算器）旨在处理上述操作流程。创建计算图并在会话中运行计算图是一项非常耗时的工作，并给调试代码的过程带来了诸多挑战。

当模型开发完成之后，将其部署到生产服务器也是一个挑战，我们可能希望将其部署在分布式环境中以获得更好的性能。此外，还可能希望在 CPU、GPU 或 TPU 上运行模型，这都需要更改代码。为了解决上述问题，我们引入了估算器。尽管估算器在 TensorFlow 2.0 之前已经存在，但是，估算器允许我们能够充分利用 TensorFlow 2.x 中引入的诸多新功能。例如，构建数据管道和模型开发之间具有明确的分离、分布式环境中的模型部署无需任何代码更改、增强的日志记录与跟踪使得调试更加容易。本章，我们将学习估算器是如何实现上述功能的，主要包含如下内容。

- ☑ 什么是 Estimators？
- ☑ 什么是预制 Estimators？
- ☑ 使用预制估算器解决分类问题。
- ☑ 使用预制估算器解决回归问题。
- ☑ 在 Keras 模型上构建自定义估算器。
- ☑ 在 tfhub 模块上构建自定义估算器。

6.1 Estimators 概述

TensorFlow 估算器是一种高级 API，它将机器学习开发的多个阶段统一为一个整体。估算器封装了用于训练、评价、预测和导出模型以供生产使用的各种 API，为现有的 TensorFlow API 提供了进一步的抽象。

6.1.1 API 接口

引入估算器后，TensorFlow 的最新 API 栈如图 6-1 所示。

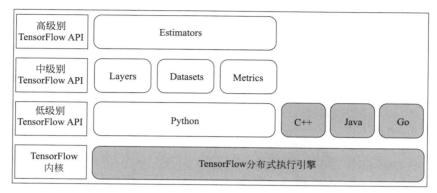

图 6-1　TensorFlow API

可以看到，截至目前，我们主要使用了中级别 API。当需要更好地控制模型开发时，低级别 API 就变得非常必要。学习了 Estimators API 之后，我们甚至不会使用中级别 API 进行模型开发。但是，已经开发的模型是否可以从估算器 API 中受益？答案是可以，那么如何使用 Estimators API 呢？幸运的是，TensorFlow 团队开发了一个接口，允许用户迁移已有模型以使用 Estimators API。TensorFlow 自行创建了一些估算器，以帮助用户快速入门，这些被称为预制估算器。不仅应用于快速入门，估算器已经过全面开发和测试，并可应用于实际项目。如果上述都不能满足我们的目的，或者想利用估算器迁移现有模型，可以使用估算器 API 开发自定义 Estimators。在学习如何使用预制 Estimators 及如何构建自定义估算器之前，我们首先详细地讨论 Estimators 的优点。

6.1.2 Estimators 的优点

以下列出了 Estimators 的主要优点。
① 为训练、评价、预测提供统一的接口。
② 通过输入函数处理数据输入。
③ 创建 checkpoint。
④ 创建摘要日志。

以上并非 Estimators 的所有点，但一定是最重要的优点，接下来将更详细讨论这些内容。

如图 6-2 所示，估算器类提供了用于训练、评价和预测的三个接口。所以，一旦我们创建了一个估算器对象，就可以在这个对象上调用训练、评价和预测方法。需要为每个方法（训练、评价、预测）传递不同的数据集，这由输入函数（input_fn）实现。本小节后面将更详细地解释这个输入函数的结构，在此只需了解输入函数的引入简化了在不同数据集下的实验即可。

训练期间创建的 checkpoint 使得我们可以回看已训练的状态，并从该状态点开始继续训练。这将节省大量训练时间，尤其是当某个 epoch 结束发生错误时。此外，这也使得调试的效率更高。训练结束后，我们可以通过 TensorBoard 来可视化模型评价期间创建的摘要日志，以快速了解模型的训练效果。

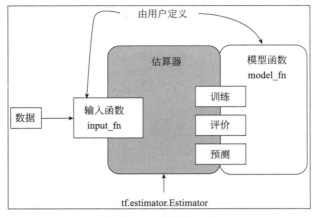

图 6-2　估算器接口

当模型训练到我们完全满意的程度之后，接下来的任务是模型部署。在 CPU、GPU、TPU、移动设备、Web 及分布式环境上部署模型，通常需要进行多次代码更改。但是如果使用估算器，则可以按模型原样或者以最少的改动在这些平台上部署训练好的模型。

介绍完估算器的优点，接下来将介绍估算器的类型。

6.1.3　Estimators 的类型

估算器可分为两类：预制估算器、自定义估算器。

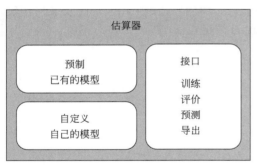

图 6-3　估算器的种类

估算器的种类如图 6-3 所示。

预制估算器就像一个盒子里的模型，模型的功能已经由 TensorFlow 团队编写。另一方面，在自定义估算器中，我们同样需要提供此模型功能。两种类型的估算器对象都将具有用于训练、评价和预测的通用接口。此外，两种类型都能够以类似的方式导出以供服务。

TensorFlow 提供了一些预制估算器，如果这些预制估算器不符合我们的目的，或者我们想利用估算器迁移现有模型，则需要创建自己的估算器类。所有估算器都是 tf.estimator.Estimator 的子类。

DNNClassifier、LinearClassifier 和 LinearRegressor 是预制估算器的几个例子。DNNClassifier 用于创建基于密集神经网络（dense neural network）的分类模型，而 LinearRegressor 用于处理线性回归问题。本章接下来的部分将介绍如何使用这两个类。

作为构建自定义估算器的一部分，我们需要将现有的 Keras 模型转换为估算器。这样做可以使我们的模型具有之前介绍的估算器的一些优点。我们首先将前一章开发的葡萄酒质量回归模型构建为自定义估算器。接下来，介绍如何基于 tfhub 模块构建自定义估算器。

要使用估算器，我们需要了解两个新概念——输入函数和特征列。输入函数基于 tf.data.dataset 创建一个数据管道，该管道将数据分批次提供给模型进行训练和评价。我们还可以创建用于预测的数据管道。DNNClassifier 项目将展示如何执行上述操作。特征列指定估算器如何解释数据。在讨论输入函数和特征列概念之前，我们首先简要介绍一下基于估算器的项目开发。

6.1.4 基于 Estimators 的项目开发流程

基于估算器的项目开发流程如下：加载数据、数据预处理、定义特征列、定义输入函数、模型实例化、模型训练、模型评价、在 TensorBorad 上判断模型性能、使用模型进行预测。

学习完之前的章节，读者肯定熟悉上述工作流程中的许多步骤，需要注意的是定义特征列和定义输入函数。接下来将介绍如何定义特征列和定义输入函数。

1. 特征列

特征列提供了一个原始数据和估算器之间的桥梁，它将各种各样的原始数据转换为估算器所需的格式。首先，使用 tf.feature_column 模块来构建特征列的列表；然后，将该列表作为估算器构造函数的输入；最后，估算器对象使用此列表解释来自输入函数的数据。整个过程如图 6-4 所示。

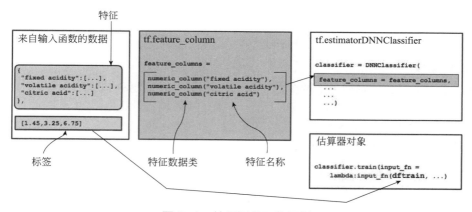

图 6-4　特征列的工作机制

如图 6-4 中间的模块所示，特征列的列表包括非挥发性酸（数值）、挥发性酸（数值）、柠檬酸（数值）等。

这些是上一章开发的葡萄酒质量评估模型的特征，列表中的每个元素都是 tf.feature_column.numeric_column 类型。以下代码片段描述了如何构建这样的列表。

```
# build numeric features array
numeric_feature = []
for col in numeric_columns:
 numeric_feature.append
   (tf.feature_column.numeric_column(key=col))
```

有时，数据集中可能包含类别字段，这些类别字段可作为特征供我们训练模型使用。以下代码片段描述了如何为类别特征构建特征列。

```
categorical_features = []
for col in categorical_columns:
 vocabulary = data[col].unique()
 cate = tf.feature_column.
       categorical_column_with_vocabulary_list
          (col, vocabulary)
 categorical_features.append
       (tf.feature_column.indicator_column(cate))
```

> 我们首先调用 categorical_column_with_vocabulary_list 方法获取词汇表，然后将指标列（indicator columns）添加到列表中。该列表作为函数参数传递给估算器的构造函数，如图 6-4 所示。如果我们的模型需要数值特征和类别特征，则需要将两者都添加到目标的特征列。图 6-4 左侧模块显示了由 Input 函数构建的数据，此数据将输入到估算器的训练、评价、预测方法中。

接下来，我们将介绍如何编写输入函数。

2. 输入函数

输入函数的目的是返回以下两个对象以供估算器模型对象使用。
① 包含相应特征数据的特征名称（键）和张量或稀疏张量（值）的字典。
② 包含一个或多个标签的张量。
其基本框架如下所示。

```
def input_fn (dataset):
    # create dictionary with feature names
    # and Tensors with corresponding data
    # create Tensor for label data
    return dictionary, label
```

可以基于此框架编写单独的函数用于训练、评价、预测。

现在，我们介绍一个输入函数的具体实验，用于本章后文中的示例，以进一步阐明输入函数背后的概念。

输入函数可按如下方式定义。

```
def input_fn(features, labels, training =
            True, batch_size = 32):
```

这里，features 和 labels 参数表示包含特征和标签数据的张量。在函数内部，通过调用 tf.data.Dataset 模块的 from_tensor_slices 函数将数据转换为张量。代码如下。

```
#converts inputs to a dataset
Dataset = tf.data.Dataset.from_tensor_slices
          ((dict(features),labels))
```

该函数的输入参数是一个包含特征和对应标签的 Python 字典。最后，函数通过 tf.data.Dataset 的 batch 方法将这些数据批量返回给函数调用者。代码如下。

```
return dataset.batch(batch_size)
```

为了进一步理解，输入函数的结构如图 6-5 所示。

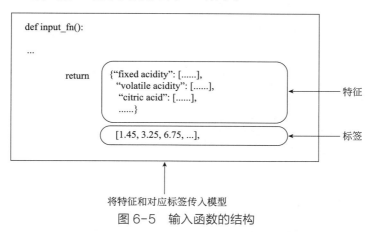

图 6-5　输入函数的结构

整个过程可能看起来有些复杂，但可以通过一个实际的示例进一步阐明。接下来，我们将实现这个过程。

6.2　设置 Estimators

我们将讨论两种类型的预制估算器——一种用于分类，另一种用于回归。首先介绍一个分类项目，该项目使用预制的 DNNClassifier。DNNClassifier 定义了一个深度神经网络模型，并将模型输入分为几个类别。该项目使用 MNIST 数据集将手写字符分类为 10 个数字（0～9）。我们要介绍的第二个项目使用名为 LinearRegressor 的预制估算器，并使用波士顿地区的 Airbnb 数据集。该数据集由 Airbnb 列出的几所房屋组成，对于列表中的每所房屋，都有一系列对应的特征，以及房屋出售或可出售的价格。使用这些房屋信息，我们将开发一个回归模型来预测新上市房屋的可能出售价格。

使用这两种不同的模型将使读者深入了解如何在自己的模型开发问题中使用预制估算器。首先，让我们从一个分类模型开始。

6.3　用于分类的 DNN 分类器

创建一个新的 Colab 项目并将其重命名为 DNNClassifier-estimator。像往常一样，使用如下语句导入 TensorFlow。

```
import tensorflow as tf
```

本项目使用 MNIST 数据集。该数据集是一个手写字符的数据集,可在 sklearn 工具包中获取。我们的任务是使用预制估算器来识别嵌入在这些图像中的数字。模型的输出为 10 个类别,对应 10 个阿拉伯数字(0 ~ 9)。

6.3.1 加载数据

使用如下代码从 sklearn 中加载 MNIST 数据。

```
from sklearn import datasets
digits = datasets.load_digits()
```

通过显示一些图像来查看加载数据的内容,以下代码将显示前四幅图像。

```
#plotting sample image
import matplotlib.pyplot as plt
plt.figure(figsize=(1,1))
fig, ax = plt.subplots(1,4)
ax[0].imshow(digits.images[0])
ax[1].imshow(digits.images[1])
ax[2].imshow(digits.images[2])
ax[3].imshow(digits.images[3])
plt.show()
```

输出结果如图 6-6 所示。

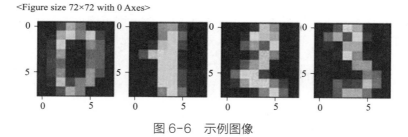

图 6-6 示例图像

可以看到,每幅图像的像素尺寸为 8×8。

6.3.2 准备数据

加载的 MNIST 图像均为彩色图像,但是我们不需要颜色分量来识别数字,只需灰度图像即可满足要求。因此,通过 reshape 函数去除图像的颜色分量。

```
n_samples = len(digits.images)
data = digits.images.reshape((n_samples, -1))
```

如果查看数据的维度,可以观察到有 1797 幅图像,每幅图像由 64 个像素值组成。使用如下代码将数据拆分为训练集和测试集。

```
from sklearn.model_selection
            import train_test_split
```

```
X_train, X_test, y_train, y_test = 
                    train_test_split(data, digits.target, test_size = 0.05, 
shuffle=False)
```

接下来，我们为估算器定义输入函数。

6.3.3 Estimators 输入函数

如前文所述，输入函数需要特定格式的数据。我们需要指定一个列列表作为估算器的输入。图像数据由 64 个像素组成，在模型训练期间，每个像素表示为一个数值列。因此，要创建由这些像素值组成的张量，我们首先为每个像素列创建一个名称。

```
# create column names for our model input function
columns = ['p_'+ str(i) for i in range(1,65)]
```

可以看到，创建的列名称分别为 p_1，p_2，…，p_64。然后，将 tf.feature_column 类中的 numeric_column 类型添加到名为 feature_columns 的数组中以构建特征列，代码如下。

```
feature_columns = [ ]
for col in columns:
  feature_columns.append (tf.feature_column.numeric_column(key = col))
```

定义输入函数如下。

```
def input_fn(features, labels, training = True, batch_size = 32):
```

第一个参数定义数据的特征，第二个参数定义数据的标签，第三个参数指定数据是用于训练还是评价。数据通过批量的方式处理——批量大小由输入函数的最后一个参数决定。

将数据集中的数据转换为张量，以便估算器进行更有效的处理，代码如下。

```
#converts inputs to a dataset
  dataset = tf.data.Dataset.from_tensor_slices ((dict(features),labels))
```

通过调用 from_tensor_slices 方法将数据转换为张量，该方法的输入为由数据的特征及其对应标签组成的 Python 字典。

如果输入数据集为训练集，则将数据集的顺序打乱，代码如下。

```
#shuffle and repeat in a training mode
  if training:
    dataset=dataset.shuffle(1000).repeat()
```

shuffle 方法用于打乱数据集，shuffle 的缓冲区大小设置为 1000。为了处理数据集太大而无法放入内存，shuffle 是批量完成的。如果缓冲区的大小大于数据集中的样本数，则会得到相同的 shuffle 结果。如果缓冲区的大小设置为 1，则不会起到打乱数据集的效果。

最后，批量返回数据，代码如下。

```
#giving inputs in batches for training
  return dataset.batch(batch_size)
```

接下来，创建一个估算器实例对象。

6.3.4 创建 Estimators 实例

使用预制估算器 DNNClassifier 来创建估算器实例对象，如下面的代码所示。

```
classifier = tf.estimator.DNNClassifier
              (hidden_units = [256, 128, 64],
                feature_columns = feature_columns,
                        optimizer = 'Adagrad',
                        n_classes = 10,
                        model_dir = 'classifier')
```

构造函数接收五个输入参数。第一个参数定义了网络架构。这里，我们在网络架构中定义了三个隐藏层；第一层包含 256 个神经元，第二层包含 128 个神经元，第三层包含 64 个神经元。第二个参数指定数据的特征列。因为我们之前创建了 feature_columns 向量，因此将其设置为默认参数。第三个参数指定要使用的优化器，这里默认设置为 Adagrad。Adagrad 是一种基于梯度的优化算法，该算法根据网络参数自动调节学习率，执行较小的更新。第四个参数（n_classes）定义了输出类别的数量。在本项目中，分类类别的数量为 10，分类类别为 0 ~ 9 之间的数字。最后一个参数 model_dir 指定了训练日志的保存路径。

接下来，将介绍模型训练过程。

6.3.5 模型训练

创建的估算器模型可以使用常规的训练方法进行训练。在开始训练之前，需要创建用于训练的输入数据集。为此，我们创建了一个包含训练数据和特征列表的 Pandas 数据流（dataframe），代码如下。

```
# create dataframes for training
import pandas as pd
dftrain = pd.DataFrame(X_train, columns = columns)
```

在之前创建的估算对象上调用 train 方法开始训练：

```
classifier.train(input_fn = lambda:input_fn
                        (dftrain,
                      y_train,
                      training = True),
                      steps = 2000)
```

该方法以我们之前定义的 input_fn 函数作为参数，而 input_fn 函数将数据特征和对应标签作为前两个参数。参数 training 设置为 True 以便对数据进行打乱，steps 设置为 2000。在机器学习中，一个 epoch 表示将训练集从头到尾遍历一次。一个 step 对应神经网络的一次前向传播和反向传播。如果我们不在数据集中创建批次 (batch)，则一个 step 即对应于一个 epoch。但是，如果将数据集拆分为批次（batches），则一个 epoch 将包含多个 setp。注意，一个 step 表示在一个 batch 数据上的单次迭代。可以使用如下公式计算在整个训练周期中总的 epoch 数量。

epochs 数量 = (batch_size × step 数量)/(训练样本数量)

在当前的示例中，batch_size 为 32，step 数量为 2000，训练样本数量为 1707，因此，完成训练需要 38 个 [(32 × 2000) / 1707] epoch。

6.3.6 模型评价

为了评价模型的性能，需要像以前一样创建一个 dataframe，使用测试数据作为输入：

dftest = pd.DataFrame(X_test, columns = columns)

调用估算器的 evaluate 方法评价模型：

```
eval_result = classifier.evaluate(
    input_fn = lambda:input_fn(dftest, y_test,
                               training = False)
)
```

其中，参数 training 被设置为 False。通过打印 eval_result 的值来查看评价结果，输出结果如图 6-7 所示。

评价完成后，可以在 TensorBoard 中加载日志文件查看各种参数，日志文件保存在 DNNClassifier 的构造函数中的 model_dir 参数指定的文件夹中。代码如下。

```
%load_ext tensorboard
%tensorboard --logdir ./classifier
```

图 6-7　模型评价结果

图 6-8 显示了日志中记录的损失值。

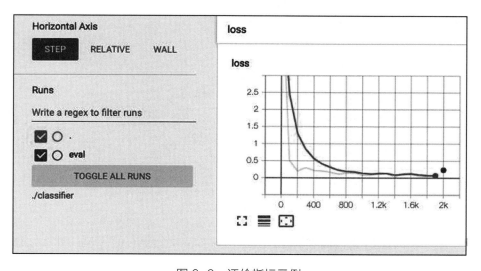

图 6-8　评价指标示例

接下来，将介绍如何使用训练好的估算器来预测未知数据。

6.3.7 预测未知数据

为了预测未知数据，首先创建一个名为 pred_input_fn 的输入函数，代码如下。

```
# An input function for prediction
def pred_input_fn(features, batch_size = 32):
    # Convert the inputs to a Dataset without labels.
    return tf.data.Dataset.from_tensor_slices
(dict(features)).batch(batch_size)
```

该函数返回仅包含特征而不包含标签的批次数据，从 dftest 中提取两个数据来测试预测功能，如下所示。

```
test = dftest.iloc[:2,:]
```

从 y_test 中提取上述两个数据对应的目标值：

```
expected = y_test[:2].tolist()
```

调用估算器对象的 predict 方法来进行实际预测：

```
pred = list(classifier.predict(
    input_fn = lambda:pred_input_fn(test))
)
```

输入函数 pred_input_fn 使用我们创建的测试数据测试了两个未知数据。
最后，使用以下循环打印这两个数据的预测类别、预测概率和实际目标值。

```
for pred_dict, expec in zip(pred, expected):
    class_id = pred_dict['class_ids'][0]
    probability = pred_dict['probabilities']
                  [class_id]
    print('predicted class {} ,
        probability of prediction {} ,
            expected label {}'.format
            (class_id,probability,expec))
```

执行上述语句的输出结果如下。

```
predicted class 8 , probability of prediction
0.9607188701629639 , expected label 8

predicted class 4 , probability of prediction
0.9926437735557556 , expected label 4
```

6.3.8 实验不同的 ANN 结构

在评估模型的性能时，可以很容易地实验不同的 ANN 架构。例如，可以更改估算器构造函数中 hidden_units 参数的值，也可以为之前的神经网络架构再添加一层。我们在现有的代码上尝试了以下配置。

```
hidden_units = [256, 128, 64, 32],
```

模型的评价结果如图 6-9 所示。

不仅是隐藏层的数量和其中神经元的数量，我们还可以在估算器实例化的代码中增加一个参数来添加 dropout，如下所示。

```
classifier = tf.estimator.DNNClassifier
            (hidden_units = [256, 128, 64, 32],
             feature_columns = feature_columns,
                 optimizer='Adagrad',
                 n_classes=10,
                 dropout = 0.2,
                 model_dir='classifier')
```

上述参数配置下的评价结果如图 6-10 所示。

```
eval_result
{'accuracy': 0.9222222,
 'average_loss': 0.21866909,
 'global_step': 2000,
 'loss': 0.2130903}
```

图 6-9　更密集神经网络结构的评价结果

```
eval_result
{'accuracy': 0.9111111,
 'average_loss': 0.44737434,
 'global_step': 2000,
 'loss': 0.44995198}
```

图 6-10　增加 dropout 之后的评价指标

可见，我们可以轻松地试验多种模型架构。此外，还可以对不同的数据集进行试验，如更改特征列列表中包含的特征数量。一旦对模型的性能感到满意，我们可以将其保存为文件，然后将保存的文件上传到生产服务器以供使用。

6.3.9　项目源码

清单 6-1 列出了本项目的完整源码，供读者快速参考。

清单 6-1 DNNClassifier 估算器完整源码

```python
import tensorflow as tf

from sklearn import datasets
digits = datasets.load_digits()

#plotting sample image
import matplotlib.pyplot as plt
plt.figure(figsize=(1,1))
fig, ax = plt.subplots(1,4)
ax[0].imshow(digits.images[0])
ax[1].imshow(digits.images[1])
ax[2].imshow(digits.images[2])
ax[3].imshow(digits.images[3])
plt.show()

# reshape the data to two dimensions
n_samples = len(digits.images)
data = digits.images.reshape((n_samples, -1))

data.shape
```

源码清单
链　接：https://pan.baidu.com/s/1NV0rimQ_8kRz22xfFHN-Cw
提取码：1218

```python
# split into training/testing
from sklearn.model_selection
    import train_test_split
        X_train, X_test, y_train, y_test = 
            train_test_split(
        data, digits.target, test_size = 0.05,
            shuffle=False)

# create column names for our model input function
  columns = ['p_'+ str(i) for i in range(1,65)]
feature_columns = []
for col in columns:
 feature_columns.append
    (tf.feature_column.numeric_column(key=col))

def input_fn(features, labels,
            training = True, batch_size = 32):
   #converts inputs to a dataset
   dataset = tf.data.Dataset.from_tensor_slices(
            (dict(features),labels))
   #shuffle and repeat in a training mode
   if training:
      dataset=dataset.shuffle(1000).repeat()

   #giving inputs in batches for training
   return dataset.batch(batch_size)

classifier = tf.estimator.DNNClassifier
            (hidden_units = [256, 128, 64],
              feature_columns = feature_columns,
                    optimizer = 'Adagrad',
                    n_classes = 10,
                    model_dir = 'classifier')

# create dataframes for training
import pandas as pd
dftrain = pd.DataFrame
            (X_train, columns = columns)

classifier.train(input_fn = 
            lambda:input_fn(dftrain,
                        y_train,
                        training = True),
                        steps = 2000)
# create dataframe for evaluation

dftest = pd.DataFrame(X_test, columns = columns)

eval_result = classifier.evaluate(
    input_fn = lambda:input_fn
                (dftest, y_test, training = False)
)

eval_result
```

```
%load_ext tensorboard
%tensorboard --logdir ./classifier

# An input function for prediction
def pred_input_fn(features, batch_size = 32):
# Convert the inputs to a Dataset without labels.
    return tf.data.Dataset.from_tensor_slices
            (dict(features)).batch(batch_size)

test = dftest.iloc[:2,:]
#1st two data points for predictions

expected = y_test[:2].tolist()
#expected labels

pred = list(classifier.predict(
    input_fn = lambda:pred_input_fn(test))
)

for pred_dict, expec in zip(pred, expected):
    class_id = pred_dict['class_ids'][0]
    probability = pred_dict['probabilities'][class_id]
    print('predicted class {} ,
        probability of prediction {} ,
    expected label {}'.
    format(class_id,probability,expec))
```

至此，我们已经介绍了如何使用预制的 DNNClassifier，接下来将介绍如何使用预制的 LinearRegressor 解决回归问题。

6.4 用于回归的 LinearRegressor

第 5 章中提到，神经网络可用于解决回归问题。为了支持这一说法，我们在 TensorFlow 库中找到了一个预制的估算器 LinearRegressor，用于回归模型的开发。本节我们将讨论如何使用这个估算器。

6.4.1 项目描述

本项目尝试解决的回归问题是估计波士顿地区的房屋售价，为此，我们使用波士顿的 Airbnb 数据集 (www.kaggle.com/airbnb/boston)。该数据集有大量特征列，必须仔细检查每个特征列是否适合作为一个特征用于模型开发。因此，在开发回归模型的过程中，我们需要对数据进行严格的预处理，以充分减少特征数量，同时在预测房屋价格时达到高准确性。

6.4.2 创建项目

像往常一样，创建一个 Colab 项目并将其重命名为 DNNRegressor-Estimator。使用如下语句导入所需的库。

```
import tensorflow as tf
```

```
import pandas as pd
import matplotlib.pyplot as plt
import numpy as np
```

6.4.3 加载数据

使用以下代码中指定的 URL 将数据下载到项目中。

```
url = 'https://raw.githubusercontent.com/Apress/artificialneural-
networks-with-tensorflow-2/main/ch06/listings.csv'
data = pd.read_csv(url)
```

使用的数据被读入 pandas 工具包的数据流中。我们可以通过调用 data 变量的 shape 方法来查看数据的大小。调用 shape 方法后可以看到，该数据集有 3585 行（总样本数量）和 95 列（特征数量）。即使对于技术娴熟的数据分析师来说，找出这 95 列特征之间的回归关系也不是一件容易的事，因此，我们意识到可以尝试使用神经网络找到解决方案。

在数据集的 95 列特征中，显然并非所有的特征都对我们训练模型有用，因此，我们需要进行数据清理来删除不需要的特征并对剩余的特征进行标准化。接下来，我们将根据上述描述来对数据进行预处理。

6.4.4 特征选择

我们要做的第一件事是列出所有特征列，通过调用 data 变量的 columns 方法来完成。执行此操作后，可以看到如图 6-11 所示的输出。

图 6-11 特征列表

我们使用 data.info() 来更好地理解该数据的结构。很容易注意到，有许多字段可能与模型训练无关。为了更好地分析可以消除哪些特征列，可以调用 describe 方法获取数据的描述信息，输出结果如图 6-12 所示。

```
# understanding data
data.describe()
```

	id	scrape_id	host_id	host_listings_count	host_total_listings_count	neighbourhood_group_cleansed	latitude
count	3.585000e+03	3.585000e+03	3.585000e+03	3585.000000	3585.000000	0.0	3585.000000
mean	8.440875e+06	2.016091e+13	2.492311e+07	58.902371	58.902371	NaN	42.340032
std	4.500787e+06	8.516813e-01	2.292781e+07	171.119663	171.119663	NaN	0.024403
min	3.353000e+03	2.016091e+13	4.240000e+03	0.000000	0.000000	NaN	42.235942
25%	4.679319e+06	2.016091e+13	6.103425e+06	1.000000	1.000000	NaN	42.329995
50%	8.577620e+06	2.016091e+13	1.928100e+07	2.000000	2.000000	NaN	42.345201
75%	1.278953e+07	2.016091e+13	3.622147e+07	7.000000	7.000000	NaN	42.354685
max	1.493346e+07	2.016091e+13	9.385411e+07	749.000000	749.000000	NaN	42.389982

图 6-12　数据描述

仔细检查数据后，我们决定共使用 17 列进行分析（本书不做介绍）。在实践中，可以使用任何已知的特征选择技术，如单因素特征选择（univariate selection）或带有热图的相关矩阵（correlation matrix with heatmap）。通过键入字段名称来选择要使用的特征列，如下面的语句所示。

```
# Selecting only few columns as our features
data = data[['property_type','room_type',
           'bathrooms','bedrooms','beds','bed_type',
           'accommodates','host_total_listings_count',
           'number_of_reviews','review_scores_value',
           'neighbourhood_cleansed','cleaning_fee',
           'minimum_nights','security_deposit',
           'host_is_superhost','instant_bookable',
           'price']]
```

特征选择之后，我们需要进行数据清洗以确保选择的特征仅包含干净的数据。

6.4.5 数据清洗

首先检查数据是否包含空值，通过调用 isnull 函数并对其输出求和来完成此操作。图 6-13 显示了包含空值的特征。

```
data.isnull().sum() #nan values

property_type                   3
room_type                       0
bathrooms                      14
bedrooms                       10
beds                            9
bed_type                        0
accommodates                    0
host_total_listings_count       0
number_of_reviews               0
review_scores_value           821
neighbourhood_cleansed          0
cleaning_fee                 1107
minimum_nights                  0
security_deposit             2243
host_is_superhost               0
instant_bookable                0
price                           0
dtype: int64
```

图 6-13　检查空值

如图 6-13 所示，共有 7 列包含空值，这些列的总和不等于零。由于 security_deposit 列在 3585 条记录中有 2243 条为空值，比例非常大，因此我们在分析中去除此列。代码如下。

```
data = data.drop('security_deposit', axis = 1)
```

我们打印一些记录来检查数据，代码如下。

```
data.head()
```

输出结果如图 6-14 所示。

图 6-14 检查数据

检查数据时，property_type 字段是用于分类的，将单独处理。此外，注意到 cleaning_fee、security_deposit 和 price 列包含一个 $ 符号，去掉 $ 符号后，将这些字段的数值转化为浮点数。另外，上述字段的数值间包含了逗号，需要去除这些逗号。为了清理上述三列数值，我们编写了一个转换函数，如下所示。

```
# function to remove $ and , characters
def transform(x):
  x = str(x)
  x = x.replace('$','')
  x = x.replace(",","")
  return float(x)
```

使用 for 循环对上述三个字段应用此转换：

```
for col in ["cleaning_fee","price"]:
  data[col] = data[col].apply(transform)
  #filling nan with mean value
  data[col].fillna(data[col].mean(),inplace = True)
```

在循环中，我们将空值替换为该字段下所有数值的平均值。对于其余的特征列，我们只将空值替换为当前列的平均值。使用了如下代码。

```
#filling nan values with mean value
for feature in ["bathrooms","bedrooms","beds",
                "review_scores_value"]:
  data[feature].fillna(data[feature].mean(),
                       inplace = True)
```

接下来查看房屋类别列——property_type。由于这是一个类别列，我们不能简单地用平均值替换空值，而是需要找到一个合适的值来替换空值。为此，我们检查此列中有哪些不同的值，如图 6-15 所示。

图 6-15 类别列中的不同值

调用 value_counts 方法来检查每个值的频率，如图 6-16 所示。

由于 Apartment 一词出现的频率最高，我们将使用该值作为此列中空值的替换文本。使用如下代码进行替换。

```
# replacing nan with Apartment
data['property_type'].fillna
('Apartment',
inplace = True)
```

6.4.6 创建数据集

使用如下代码从预处理的数据中提取特征和标签。

```
feature = data.drop('price', axis = 1)
#input data
target = data['price']
```

图 6-16 property_type 值的频率分布

想预测新签约房屋的价格，首先看一下数据集中的价格分布。使用以下代码绘制价格的直方图。

```
# price value histogram
data['price'].plot(kind='hist',grid = True)
plt.title('price distribution')
plt.xlabel('price')
```

输出的直方图如图 6-17 所示。

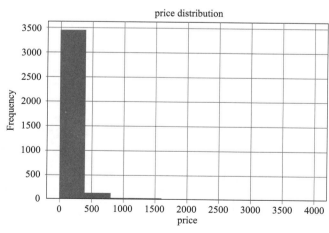

图 6-17 原始价格数据分布

如图 6-17 所示，大多数价格数值都分布在较低范围内。为了更好地训练模型，这些价格需要有更好的数据分布，而不是现在的偏态分布。可以通过取价格的对数来完成，如以下代码所示。

```
target = np.log(data.price) #output data
target.hist()
plt.title
    ('price distribution after log transformation')
```

变换后的数据分布如图 6-18 所示。

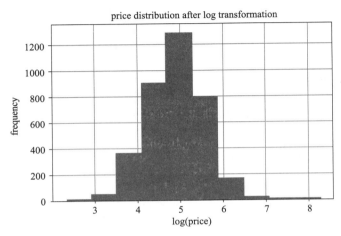

图 6-18　变换后的价格数据分布

至此，我们完成了数据预处理，接下来创建训练集和测试集。为了创建训练集和测试集，使用 sklearn 的 train_test_split 方法：

```
#splitting input and output data for training and testing
from sklearn.model_selection import train_test_split
xtrain,xtest,ytrain,ytest=train_test_split
    (feature, target, test_size = 0.2,
     random_state = 42)
```

可以看到，我们保留了 20% 的数据用于测试。接下来我们将创建特征列，用于创建估算器实例。

6.4.7　建立特征列

在上一个项目中，我们创建了一个特征列的列表，该列表中的所有列的内容都为数值。当前的数据集除了数值之外，还有类别特征。首先，我们构建数值列表，代码如下。

```
# selecting numerical feature columns
numeric_columns = feature.select_dtypes
    (include = np.number).columns.tolist()
numeric_columns
```

我们从 feature 列表中选择出所有类型为 np.number 的列，并为其创建一个名为 numeric_columns 的新列表。打印这个新列表，可以看到如图 6-19 所示的结果。

```
['bathrooms',
 'bedrooms',
 'beds',
 'accommodates',
 'host_total_listings_count',
 'number_of_reviews',
 'review_scores_value',
 'cleaning_fee',
 'minimum_nights']
```

图 6-19　数值特征列表

可以看到，特征列表中所有数值特征列的名称都已添加到新列表中。接下来，我们将把这个列表转换成估算器的输入函数所需的格式。输入函数要求列表中应包含特征列及其数据类型（tf.feature_column），我们使用 for 循环构建这个列表，如下所示。

```
# build numeric features array
numeric_feature = [ ]

for col in numeric_columns:
 numeric_feature.append
    (tf.feature_column.numeric_column(key=col))
numeric_feature
```

打印这个列表，可以看到如图 6-20 所示的输出。

```
[NumericColumn(key='bathrooms', shape=(1,), default_value=None, dtype=tf.float32, normalizer_fn=None),
 NumericColumn(key='bedrooms', shape=(1,), default_value=None, dtype=tf.float32, normalizer_fn=None),
 NumericColumn(key='beds', shape=(1,), default_value=None, dtype=tf.float32, normalizer_fn=None),
 NumericColumn(key='accommodates', shape=(1,), default_value=None, dtype=tf.float32, normalizer_fn=None),
 NumericColumn(key='host_total_listings_count', shape=(1,), default_value=None, dtype=tf.float32, normalizer_fn=None),
 NumericColumn(key='number_of_reviews', shape=(1,), default_value=None, dtype=tf.float32, normalizer_fn=None),
 NumericColumn(key='review_scores_value', shape=(1,), default_value=None, dtype=tf.float32, normalizer_fn=None),
 NumericColumn(key='cleaning_fee', shape=(1,), default_value=None, dtype=tf.float32, normalizer_fn=None),
 NumericColumn(key='minimum_nights', shape=(1,), default_value=None, dtype=tf.float32, normalizer_fn=None),
 NumericColumn(key='security_deposit', shape=(1,), default_value=None, dtype=tf.float32, normalizer_fn=None)]
```

图 6-20　特征列

同样地，我们将在特征列列表中创建一个类别特征列表。我们选择并构建具有类别属性的类别列表，如下所示。

```
#selecting categorical feature columns
categorical_columns = feature.select_dtypes
        (exclude=np.number).columns.tolist()
categorical_columns
```

打印类别列表，可以得到如图 6-21 所示的结果。

可以看到，有六个特征列具有类别值。我们使用以下 for 循环构建类别特征的特征列。

```
categorical_features = [ ]
for col in categorical_columns:
 vocabulary = data[col].unique()
 cate = tf.feature_column.
```

```
['property_type',
 'room_type',
 'bed_type',
 'neighbourhood_cleansed',
 'host_is_superhost',
 'instant_bookable']
```

图 6-21　类别列表

```
            categorical_column_with_vocabulary_list
            (col, vocabulary)
    categorical_features.append
            (tf.feature_column.indicator_column(cate))
```

首先提取类别特征中的不同值，然后构建词汇表，最后将其添加到 categorical_features 列表中。打印这个列表，可以看到如图 6-22 所示的结果。

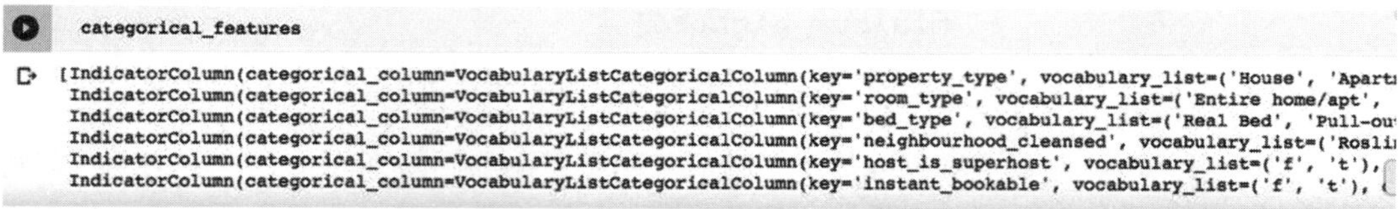

```
categorical_features

[IndicatorColumn(categorical_column=VocabularyListCategoricalColumn(key='property_type', vocabulary_list=('House', 'Apart
 IndicatorColumn(categorical_column=VocabularyListCategoricalColumn(key='room_type', vocabulary_list=('Entire home/apt',
 IndicatorColumn(categorical_column=VocabularyListCategoricalColumn(key='bed_type', vocabulary_list=('Real Bed', 'Pull-ou
 IndicatorColumn(categorical_column=VocabularyListCategoricalColumn(key='neighbourhood_cleansed', vocabulary_list=('Rosli
 IndicatorColumn(categorical_column=VocabularyListCategoricalColumn(key='host_is_superhost', vocabulary_list=('f', 't'),
 IndicatorColumn(categorical_column=VocabularyListCategoricalColumn(key='instant_bookable', vocabulary_list=('f', 't'),
```

图 6-22　类别特征的特征列

最后，我们将数值特征和类别特征组合到一个列表中，作为参数传递到接下来定义的输入函数中。代码如下。

```
# combining both features as our final features list
features = categorical_features +
            numeric_feature
```

6.4.8　定义输入函数

输入函数的定义如下。

```
# An input function for training and evaluation
def input_fn(features,labels,training = True,batch_size = 32):
 #converts inputs to a dataset
 dataset=tf.data.Dataset.from_tensor_slices
            ((dict(features), labels))
 #shuffle and repeat in a training mode
 if training:
   dataset=dataset.shuffle(10000).repeat()
 # return batches of data
 return dataset.batch(batch_size)
```

如上一个项目所示，该函数将特征和标签作为前两个参数。training 参数决定数据是否用于训练，如果是，则对数据进行清洗。from_tensor_slices 函数为模型的数据管道创建了一个张量。输入函数最终返回批量数据。

6.4.9　创建 Estimators 实例对象

使用如下代码创建估算器实例对象。

```
linear_regressor = tf.estimator.LinearRegressor(
   feature_columns = features,
   model_dir = "linear_regressor")
```

可以看到，我们使用了 LinearRegressor 类作为回归问题的预制估算器。 第一个参数是特征列，指定了组合的数值特征和类别特征列表。第二个参数为维护日志的保存路径。

6.4.10 模型训练

调用模型的 train 方法来训练模型，代码如下。

```
linear_regressor.train(input_fn = lambda:input_fn(xtrain,
                       ytrain,
                       training = True),
                       steps = 2000)
```

输入函数以 xtrain 中的特征数据和 ytrain 中的标签值为函数参数。training 参数设置为 True 用于启用数据清洗，steps 参数决定了训练阶段的 epoch 数量。

6.4.11 模型评估

通过调用模型的 evaluate 方法来评价模型性能，代码如下。

```
linear_regressor.evaluate(
    input_fn = lambda:input_fn
            (xtest, ytest, training = False)
)
```

输入函数使用 xtest 和 ytest 进行评价，评价的输出结果如下。

```
{'average_loss': 0.18083459,
 'global_step': 2000,
 'label/mean': 4.9370537,
 'loss': 0.1811692,
 'prediction/mean': 4.956979}
```

评价完成后，可以通过加载 TensorBoard 来查看各种评价指标：

```
%load_ext tensorboard
%tensorboard --logdir ./linear_regressor
```

TensorBoard 中显示的损失曲线如图 6-23 所示。

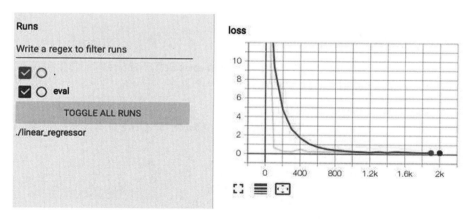

图 6-23　损失值指标

6.4.12 项目源码

清单 6-2 列出了本项目的完整源码，供读者快速参考。

清单 6-2 LinearRegressor-Estimator 完整源码

源码清单
链　接：https://pan.baidu.com/s/1NV0rimQ_
8kRz22xfFHN-Cw
提取码：1218

```python
import tensorflow as tf
import pandas as pd
import matplotlib.pyplot as plt
import numpy as np
url = 'https://raw.githubusercontent.com/Apress/artificialneural-networks-with-tensorflow-2/main/ch06/listings.csv'
data = pd.read_csv(url)

# listing column names
data.columns

# understanding data
data.describe()

# Selecting only few columns as our features
data = data[['property_type','room_type',
            'bathrooms','bedrooms','beds','bed_type',
            'accommodates','host_total_listings_count',
            'number_of_reviews','review_scores_value',
            'neighbourhood_cleansed','cleaning_fee',
            'minimum_nights','security_deposit',
            'host_is_superhost','instant_bookable',
            'price']]

data.isnull().sum() #nan values

data.head()

# function to remove $ and , characters
def transform(x):
  x = str(x)
  x = x.replace('$','')
  x = x.replace(",","")
  return float(x)
for col in ["cleaning_fee","security_deposit","price"]:
    data[col] = data[col].apply(transform)
# apply function
    #filling nan with mean value
    data[col].fillna(data[col].mean(),inplace = True)

#filling nan values with mean value
for feature in ["bathrooms","bedrooms","beds",
                "review_scores_value"]:
  data[feature].fillna(data[feature].mean(),
                inplace = True)
```

```python
data['property_type'].unique()

# get frequency of all unique values
data['property_type'].value_counts()

# replacing nan with Apartment
data['property_type'].fillna
            ('Apartment', inplace = True)

feature = data.drop('price', axis = 1)
#input data
target = data['price']

# price value histogram
data['price'].plot(kind='hist',grid = True)
plt.title('price distribution')
plt.xlabel('price')

# make a log transformation to remove skew.
target = np.log(data.price) #output data
target.hist()
plt.title('price distribution after log transformation')

#splitting input and output data for training and testing
from sklearn.model_selection
            import train_test_split
            xtrain,xtest,ytrain,ytest =
            train_test_split
            (feature, target, test_size = 0.2,
            random_state = 42)

# selecting numerical feature columns
numeric_columns = feature.select_dtypes
            (include = np.number).columns.tolist()
numeric_columns

# build numeric features array
numeric_feature = []
for col in numeric_columns:
  numeric_feature.append
        (tf.feature_column.numeric_column(key=col))
numeric_feature

#selecting categorical feature columns
categorical_columns = feature.select_dtypes
        (exclude=np.number).columns.tolist()
categorical_columns

categorical_features = []
for col in categorical_columns:
```

```python
    vocabulary = data[col].unique()
    cate = tf.feature_column.
            categorical_column_with_vocabulary_list
            (col, vocabulary)
categorical_features.append
        (tf.feature_column.indicator_column(cate))
categorical_features

# combining both features as our final features list
features = categorical_features +
            numeric_feature
# An input function for training and evaluation
def input_fn(features,labels,training = True,batch_size = 32):
    #converts inputs to a dataset
    dataset=tf.data.Dataset.from_tensor_slices(
                (dict(features), labels))

    #shuffle and repeat in a training mode
    if training:
        dataset=dataset.shuffle(10000).repeat()

    # return batches of data
    return dataset.batch(batch_size)

linear_regressor = tf.estimator.LinearRegressor(
    feature_columns = features,
    model_dir = "linear_regressor")

linear_regressor.train(input_fn = lambda:input_fn(xtrain,
                ytrain,
                training = True),
                steps = 2000)

linear_regressor.evaluate(
    input_fn = lambda:input_fn
            (xtest, ytest, training = False)
)

%load_ext tensorboard
%tensorboard --logdir ./linear_regressor
```

本项目演示了如何使用预制的 LinearRegressor 来解决线性回归问题,接下来我们将介绍如何构建自定义估算器。

6.5 自定义 Estimators

第 5 章我们开发了一个回归模型用来估计葡萄酒的质量,我们使用了三种不同的架构——小型、中型和大型——来比较它们的结果。本节我们将展示如何将这些已有的 Keras 模型转换为自定义估算器,以利用估算器提供的功能。

6.5.1 创建项目

创建一个新的 Colab 项目并将其重命名为 ModelToEstimator。使用如下代码导入需要的工具包。

```
import tensorflow as tf
import pandas as pd
from sklearn.model_selection
        import train_test_split
from sklearn.preprocessing
        import StandardScaler
```

6.5.2 加载数据

我们继续使用之前的白葡萄酒质量数据集。使用以下代码将数据从 UCI 机器学习存储库下载到项目中。

```
data_url='https://raw.githubusercontent.com/Apress/artificialneural-
networks-with-tensorflow-2/main/Ch05/winequality-
white.csv'
data=pd.read_csv(data_url,delimiter=';')
```

由于我们已经熟悉这些数据，因此这里不再描述该数据集，也不再描述数据预处理的过程，而是直接介绍特征选择。

6.5.3 创建数据集

我们从第 5 章知道，数据集中的葡萄酒质量字段用作标签，其余字段用作特征。使用以下代码将特征和标签分离开来。

```
x = data.iloc[:,:-1]
y = data.iloc[:,-1]
```

将所有选择的数字特征缩放为正态分布：

```
sc = StandardScaler()
x = sc.fit_transform(x)
```

使用 train_test_split 方法划分训练集和测试集：

```
xtrain, xtest, ytrain, ytest =
    train_test_split(x, y, test_size =
                        0.3,random_state = 20)
```

在变量中捕获 xtrain 的维度，以供我们定义 Keras Sequential 模型：

```
input_shape = xtrain.shape[1]
```

6.5.4 定义模型

我们采用第 5 章案例中的小模型，定义如下。

```
small_model = tf.keras.models.Sequential([
```

```
        tf.keras.layers.Dense
            (64,activation = 'relu',
            input_shape = (input_shape,)),
        tf.keras.layers.Dense(1)
])
```

模型的最后一层为单个神经元，该神经元输出一个浮点值，表示葡萄酒的质量。

> 注意
>
> 与第 5 章讨论的葡萄酒质量问题一样，我们将此问题视为回归问题。

调用模型的 compile 方法以编译模型：

```
small_model.compile
            (loss = 'mse', optimizer = 'adam')
```

由于该问题为回归问题，因此使用 MSE 作为损失函数。我们将在估算器的实例化中使用这个编译完成的模型。在此之前，首先定义输入函数。

6.5.5 定义输入函数

与之前的示例相同，输入函数的定义代码如下。

```
def input_fn(features, labels,
            training = True, batch_size = 32):
 #converts inputs to a dataset
 dataset = tf.data.Dataset.from_tensor_slices
            (({'dense_input':features},labels))
   #shuffle and repeat in a training mode

if training:
    dataset = dataset.shuffle(1000).repeat()

#giving inputs in batch for training
return dataset.batch(batch_size)
```

可以看到，输入函数的定义和之前的示例完全一致，因此这里不再解释。接下来，我们将已定义的模型转换为估算器。

6.5.6 将模型转换为 Estimator

为了将现有的 Keras 模型转换为估算器实例对象，TensorFlow 库提供了一个名为 model_to_estimator 的函数。函数调用如下面的代码所示。

```
keras_small_estimator = tf.keras.estimator.model_to_estimator(
    keras_model = small_model,
        model_dir = 'keras_small_classifier')
```

该函数有两个参数：第一个参数指定之前已编译好的 Keras 模型，第二个参数指定在训练期间维护日志的保存路径。该函数返回一个估算器实例，并可以像之前的示例一样用来训练、评价和预测。

6.5.7 模型训练

与之前的示例一样，通过调用模型的 train 方法训练该模型：

```
keras_small_estimator.train
            (input_fn = lambda:input_fn
            (xtrain, ytrain), steps = 2000)
```

对于训练过程，输入函数以 xtrain 中的特征数据和 ytrain 中的标签作为输入参数。steps 参数设置为 2000，这决定了训练 epoch 的数量；training 参数采用默认值 True。

6.5.8 模型评价

模型训练完成之后，可以通过调用估算器实例的 evaluate 方法来评估模型的性能：

```
eval_small_result =
            keras_small_estimator.evaluate(
    input_fn = lambda:input_fn
            (xtest, ytest, training = False),
            steps=1000)
print('Eval result: {}'.format
            (eval_small_result))
```

这次，使用之前创建的测试数据集来评价模型，training 参数设置为 False。模型评价结束后，评价结果将打印在控制台上。我们可以使用 TensorBoard 查看生成的各种指标：

```
%load_ext tensorboard
%tensorboard --logdir ./keras_small_classifier
```

> 我们从之前指定的日志路径中获取评价指标。

6.5.9 项目源码

清单 6-3 列出了本项目的完整源码，供读者快速参考。

清单 6-3 ModelToEstimator 完整源码

```
import tensorflow as tf
import pandas as pd
from sklearn.model_selection
```

源码清单

链　接：https://pan.baidu.com/s/1NV0rimQ_8kRz22xfFHN-Cw

提取码：1218

```python
              import train_test_split
from sklearn.preprocessing 
          import StandardScaler

data_url='https://raw.githubusercontent.com/Apress/artificialneural-
networks-with-tensorflow-2/main/Ch05/winequality-
white.csv'
data=pd.read_csv(data_url,delimiter=';')

x = data.iloc[:,:-1]
y = data.iloc[:,-1]

sc = StandardScaler()
x = sc.fit_transform(x)

xtrain, xtest, ytrain, 
         ytest = train_test_split
           (x, y, test_size = 0.3,random_state = 20)

input_shape = xtrain.shape[1]

small_model = tf.keras.models.Sequential([
              tf.keras.layers.Dense
                 (64,activation = 'relu', 
                   input_shape = (input_shape,)),
              tf.keras.layers.Dense(1)
])

small_model.compile
         (loss = 'mse', optimizer = 'adam')

def input_fn(features, labels, training = 
             True, batch_size = 32):
  #converts inputs to a dataset
  dataset = tf.data.Dataset.from_tensor_slices
          (({'dense_input':features},labels))

  #shuffle and repeat in a training mode
  if training:
    dataset = dataset.shuffle(1000).repeat()

  #giving inputs in batch for training
  return dataset.batch(batch_size)

keras_small_estimator = tf.keras.estimator.model_to_estimator(
     keras_model = small_model, model_dir = 
                  'keras_small_classifier')

keras_small_estimator.train
     (input_fn = lambda:input_fn
               (xtrain, ytrain), steps = 2000)
```

```
eval_small_result = 
        keras_small_estimator.evaluate(
    input_fn = lambda:input_fn
            (xtest, ytest, training = False),
               steps=1000)
print('Eval result: {}'.format
      (eval_small_result))

%load_ext tensorboard
%tensorboard --logdir ./keras_small_classifier
```

6.6 为预训练模型定义 Estimators

在第 4 章，我们学习了使用 TensorFlow Hub 的预训练模型，并了解了如何在自己的模型中重用这些预训练模型。当然，在扩展预训练模型架构的同时还可以将这些模型转换为估算器。本节，我们将介绍如何扩展 VGG16 模型，并在扩展的模型上创建自定义估算器。VGG16 是最先进的深度学习模型，经过训练可以将图像分为 1000 个类别。假设我们只需要将输入数据分为两类——猫和狗，所以我们只需要一个二元分类器。为此，我们将输出类别从多元更改为二元。本项目展示了如何使用具有单个神经元的密集层替换 VGG16 现有的输出层。

6.6.1 创建项目

创建一个新的 Colab 项目并将其重命名为 tfhub-custom-etimator。使用如下代码导入 TensorFlow。

```
import tensorflow as tf
```

6.6.2 导入 VGG16

使用如下代码将 VGG16 预训练模型导入项目。

```
keras_Vgg16 = tf.keras.applications.VGG16(
    input_shape=(160, 160, 3), include_top=False)
```

将输出数据的维度设置为原始 VGG16 模型的输入维度。include_top 参数设置为 False 表明去除掉模型的最顶层。显然，我们不需要为这个已经训练好的模型重新生成随机权重，因此，将可训练参数设置为 False：

```
keras_Vgg16.trainable = False
```

6.6.3 创建自定义模型

在加载的 VGG16 模型基础上构建自定义模型，代码如下。

```
estimator_model = tf.keras.Sequential([
    keras_Vgg16,
    tf.keras.layers.GlobalAveragePooling2D(),
    tf.keras.layers.Dense(256),
```

```
tf.keras.layers.Dense(1)
])
```

我们以 VGG16 为基础层构建了一个 Sequential 模型。在此基础上,添加了一个池化层和两个密集层,最后一个密集层为二元输出层。

可以打印两个模型的摘要信息以查看我们所做的更改。首先使用如下代码打印原始模型的摘要信息。

```
keras_Vgg16.summary()
```

输出结果如图 6-24 所示。

```
Layer (type)                    Output Shape              Param #
=================================================================
input_1 (InputLayer)            [(None, 160, 160, 3)]     0
block1_conv1 (Conv2D)           (None, 160, 160, 64)      1792
block1_conv2 (Conv2D)           (None, 160, 160, 64)      36928
block1_pool (MaxPooling2D)      (None, 80, 80, 64)        0
block2_conv1 (Conv2D)           (None, 80, 80, 128)       73856
block2_conv2 (Conv2D)           (None, 80, 80, 128)       147584
block2_pool (MaxPooling2D)      (None, 40, 40, 128)       0
block3_conv1 (Conv2D)           (None, 40, 40, 256)       295168
block3_conv2 (Conv2D)           (None, 40, 40, 256)       590080
block3_conv3 (Conv2D)           (None, 40, 40, 256)       590080
block3_pool (MaxPooling2D)      (None, 20, 20, 256)       0
block4_conv1 (Conv2D)           (None, 20, 20, 512)       1180160
block4_conv2 (Conv2D)           (None, 20, 20, 512)       2359808
block4_conv3 (Conv2D)           (None, 20, 20, 512)       2359808
block4_pool (MaxPooling2D)      (None, 10, 10, 512)       0
block5_conv1 (Conv2D)           (None, 10, 10, 512)       2359808
block5_conv2 (Conv2D)           (None, 10, 10, 512)       2359808
block5_conv3 (Conv2D)           (None, 10, 10, 512)       2359808
block5_pool (MaxPooling2D)      (None, 5, 5, 512)         0
=================================================================
Total params: 14,714,688
Trainable params: 0
Non-trainable params: 14,714,688
```

图 6-24 VGG16 模型摘要信息

从图 6-24 可知,VGG16 模型有多个隐藏层,且可训练参数的总量超过了 1400 万。可以想象训练此模型所消耗的时间和资源。显然,当在自己的应用中使用这个模型时,我们不希望重新训练这个模型。

使用如下代码打印新构建模型的摘要信息。

```
estimator_model.summary()
```

输出结果如图 6-25 所示。

```
Model: "sequential"
_____
Layer (type)                 Output Shape              Param #
=================================================================
vgg16 (Model)                (None, 5, 5, 512)         14714688
_____
global_average_pooling2d (Gl (None, 512)               0
_____
dense (Dense)                (None, 256)               131328
_____
dense_1 (Dense)              (None, 1)                 257
=================================================================
Total params: 14,846,273
Trainable params: 131,585
Non-trainable params: 14,714,688
_____
```

图 6-25　扩展模型的摘要信息

可以看到，新构建的模型只有大约 10 万个可训练参数。

6.6.4　编译模型

使用所需的优化器、损失函数和评价指标编译模型，代码如下。

```
# Compile the model
estimator_model.compile(
    optimizer = 'adam',
    loss=tf.keras.losses.BinaryCrossentropy
            (from_logits = True),
    metrics = ['accuracy'])
```

6.6.5　创建 Estimator

使用 model_to_estimator 方法创建估算器，代码如下。

```
est_vgg16 = tf.keras.estimator.model_to_estimator
            (keras_model = estimator_model)
```

在训练估算器之前，我们需要对输入数据做一些处理。

6.6.6　处理数据

创建模型时，我们知道该模型需要维度为 160 像素 ×160 像素的图像，因此，使用 TensorFlow 中提供的预处理函数对图像进行预处理，代码如下。

```
IMG_SIZE = 160
import tensorflow_datasets as tfds
def preprocess(image, label):
```

```
    image = tf.cast(image, tf.float32)
    image = tf.image.resize
            (image, (IMG_SIZE, IMG_SIZE))
    return image, label
```

以下输入函数为犬和猫的训练数据定义了数据管道，这些训练数据可在 TensorFlow 的内置数据集中获取，代码如下。

```
def train_input_fn(batch_size):
  data = tfds.load('cats_vs_dogs',
          as_supervised=True)
  train_data = data['train']
  train_data = train_data.map(preprocess).shuffle(500).batch
                    (batch_size)
  return train_data
```

6.6.7 训练、评价

使用估算器实例的 train 方法训练该估算器，代码如下。

```
est_vgg16.train(input_fn =
      lambda: train_input_fn(32), steps = 500)
```

模型训练好之后，可以评价模型的性能，代码如下。

```
est_vgg16.evaluate(input_fn = lambda: train_input_fn(32),
steps=10)
```

> **注意**
> 由于没有单独的测试集可供使用，我们这里使用了不同的 steps 在训练集上对模型进行评价，结果如下。

```
{'accuracy': 0.96875, 'global_step': 500, 'loss': 0.27651623}
```

6.6.8 项目源码

清单 6-4 列出了本项目的完整源码，供读者快速参考。

清单 6-4 VGG16-custom-estimator 完整源码

源码清单
链　接：https://pan.baidu.com/s/1NV0rimQ_8kRz22xfFHN-Cw
提取码：1218

```
import tensorflow as tf

keras_Vgg16 = tf.keras.applications.VGG16(
    input_shape=(160, 160, 3), include_top=False)
keras_Vgg16.trainable = False
```

```python
estimator_model = tf.keras.Sequential([
    keras_Vgg16,
    tf.keras.layers.GlobalAveragePooling2D(),
    tf.keras.layers.Dense(256),
    tf.keras.layers.Dense(1)
])

keras_Vgg16.summary()

estimator_model.summary()
# Compile the model
estimator_model.compile(
    optimizer = 'adam',
    loss=tf.keras.losses.BinaryCrossentropy
            (from_logits = True),
    metrics = ['accuracy'])

est_vgg16 = tf.keras.estimator.model_to_estimator
            (keras_model = estimator_model)

IMG_SIZE = 160
import tensorflow_datasets as tfds
def preprocess(image, label):
 image = tf.cast(image, tf.float32)
 image = tf.image.resize
            (image, (IMG_SIZE, IMG_SIZE))
 return image, label

def train_input_fn(batch_size):
 data = tfds.load('cats_vs_dogs',
        as_supervised = True)
 train_data = data['train']
 train_data = train_data.map(preprocess).shuffle(500).batch
            (batch_size)
 return train_data

est_vgg16.train(input_fn = lambda: train_input_fn(32),
steps=500)

est_vgg16.evaluate(input_fn = lambda: train_input_fn(32),
steps=10)
```

综上所示，本项目通过一个简单的例子展示了如何将优秀的预训练模型应用于自定义模型，并转换为估算器。

总结

估算器通过为训练、评价和预测提供统一的接口来促进模型开发，同时，它将数据管道与模型的开发分离开来,从而允许我们轻松地试验不同数据集。估算器可分为预制估算器和自定义估算器。本章,我们学习了使用预制估算器解决分类和回归问题。自定义估算器用于将现有模型迁移到估算器接口中,以利用估算器的一些优势。可以在分布式环境甚至 CPU、GPU、TPU 上训练基于估算器的模型。使用估算器开发模型后,即使在分布式环境中也可以轻松部署,而无须更改任何代码。此外,我们还学习了如何基于优秀的预训练模型(如 VGG16)构建自定义模型。在开发过程中,我们建议首先尽可能使用预训练模型,如果不存在符合要求的模型,则考虑创建自定义估算器(模型)。

第 7 章
文本生成

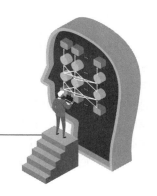

到目前为止,我们已经使用了多种普通神经网络,即无法记住过去信息的神经网络。这些网络接收一个固定大小的向量作为输入并产生一个固定大小的输出。以图像分类为例,模型的输入为图像,而输出为模型已训练(已知)的类别之一。现在,考虑需要用到之前预测知识的模型预测情况。举个例子,假设你正在看电影,你的大脑一直在猜测下一个场景是什么。你的猜测不仅取决于当前发生的事情,还取决于 15min 前甚至 1h 前发生的事情。就工作方式而言,普通神经网络没有记忆功能来记忆过去发生的事情,并将其应用于当前的预测。再举一个例子,某人告诉你说他来自印度。过了一段时间,要求你猜测他说什么语言。你猜测他说印度语的可能性在很大程度上取决于是否还记得他来自印度的事实。到目前为止,我们所学过的神经网络无法解决此类问题。为此,我们引入循环神经网络(recurrent neural networks, RNN)及其特殊版本长短时记忆(long short-term memory, LSTM)的概念。本章,我们将分别介绍 RNN 和 LSTM,然后介绍一个文本生成的示例。所有后续章节也将使用这些网络来解决不同领域的问题,因此,充分理解本章的内容将有助于更好地理解其余部分。

在本章中,我们将训练一个 RNN 字符级语言模型。首先提供一段文本作为模型的输入,然后训练模型以理解文本中字符的排序,最后让该模型预测用户定义的字符序列中某个字符出现的概率。简而言之,如果让你预测字符序列"hell"后面的字符,你很可能会给出答案"o",因为字符"o"出现的概率最高。现在,我们将训练神经网络来执行此类预测。

具体来说,本章将讨论两个文本生成的应用程序。第一个应用是一个简单的示例,我们可以了解文本生成过程的细节。在这个简单的应用中,将训练一个模型,使其根据现有姓名生成新的婴儿姓名。第二个应用是一个更大胆的尝试,在让模型从一本知名小说中学习语义信息和写作风格后,我们将尝试创作一部新的小说。

接下来,我们首先介绍什么是循环神经网络。

7.1 循环神经网络

与图像分类问题中使用的 CNN 一样，RNN 已经存在了几十年。近些年不断增长的计算资源和计算能力，使得研究人员可以充分发挥这些神经网络的潜力。RNN 现在已被用作强大的语言模型，如将语音转换为文本、执行语言翻译和生成手写文本、用于时间序列的预测等。我们也在计算机视觉应用中发现了 RNN，如执行帧级视频分类、图像字幕生成、视觉问答等。目前，有众多应用程序使用了 RNN 及其变体。在本书接下来的部分，我们将学习使用 RNN 来开发各种类型的应用程序。

RNN 允许我们对向量序列进行操作，这些向量可以是输入、输出或者两者均为向量，这三种情况如图 7-1 所示。

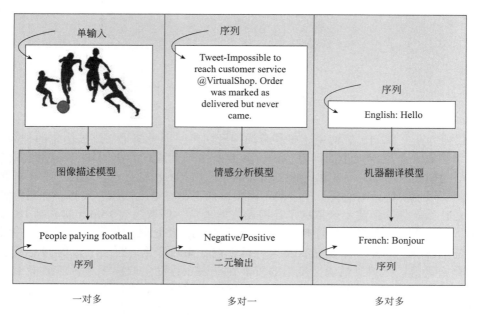

图 7-1 具有各种输入、输出的神经网络模型

如图 7-1 所示，对于图像描述模型，它需要单个输入（一幅图像），然后为图像生成一个标题（一个单词或字符序列）；在情感分析模型中，模型的输入为一条推文（由一系列单词或字符组成），模型的输出为二元的正值或负值；在机器翻译模型中，模型的输入为一个英语单词或句子（单词或序列），输出为对应的法语译文（单词或序列）。

了解了 RNN 的优势之后，我们将介绍一个简单的 RNN 架构。

7.1.1 朴素 RNN

描述 RNN 最好的方式是借助一幅网络架构图，如图 7-2 所示。

对于每个节点来说，X_i 为输入向量，h_i 为输出向量。整个网络中除第一个节点外，其余每个节点都将前一个节点的输出作为额外输入。此外，整个网络由大量节点组成，其中每个节点的输出不仅受其输入的影响，还受之前节点预测结果的影响。因此，RNN 是一种旨在识别序列数据中的模式的人工神经网络。简而言之，RNN 不仅接收当前的输入，而且还接收网络之前感知到的内容作为输入。因此，RNN 网络需要两个输入，即当前值和过去值。那么，问题是我们需要让 RNN 记忆多长时间的过去信息呢？当训练深度过大的 RNN 时，我们将面临所谓的梯度消失问题。梯度消失和梯度爆炸是训练大规模深度神经网络时常见的问题。在讨论梯度消失问题的解决方案之前，我们简要描述一下梯度消失和梯度爆炸的含义。

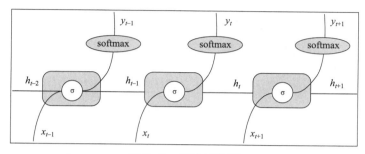

图 7-2　网络架构图

7.1.2　梯度消失和梯度爆炸

在神经网络模型的训练期间，参数梯度会在神经网络中反向传播，直到网络的初始层。在反向传播过程中，梯度信息每经过一层网络，都会进行一次矩阵乘法运算。对于深度网络，由于反向传播链很长，梯度会以指数方式缩小，不断接近远低于 1 的值，最终梯度信息消失，致使模型停止了学习，这被称为梯度消失问题。另一方面，如果梯度信息中具有大于 1 的较大值，这些值会随着反向传播在传播链中不断放大，最终数值过大导致模型崩溃，这被称为梯度爆炸问题。

从上述讨论中，我们可以理解为什么 RNN 容易出现梯度消失问题。为了克服梯度消失问题，提出了一种称为 LSTM 的新模型。接下来，我们将对 LSTM 进行讨论。

7.1.3　LSTM（一个特例）

LSTM 是一种特殊的 RNN。该模型能够学习长期依赖关系，同时避免出现简单 RNN 模型中的梯度消失问题。LSTM 于 1997 年由 Hochreiter 和 Schmidhuber 提出，随后许多人对其进行了改进和推广。LSTM 现在已被广泛用于解决各种各样的问题。

提出 LSTM 的主要动机是模型应该记住一段长时间的信息，这应该是模型的默认属性，而不是需要努力解决的问题。

LSTM 与其他循环神经网络一样，是具有重复模块的链式结构，只是重复模块的内部架构与标准的 RNN 不同。图 7-2 已经描述了一个标准的 RNN 模块及其模块链。RNN 中每个模块都是一个简单的架构与 RNN 不同，LSTM 模块具有更复杂的结构，如图 7-3 所示。

可以看到，每个重复模块有四个网络层，网络层之间以特殊的方式进行交互。在 LSTM 中，信息流通过一种称为细胞状态的机制进行传递，我们可以借助以下四种状态来解释整个流程：遗忘门、输入门、更新门、输出门。

接下来，我们逐一介绍上述状态。

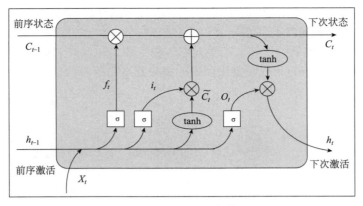

图 7-3 LSTM 架构

1. 遗忘门

遗忘门如图 7-4 中的高亮部分所示。

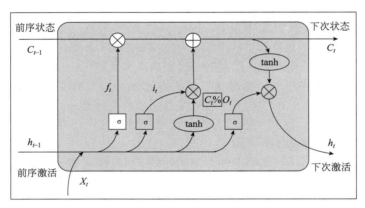

图 7-4 遗忘门

遗忘门使用一个 sigmoid 函数 (σ) 来决定舍弃或保留哪些信息。该门控的输入是前一个隐藏状态 (h_{t-1}) 和当前的输入 (X_t),输出为二元输出。遗忘门的输出为 False 时,网络将忘记之前的信息;输出为 True 时,网络将保留细胞状态 C_{t-1} 中的所有信息。数学上,可以使用如下公式表示。

$$f_t = \sigma\left(W_f\left[h_{t-1}, x_t\right] + b_f\right)$$

2. 输入门

输入门如图 7-5 中的高亮部分所示。

sigmoid 输入层决定了网络将更新哪些值,而 tanh 激活层创建了一个新的候选值 $\widetilde{C}_t$。数学上,可以使用如下公式表示。

$$i_t = \sigma\left(W_i\left[h_{t-1}, x_t\right] + b_i\right)$$

$$\widetilde{C}_t = \tanh\left(W_c\left[h_{t-1}, x_t\right] + b_c\right)$$

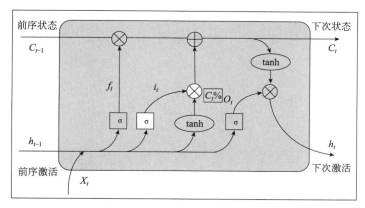

图 7-5 输入门

3. 更新门

更新门如图 7-6 中的高亮部分所示。

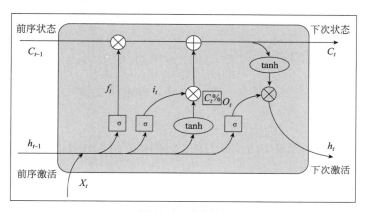

图 7-6 更新门

可以看到，新的细胞状态 C_t 等于 $t-1$ 时刻的细胞状态 C_{t-1} 乘以 f_t 与 i_t C_t 的结果之和。数学上，可以表示为如下公式。

$$C_t = f_t C_{t-1} + i_t \widetilde{C}_t$$

4. 输出门

输出门如图 7-7 中的高亮部分所示。

首先，sigmoid 层决定了模型要保留细胞状态的哪一部分。然后，通过 tanh 激活函数输入细胞状态，并将其与 sigmoid 层的输出相乘以获得最终输出。数学上，表示为如下公式。

$$o_t = \sigma\left(W_o\left[h_{t-1}, x_t\right] + b_o\right)$$

$$h_t = o_t \tanh(C_t)$$

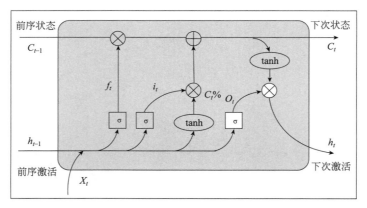

图 7-7　输出门

综上所述，通过对 RNN 模块的增强，LSTM 实现了长短期记忆并克服了在简单 RNN 中观察到的梯度消失问题。

由于具有长期记忆的能力，LSTM 被广泛应用于许多领域。本章，我们将重点介绍 LSTM 在语言模型中的应用。在语言模型中，依然有无数的应用程序，其中一些列举如下。

① 打字时预测下一个字符或单词。
② 单词或句子填充。
③ 学习大量文本的句法和语义。例如，学习莎士比亚和阿加莎·克里斯蒂的风格，并根据学习到的风格创作新故事。
④ 机器翻译。

简要介绍完所有相关理论之后，接下来，我们将介绍 LSTM 在文本生成中的应用。

7.2　文本生成

语言文字是作者向读者传达语法和语义的一系列字符，通常每个作者都有自己的写作风格。我们在文本生成中的任务是创建一个新文本，新文本遵循作者的写作风格，并且具有语法和语义意义。例如，根据莎士比亚的写作风格写一部新小说，或者根据之前判决的法律案件及此类文件的风格创作新的法律文件。在理解了其他 Python 程序的语法和代码结构后，你甚至可能还想生成新的 Python 源码。可能性永无止境，其中大部分需要在计算资源和计算时间方面进行大量处理。在本章接下来的示例中，我们首先通过一个简单的示例介绍如何生成婴儿的姓名，然后，介绍一个基于知名小说文本生成新文本的示例。

接下来，我们描述一些文本生成的相关理论。

7.2.1　模型训练

和字符级别的预测一样，我们需要将整个文本拆分为字符组，这些字符组被称为序列（sequences）。给定训练集中的一个输入文本为："Investors so much about their startup hubs. As a lot of mind I don't know the more airborning case of the European of the schedule, and from such sites …"。

把这个输入文本拆分为每 25 个字符一组的序列，如图 7-8 所示。

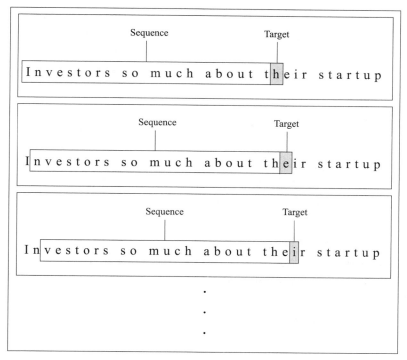

图 7-8　将输入文本拆分为序列

可以看到，前 25 个字符将按照它们出现的顺序组合在一起以创建第一个序列。此序列的最后为字母 h。因此，在训练网络的过程中，对于这个给定的字符序列，我们告诉模型下一个字符应该是 h。然后，将窗口向右移动一个字符，且对于新序列，我们告诉模型该序列的下一个字符是 e。接下来，将窗口继续向右移动一个字符，如图 7-8 中最后一个序列所示，下一个字符是 i。我们继续使用整个语料库训练模型，并告诉模型输入文本中任何给定的 25 个字符序列的下一个字符是什么。至此，相信读者可以理解为什么不能将简单的 DNN 用于此类程序，因为这类程序需要具有长期记忆的功能来记住当前序列及其后面的字符。

模型训练完成，我们认为它已经学习了文本的写作风格、语法和语义。显然，学习的语料库越多，模型对语料库的理解就越好。这类似于一个人在阅读了阿加莎·克里斯蒂写的几部小说后，便开始了解她的写作风格。

接下来，我们介绍如何使用训练的模型以所学的风格生成新文本。

7.2.2　预测

我们遵循在训练期间所做的类似的步骤，由预定义序列组成的种子序列开始。假设我们定义了一个含有 25 个字符的序列，与在模型训练期间所做的相同。用于预测的序列大小不必与训练期间使用的序列大小相同。根据给定的 25 个字符的序列，我们要求模型预测第 26 个字符，并将预测结果存储起来作为预测文本的一部分。接下来，将窗口向右移动一个字符，这次将刚刚预测的字符作为新序列中的最后一个字符（第 25 个字符）。对于这个新创建的序列，我们要求模型预测下一个字符。新的预测字符将被继续添加到下一个序列，以此类推。同样地，给定一个输入种子序列，我们可以使用模型预测任意数量的字符。如果模型对语料库中的序列学习得足够好，它就会产生有意义的文本。本章，我们将在大型语料库示例中逐步展示长时间训练过程中的结果。

7.2.3 模型定义

用于训练的神经网络模型主要由多个 LSTM 层组成，如图 7-9 所示。

在图 7-9 中，每个 LSTM 层都由大量节点组成，节点的数量越多，模型的长期记忆能力就越强。但是，这意味着要训练更多的网络权重。通过构建更深的网络模型，即增加更多的 LSTM 层数，该模型能够更好地学习复杂的数据。对于每个 LSTM 层，需要设置 "return_sequences" 参数为 True，使得其输出能够连接到下一个 LSTM 层。对于最后一个 LSTM 层，此参数设置为 False。最后一个 LSTM 层的输出将传入到密集层，然后根据 softmax 函数输出的每个可能字符的概率进行字符分类。

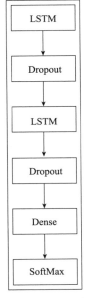

图 7-9　用于文本生成的神经网络

以上介绍了文本生成的简单原理，接下来我们通过一个实际示例来更好地理解这些理论，这个示例的目标是根据一组给定的名字生成新的婴儿名字。

7.3　生成新生儿名字

本项目使用著名的 Andrej Karpathy 博客中关于文本生成的数据集——循环神经网络的不合理有效性 (*The Unreasonable Effectiveness of Recurrent Neural Networks*, http://karpathy.github.io/2015/05/21/rnn-effectiveness/)。该数据集由一些已知的新生儿名字组成，名字之间由换行符分隔。我们将使用这些名字来训练 LSTM 模型，以使模型记住数据集中定义的序列。训练结束后，我们使用该模型根据学到的语义知识预测一些新名字。

7.3.1　创建项目

创建一个新的 Colab 项目并将其重命名为 TextGenerationBabyNames。使用如下代码导入所需的模块。

```
import sys
import re
import requests
import numpy as np
import tensorflow as tf
from tensorflow.keras import Sequential
from tensorflow.keras.callbacks
            import ModelCheckpoint
from tensorflow.keras.layers
            import Dense, Activation,
                Dropout, LSTM
```

7.3.2 下载文本

使用以下命令对指定的 URL 发出 HTTP 请求来下载数据。

r = requests.get('https://cs.stanford.edu/people/karpathy/namesGenUnique.txt')

传输完成后,将 HTTP 响应中的文本复制到局部变量中以提取数据集,代码如下。

raw_txt = r.text

调用 len 方法查看 raw_txt 变量中的数据长度:

len(raw_txt)

得到输出结果为 52127,表示数据集中共有 52127 个字符。要查看数据中的内容,只需在终端上打印 raw_txt:

raw_txt

输出结果如图 7-10 所示。

'jka\nDillie\nRyine\nCherita\nDasher\nChailine\nFrennide\nGremaley\nPatj\nH

图 7-10 数据文件中的内容

可以看到一个包含由换行符 (\n) 分隔的名字的长列表。我们可以使用 print 语句将前几个名字按行打印出来:

print(raw_txt[:100])

输出结果如下。

```
jka
Dillie
Ryine
Cherita
Dasher
Chailine
Frennide

Gremaley
Patj
Handi
Gully
Wennie
Ferentra
Jixandli
```

输出结果显示了数据集中的前 100 个字符,可以看到,这些名字的长度是不同的。

7.3.3 处理文本

为了将数据输入到模型中进行训练,我们首先需要去除 \n 字符。使用以下命令将其

替换为空格:

```
raw_txt = raw_txt.replace('\n' , ' ')
```

在 raw_txt 上创建一个 Pyhton 集合来从 raw_txt 中提取出所有不相同的字符:

```
set(raw_txt)
```

部分结果如图 7-11 所示。

可以看到,该集合包含空格、破折号(——)、点 (.)、冒号 (:) 和数字 (0～9) 等字符。在训练模型之前,这些字符应该从集合中删除,因为这些字符在模型生成的新名字中没有用。使用正则表达式删除这些字符:

```
raw_txt = re.sub('[-.0-9:]' , '' , raw_txt)
```

此外,我们不需要在生成的婴儿名字中同时存在大写和小写字母,因此,调用 lower 方法将所有字母转换为小写:

```
raw_txt1 = raw_txt.lower()
set(raw_txt1)
```

再次打印该集合,注意到现在 raw_text1 变量中只包含小写字母和空格字符。

读者可能想要知道,如果最后我们只想要一组小写字母和空格字符,为什么要对整个数据集进行处理? 为了生成婴儿名字,我们的目标字符集只包含字母和空格以分隔名字。但是,在更高级的文本生成应用程序中,如生成数学方程式文档、法律文档、科学摘要文档等,所需的目标字符集会更大,包含了各种字符。但是,随着目标字符集的增大,模型的训练周期也会呈指数增长。因此,一般情况下,我们会从原文中去掉一些不需要的字符。这种去除无关字符的操作可以使得模型的训练更快。

现在,打印新集合的大小,代码如下。

```
len1 = len(set(raw_txt1))
print (len1)
```

可以在终端中看到输出结果为 27。

由于神经网络模型只能进行数值运算,不能直接操作字符,因此我们需要将字母映射为不同的数字。此外,模型的预测结果会以数字形式输出,必须将这些数字映射回字母才能有意义。因此,我们创建了两个数组用于数字与字符之间的映射,代码如下。

```
chars = sorted(list(set(raw_txt1)))

arr = np.arange(0, len1)

char_to_ix = {}
ix_to_char = {}
for i in range(len1):
    char_to_ix[chars[i]] = arr[i]
    ix_to_char[arr[i]] = chars[i]
```

char_to_ix 数组提供了一个从集合中的字符到整数之间的映射,而 ix_to_char 提供了一个从整数到字符之间的反向映射。打印 ix_to_char 数组,可以看到如图 7-12 所示的部分输出。

```
{' ',
 '_',
 '.',
 '0',
 '1',
 '2',
 '3',
 '4',
 '5',
 '6',
 '7',
 '8',
 '9',
 ':',
 'A',
 'B',
 'C',
 'D',
```

图 7-11　不相同的部分字符集合

```
ix_to_char

{0: ' ',
 1: 'a',
 2: 'b',
 3: 'c',
 4: 'd',
 5: 'e',
 6: 'f',
 7: 'g',
 8: 'h',
 9: 'i',
 10: 'j',
 11: 'k',
 12: 'l',
 13: 'm',
```

图 7-12　预处理之后的部分不相同字符集合

接下来，使用如下代码片段创建输入和输出序列。

```
maxlen = 5
x_data = []
y_data = []
for i in range(0, len(raw_txt1) - maxlen, 1):
    in_seq = raw_txt1[i: i + maxlen]
    out_seq = raw_txt1[i + maxlen]

    x_data.append([char_to_ix[char]
                        for char in in_seq])
    y_data.append([char_to_ix[out_seq]])
    nb_chars = len(x_data)
    print('Text corpus: {}'.format(nb_chars))
    print('Sequences # ', int(len(x_data) / maxlen))
```

注意

我们定义的序列长度为 5，所以前 5 个字符将被输入网络模型，第 6 个字符将作为真实标签。在下一个循环中，从 2 ~ 6 的字符将作为输入序列，第 7 个字符作为真实标签，以此类推。因此，在 for 循环中，我们创建 x_data 来训练模型，y_data 为模型训练时使用的真实标签。

上述代码片段的输出如下。

```
Text corpus: 52038
Sequences # 10407
```

可以看到，该数据集由 52038 个字符组成，分为 10407 个序列，每个序列的长度为 5。接下来，我们将数据转换为 numpy 数组以输入到模型中，并将训练数据归一化为 0～1 之间的数值。代码如下。

```
x = np.reshape(x_data , (nb_chars , maxlen , 1))
x = x/float(len(chars))
```

将训练标签序列转换为类别列，代码如下。

```
y = tf.keras.utils.to_categorical(y_data)
y[:1]
```

在上述语句中，当我们对标签数据 y_data 转换完，并打印转换后的数据（y）中的一项时，得到如图 7-13 所示的输出。

```
array([[0., 0., 0., 0., 0., 0., 0., 0., 0., 1., 0., 0., 0., 0., 0.,
        0., 0., 0., 0., 0., 0., 0., 0., 0., 0.]], dtype=float32)
```

图 7-13　转换后的标签数据示例

可以看到，在此数组中，其中一个值为 1，其余值均为 0。数值 1 对应于 char_to_ix 数组中该索引处对应的字符。

打印 x_data 中的维度：

```
x.shape
```

输出结果如下。

```
(52038, 5, 1)
```

结果表明输入数据中有 52038 个序列，每个序列的长度为 5。我们也可以通过调用 y.shape 来查看 y 的维度：

```
y.shape
```

输出结果如下。

```
(52038, 27)
```

可以看到，模型输出一共有 27 个类别。

7.3.4　定义模型

定义模型的代码如下。

```
model = tf.keras.Sequential([
            tf.keras.layers.LSTM(256,
                input_shape = (maxlen, 1),
                    return_sequences = True),
```

```
                tf.keras.layers.LSTM(256,
                        return_sequences = True),
                tf.keras.layers.Dropout(0.2),
                tf.keras.layers.LSTM(64),
                tf.keras.layers.Dropout(0.2),
                tf.keras.layers.Dense(len(y[1]),
                        activation='softmax')
])
```

模型的摘要信息如下。

```
Model: "sequential"
_____
Layer (type)                 Output Shape              Param #
=================================================================
lstm (LSTM)                  (None, 5, 256)            264192
_____
lstm_1 (LSTM)                (None, 5, 256)            525312
_____
dropout (Dropout)            (None, 5, 256)            0
_____
lstm_2 (LSTM)                (None, 64)                82176
_____
dropout_1 (Dropout)          (None, 64)                0
_____
dense (Dense)                (None, 27)                1755
=================================================================
Total params: 873,435
Trainable params: 873,435
Non-trainable params: 0
```

可以看到，该模型有三个 LSTM 层，每个 LSTM 层由 500 个节点组成。每个 LSTM 层后衔接一个具有 20% dropout 的 dropout 层。模型的最后一层是一个密集层，包含 27 个节点，最后由一个 softmax 层进行分类。

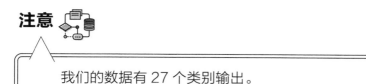

我们的数据有 27 个类别输出。

该模型的可视化结构如图 7-14 所示。

7.3.5 编译

使用 categorical cross-entropy 损失函数和 Adam 优化器编译模型，代码如下。

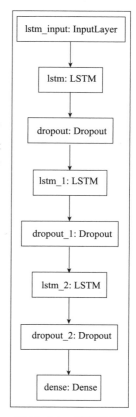

图 7-14 模型架构

```
model.compile(loss = 'categorical_crossentropy',
                       optimizer = 'adam')
```

> 对于这类语言模型的问题，没有测试集可供使用。我们对整个数据集进行建模以预测序列中每个分类字符的概率。模型预测下一个字符的正确率对我们来说并不重要，相反，我们感兴趣的是最小化模型的损失函数。因此，我们试图在模型的泛化性能和过拟合之间取得平衡。

7.3.6 创建 checkpoints

训练 LSTM 网络通常需要很长时间。由于网络本身的性质，每个 epoch 之后的损失值可能会增大，也可能会减小。通常情况下，损失值最小对应的模型的性能最优，可以得到最好的预测结果，因此，我们需要捕获损失值最小的 epoch 对应的模型权重。通过使用 ModelCheckPoint 方法并设置回调函数完成，代码如下。

```
filepath = "model_weights_babynames.hdf5"
checkpoint = ModelCheckpoint(filepath,
            monitor = 'loss', verbose = 1,
            save_best_only = True, mode = 'min')
model_callbacks = [checkpoint]
```

我们在第 4 章的项目中使用过早停（early stopping）这样的回调函数，本小节介绍的是另一种类型的回调函数，该函数会在每个 epoch 之后调用。

7.3.7 训练

调用模型的 fit 方法训练模型：

```
model.fit(x,y, epochs = 300, batch_size = 32 ,
          callbacks = model_callbacks)
```

我们将 epoch 设置为 10，batch_size 设置为 32 个序列，稍后我们还将展示增大这两个数值的效果。当在 GPU 上运行这段代码时，训练时间大约为每个 epoch 花费 6s，这对我们来说是可接受的。这里，我们关心训练时间的原因是，即使利用一组 GUP 进行分布式训练，在大型语料库上训练具有大量分类类别的 LSTM 模型也需要几个小时才能完成。

训练完成后，接下来，让我们尝试做一些预测。

7.3.8 预测

首先，需要创建一个输入序列。我们从原始数据集上定义了一个输入序列，使用如下代码生成输入序列。

```
pattern = []
```

```
seed = 'handi'
for i in seed:
    value = char_to_ix[i]
    pattern.append(value)
```

我们使用的输入序列为"handi",其长度为 5。使用之前创建的 char_to_ix 数组将种子序列中的每个字符转换为对应的整数值。

接下来,使用一个 for 循环来预测种子序列"handi"之后的 100 个字符。作为对比,我们首先打印种子序列,并将字符表中的字符数量设置为 n_vocab 变量。

```
print(seed)
n_vocab = len(chars)
```

设置预测的循环次数为 100:

```
for i in range(100):
```

使用以下两个语句改变输入数据的维度,并对变换后的数据进行归一化:

```
X = np.reshape(pattern, (1, len(pattern) , 1))
X = X/float(n_vocab)
```

将数据输入到模型,并使其预测给定输入模式的下一个字符:

```
int_prediction = model.predict(X , verbose = 0)
```

提取最大预测概率对应字符的索引,并使用之前创建的 ix_to_char 数组将预测的索引值转换为字符:

```
index = np.argmax(int_prediction)
prediction = ix_to_char[index]
```

将预测的字符打印在终端上:

```
sys.stdout.write(prediction)
```

将预测的字符追加到我们的种子序列中,并通过提取最后 5 个字符(输入序列的长度)重新创建一个新序列,然后将创建的新序列输入模型进行下一次预测。以此类推,从定义的种子序列开始,我们使用模型进行 100 次预测。代码如下。

```
pattern.append(index)
pattern = pattern[1:len(pattern)]
```

清单 7-1 列出了生成 100 个字符的完整 for 循环代码,供读者快速参考。

清单 7-1 用于生成 100 个字符的循环代码

源码清单
链　接:https://pan.baidu.com/s/1NV0rimQ_8kRz22xfFHN-Cw
提取码:1218

```
print(seed)
n_vocab = len(chars)
for i in range(100):
    X = np.reshape(pattern , (1, len(pattern) , 1))
    X = X/float(n_vocab)
    int_prediction = model.predict(X , verbose = 0)
```

```
index = np.argmax(int_prediction)
prediction = ix_to_char[index]
sys.stdout.write(prediction)
pattern.append(index)
pattern = pattern[1:len(pattern)]
```

7.3.9　项目源码 –TextGenerationBabyNames

清单 7-2 列出了生成婴儿名字项目的完整源码。

清单 7-2 TextGenerationBabyNames 完整源码

源码清单
链　接：https://pan.baidu.com/s/1NV0rimQ_8kRz22xfFHN-Cw
提取码：1218

```
import sys
import re
import requests
import numpy as np
import tensorflow as tf
from tensorflow.keras import Sequential
from tensorflow.keras.callbacks
           import ModelCheckpoint
from tensorflow.keras.layers
           import Dense, Activation,
                  Dropout, LSTM

r = requests.get('https://cs.stanford.edu/people/karpathy/
namesGenUnique.txt')

raw_txt = r.text
len(raw_txt)

raw_txt

print(raw_txt[:100])

set(raw_txt)

len(set(raw_txt))
raw_txt = raw_txt.replace('\n' , ' ')

set(raw_txt)

len(set(raw_txt))

raw_txt = re.sub('[-.0-9:]' , '' , raw_txt)
len(set(raw_txt))

raw_txt1 = raw_txt.lower()
set(raw_txt1)

len1 = len(set(raw_txt1))
```

```python
print (len1)

len1
chars = sorted(list(set(raw_txt1)))

arr = np.arange(0, len1)

char_to_ix = {}
ix_to_char = {}
for i in range(len1):
    char_to_ix[chars[i]] = arr[i]
    ix_to_char[arr[i]] = chars[i]

char_to_ix

ix_to_char

#print("Total length of file : {}".format(len(raw_txt1)))

maxlen = 5
x_data = []
y_data = []
for i in range(0, len(raw_txt1) - maxlen, 1):
    in_seq = raw_txt1[i: i + maxlen]
    out_seq = raw_txt1[i + maxlen]

    x_data.append([char_to_ix[char]
                    for char in in_seq])
    y_data.append([char_to_ix[out_seq]])
nb_chars = len(x_data)
print('Text corpus: {}'.format(nb_chars))
print('Sequences # ', int(len(x_data) / maxlen))

#y_data[:5]

#x_data[1][:]

x = np.reshape(x_data , (nb_chars , maxlen , 1))
x = x/float(len(chars))
y = tf.keras.utils.to_categorical(y_data)
y[:1]

x.shape

y.shape

model = tf.keras.Sequential([
            tf.keras.layers.LSTM(256,
                    input_shape = (maxlen, 1),
                        return_sequences = True),
```

```python
                tf.keras.layers.LSTM(256,
                        return_sequences = True),
                tf.keras.layers.Dropout(0.2),
                tf.keras.layers.LSTM(64),
                tf.keras.layers.Dropout(0.2),
                tf.keras.layers.Dense(len(y[1]),
                        activation='softmax')
])

model.summary()

model.compile(loss = 'categorical_crossentropy',
                        optimizer = 'adam')

filepath = "model_weights_babynames.hdf5"
checkpoint = ModelCheckpoint(filepath,
            monitor = 'loss', verbose = 1,
            save_best_only = True, mode = 'min')
model_callbacks = [checkpoint]

model.fit(x,y, epochs = 300, batch_size = 32 ,
                callbacks = model_callbacks)

pattern = []
seed = 'handi'
for i in seed:
    value = char_to_ix[i]
    pattern.append(value)

print(seed)
n_vocab = len(chars)
for i in range(100):
    X = np.reshape(pattern , (1, len(pattern) , 1))
    X = X/float(n_vocab)
    int_prediction = model.predict(X , verbose = 0)
    index = np.argmax(int_prediction)
    prediction = ix_to_char[index]
    sys.stdout.write(prediction)
    pattern.append(index)
    pattern = pattern[1:len(pattern)]

from google.colab import drive
drive.mount('/content/drive')

cd 'My Drive'

model.save('baby_names_model.h5')

from tensorflow.keras.models import load_model
saved_model = load_model('baby_names_model.h5')
```

```
pattern = []
seed = 'bgajm'
for i in seed:
    value = char_to_ix[i]
    pattern.append(value)

print(seed)
n_vocab = len(chars)
for i in range(100):
    X = np.reshape(pattern , (1, len(pattern) , 1))
    X = X/float(n_vocab)
    int_prediction = saved_model.predict
                    (X , verbose = 0)
    index = np.argmax(int_prediction)
    prediction = ix_to_char[index]
    sys.stdout.write(prediction)
    pattern.append(index)
    pattern = pattern[1:len(pattern)]
```

运行上述代码，可以得到如图 7-15 所示的输出结果。

handi
saree carie carie carie carie carie carie carie carie carie carie

图 7-15 以"handi"为种子序列生成的婴儿名字

可以看到，网络模型生成了一些有意义的名字。接下来，我们增加 epoch 的数量以查看预测结果是否有所改进。

使用 epoch 为 50、batch_size 为 128、种子序列为"handi"训练模型，可以得到如图 7-16 所示的预测结果。

handi
a craskie llonpre trestor allussea wande cherita merylee gilphon salda gerrek

图 7-16 50 个 epoch 之后的婴儿名字

这次的预测结果可能更好些，当然这取决于个人对名字的选择，但起码生成的名字不再重复了。

需要注意，通过提升训练 epoch 的数量、增加 LSTM 层中的节点数、增加 LSTM 的层数、调整序列的长度等操作，都可能会提高预测的质量。事实上，对于来自原始文本的已知种子序列，经过完全训练的大规模网络模型，在理论上能够产生与原始文本相同的输出。这好比一个拥有超强记忆的人，能够以与原始文本相同的顺序复制出原始文本。因此，LSTM 神经网络可以有效地用于生成与原始文本相似的高质量文本。

就生成婴儿名字的应用而言，使用神经网络生成数据集中已经存在的名字是没有意义的。

要生成数据集中不存在的名字并且听起来与现有名字相似，我们需要使用原始文本中不存在的输入序列作为种子序列，为此，通常会生成一个随机种子。笔者尝试了一个随机种子"bgajm"，得到的输出结果如图 7-17 所示。

```
bgajm
 sherine margoria solly adisa naril rurodore je viyette alady goyne jasel minco
```

图 7-17　根据随机种子序列"bgajm"生成的名字

在继续讨论更复杂的文本生成问题之前，我们首先介绍神经网络训练时间对 batch_size 的一个更重要的依赖。如果 batch_size 较小，则需要更多时间来覆盖整个语料库中的所有字符，从而导致训练时间增加；而增加 batch_size 的大小则需要额外的系统内存资源。因此，需要在 batch_size 和分配的训练资源之间取得折中，同时获得最佳的网络训练时间。考虑到 LSTM 的训练时间较长，我们建议在网络训练完成后保存模型，然后再将其应用于不同的种子序列。接下来，我们将展示如何保存和重用模型以进行后续预测。

7.3.10　保存、重用模型

我们将训练好的模型保存到 Google Drive。首先装载磁盘驱动器，代码如下：

```
from google.colab import drive
drive.mount('/content/drive')
```

在磁盘装载过程中，会要求用户输入授权码。

装载完磁盘驱动器后，切换到要保存模型的路径下：

```
cd 'drive/My Drive/TextGenerationDemo'
```

调用模型的 save 方法将其保存为所需的文件名：

```
model.save('baby_names_model.h5')
```

可以随时调用 load_model 方法重新加载已保存的模型：

```
from tensorflow.keras.models import load_model
saved_model = load_model('baby_names_model.h5')
```

加载的模型在变量 saved_model 中，并可用于模型预测或进一步训练。

我们使用了一个简单的示例对文本生成进行了介绍，接下来我们继续介绍一个更现实的示例，该示例使用了一个更大的文本语料库。

7.4　高级文本生成

在本项目中，我们使用列夫·托尔斯泰 (Leo Tolstoy) 的著名小说《战争与和平》(https://cs.stanford.edu/people/karpathy/char-rnn/warpeace_input.txt) 中的文本。这部小说使用了许多特殊字符，如问号、感叹号、引号和句号。在之前的示例中，我们去除了这些字符，因为只想生成婴儿名称。本项目不会删除这些特殊字符，因为我们想创作另一部小说，或者至少生成一个包含所有字符的段落。

在如此庞大的语料库上训练复杂的 LSTM 模型所需的时间非常长，因此，我们需要定期存储模型的训练状态。本项目将展示如何在每个 epoch 保存模型的状态。这样，如果在训练期间断开了连接，我们可以从已知的断点（checkpoint）处继续训练。此外，我们还将创建另一个测试模型性能的工具，使得它在每个 epoch 结束时对一个固定的种子进行预测，将预测结果存储到 Google Drive 中的一个文件中。这样，我们可以一直在后台训练模型，并通过查看磁盘上预测文件的内容来定期查看模型性能。

7.4.1 创建项目

创建一个新的 Colab 项目并将其重命名为 LargeCorpusTextGeneration。使用如下代码导入所需的库。

```
import sys
import requests
import numpy as np
import tensorflow as tf
from tensorflow.keras import Sequential
from tensorflow.keras.callbacks
            import ModelCheckpoint
from tensorflow.keras.layers
            import Dense, Activation,
                   Dropout, LSTM
```

我们将在训练期间的每个 epoch 结束时保存 checkpoint 数据和模型的预测结果。装载 Google Drive 并指向相应的路径以保存上述数据：

```
from google.colab import drive
drive.mount('/content/drive')
```

在磁盘装载过程中，会要求用户输入授权码。装载完磁盘驱动器后，设置用于存储数据的路径如下。

```
cd '/content/drive/My Drive/TextGenerationDemo'
```

7.4.2 加载文本

通过发出以下 HTTP 请求将小说文本加载到项目中：

```
r = requests.get("https://cs.stanford.edu/people/karpathy/char-rnn/warpeace_input.txt")
```

将小说文本从响应对象读入局部变量：

```
raw_txt = r.text
```

得到文本中不相同字符的列表，并打印语料库的大小和输出类别的数量：

```
chars = sorted(list(set(raw_txt)))
print("Corpus: {}".format(len(raw_txt)))
print("Categories: {}".format(len(chars)))
```

可以看到如下输出结果。

```
Corpus: 3258246
Categories: 87
```

这部小说有超过 300 万个字符，文本中包含 87 个不相同的字符。这两个数字都比生成婴儿名字模型中的数字大得多。

7.4.3 处理数据

与前一个示例相同，我们需要将所有不相同的字符映射为整数以供模型处理，还需要提供对应的反向映射来解释模型的输出。构建以下两个数组实现该功能。

```
ix_to_char = {ix:char for ix,
              char in enumerate(chars)}
char_to_ix = {char:ix for ix,
              char in enumerate(chars)}
```

使用以下代码段将整个文本拆分为序列。

```
maxlen = 10
x_data = []
y_data = []
for i in range(0, len(raw_txt) - maxlen, 1):
    in_seq = raw_txt[i: i + maxlen]
    out_seq = raw_txt[i + maxlen]
    x_data.append([char_to_ix[char]
                   for char in in_seq])
    y_data.append([char_to_ix[out_seq]])
nb_chars = len(x_data)
print('Number of sequences:',
      int(len(x_data)/maxlen))
```

此代码与前一个示例中的代码相似。当序列大小为 10 时，创建的序列数为 325823。上述代码的输出如下。

```
Number of sequences: 325823
```

更改数据的维度，然后对数据进行归一化处理，使其满足网络输入数据的要求：

```
# scale and transform data
x = np.reshape(x_data , (nb_chars , maxlen , 1))
n_vocab = len(chars)
x = x/float(n_vocab)
```

调用 to_categorical 方法将输出的结果转换为对应的类别：

```
y = tf.keras.utils.to_categorical(y_data)
```

使用 print 语句查看输入和输出的大小：

```
print("The shape of x_training data : " ,x.shape)
print("The shape of y_training data : " ,y.shape)
```

输出结果如下。

```
The shape of x_training data : (3258236, 10, 1)
The shape of y_training data : (3258236, 86)
```

可以看到，训练数据中有大量的字符序列。网络的输入序列长度为 10，输出为 86 个类别。

7.4.4 定义模型

使用如下代码定义模型。

```
Model = tf.keras.Sequential([
            tf.keras.layers.LSTM(800 ,
            input_shape = (len(x[1]) , 1) ,
                return_sequences = True),
            tf.keras.layers.Dropout(0.2),
                tf.keras.layers.LSTM(800,
                return_sequences = True),
            tf.keras.layers.Dropout(0.2),
            tf.keras.layers.LSTM(800),
            tf.keras.layers.Dropout(0.2),
            tf.keras.layers.Dense(len(y[1]),
                activation = 'softmax')
])
```

该模型的整体结构与前一个示例相同，考虑到输入文本语料库的大小，我们增加了每一个网络层的节点数量。

使用交叉熵损失和 Adam 优化器编译模型，代码如下。

```
Model.compile(loss = '
            categorical_crossentropy' ,
            optimizer = 'adam')
```

接下来，将介绍该项目最重要的部分，即在每个 epoch 结束后保存模型的状态及其预测结果。

7.4.5 创建 checkpoints

要创建 checkpoints，我们需要为训练方法创建自定义回调函数，首先设置一个 checkpoint 文件名：

```
filepath = "model_weights_saved.hdf5"
```

使用 ModelCheckpoint 方法创建回调函数：

```
checkpoint = ModelCheckpoint(filepath,
            monitor = 'loss', verbose = 1,
            save_best_only = True, mode = 'min')
```

在回调函数中，我们将监控损失值并在损失值最小的时刻保存模型的权重。

创建一个变量来列出回调的数量，在本示例中只有一个。代码如下。

```
model_callbacks = [checkpoint]
```

7.4.6 自定义回调类

我们创建另一个回调函数来存储模型的预测结果。该函数在每个 epoch 结束后将预测结果存储在一个文本文件中。创建一个全局变量来跟踪 epoch 数量：

```
epoch_number = 0
```

声明一个用于存储预测结果的文件名，代码如下。

```
filename = 'predictions.txt'
```

如果文件已经存在，则覆盖掉文件的内容：

```
file = open(filename , 'w')
file.truncate()
file.close()
```

声明自定义类：

```
class CustomCallback(tf.keras.callbacks.Callback):
```

定义一个名为 on_epoch_end 的方法：

```
def on_epoch_end(self , epoch , logs = None):
```

该方法将在每个 epoch 结束时调用。在方法体中，首先增加全局 epoch 计数：

```
global epoch_number
epoch_number = epoch_number + 1
```

以追加模式打开预测结果文件用于添加预测结果：

```
filename = 'predictions.txt'
file = open(filename , 'a')
```

声明种子序列：

```
seed = "looking fo"
```

通过一个简单的 for 循环将种子序列转换为数值：

```
pattern = []
for i in seed:
    value = char_to_ix[i]
    pattern.append(value)
```

首先将 epoch 数写入文件：

```
file.seek(0)
file.write("\n\n Epoch number :
          {}\n\n".format(epoch_number))
```

创建一个循环进行 100 次预测：

```
for i in range(100):
```

改变输入数据的维度，并对数据进行归一化处理：

```
X = np.reshape(pattern ,
               (1, len(pattern) , 1))
X = X/float(n_vocab)
```

调用模型的 predict 方法对给定种子序列进行预测：

```
int_prediction = Model.predict(X ,
                      verbose = 0)
```

选取最大输出概率对应的字符并将其复制到预测变量中：

```
index = np.argmax(int_prediction)
prediction = ix_to_char[index]
```

将预测的字符写入文件：

```
file.write(prediction)
```

将预测的字符添加到输入序列中并提取最后 10 个字符（种子的大小）以创建新的输入序列：

```
pattern.append(index)
pattern = pattern[1:len(pattern)]
```

以此类推，进行下一次预测。当预测 100 次以后，关闭文件：

```
file.close()
```

7.4.7 模型训练

创建完两个回调函数后，使用以下语句训练模型。

```
Model.fit(x, y , batch_size = 200,
              epochs = 10 ,
              callbacks = [CustomCallback() ,
              model_callbacks])
```

我们定了一个足够大的 batch_size，为 200 个序列，之前定义的两个回调函数在 fit 方法中指定。自定义的回调函数用于存储预测结果，模型的回调函数用于存储模型的状态。

当运行模型训练时，在 GPU 上运算的每个 epoch 大约需要 550s。幸运的是，训练和预测结果会在每个 epoch 结束时保存。因此，我们不用担心网络超时或断开连接，因为我们能够看到最后一次连接断开时的模型性能，并且可以从最后一个断点处继续训练。代码将在"结果"小节之后讨论。

7.4.8 结果

第 1 个、第 5 个和第 10 个 epoch 后的预测结果如下。

```
Epoch number : 1
```

r the soldiers were all the same time the soldiers were all the same time the soldiers were all the

Predictions after 5 epochs:

Epoch number : 5
r the first time to the countess was a serious and the servants and the servants and the servants an

Predictions after 10 epochs:

Epoch number : 10

r the first time they had been sent to the countess was a still more than the countess was a still m

可以看到，随着 epoch 数量的增加，模型的性能在不断提升。

7.4.9 断点续训练

使用以下代码从最后一个已知的 checkpoint 继续训练：

```
try:
    Model.load_weights(filepath)
except Exception as error:
    print("Error loading in model : {}".format(error))
```

我们只需从存储的 checkpoint 文件中加载权重，然后调用模型的 fit 方法即可继续训练：

```
Model.fit(x, y , batch_size = 200, epochs = 10 ,
          callbacks = [CustomCallback() ,
          model_callbacks])
```

注意

> epoch 数值将从全局 epoch 的最后一个值开始。

以下是笔者通过断点继续训练进行了 50 个 epoch 的一些预测结果。

Epoch number : 20

r the first time to the countess was a small conversation with the state of a strange and the counte

Epoch number : 30

r the first time the soldiers who were all the same time he had seen and was about to see the counte

Epoch number : 40

r the first time the staff officer who had been at the same time he had seen him to the countess was

Epoch number : 50

r the first time the streets of the countess was a man of his soul and the same time he had seen and

可以看到，随着 epoch 数量的增加，模型的性能在继续提升。

7.4.10 过程观察

为了进行一些实验并减少训练时间，我们将每个 LSTM 层中的节点数量从 800 减少到 100。这将训练时间从每个 epoch 大约 10 min 缩短为 2 min。运行 100 个 epoch 之后，部分结果如下。

Epoch number : 1

and the same and the same and the same and the same and the same and the same and the same and the

Epoch number : 25

and the service and the service and the service and the service and the service and the service and

Epoch number : 50

and the same to the same to the same to the same to the same to the same to the same to the same to

Epoch number : 75

and the strength to the same and the strength to the same and the strength to the same and the stre

Epoch number : 100

and the same and the same and the same and the same and the same and the same and the same and the

可以看到，尽管 epoch 数量很大，但模型却不再学习了。因此，我们可以得出结论，理解大规模文本语料库时，需要更多的内存空间，同时需要更多的 LSTM 节点数量。

我们将每个 LSTM 层中的节点数量增加到 500 来进行另一个实验，训练时间增加到每个 epoch 约 11min。结果如下。

```
Epoch number : 5

the same time to the same time to the same time to the same
time to the same time to the same time

Epoch number : 10

the countess was a strange and the same things and the same
things and the same things and the same

Epoch number : 15

the countess and the same time the soldiers and the same time
the soldiers and the same time the so

Epoch number : 20

, and the same time the countess was still the staff of the
countess was still the staff of the coun
```

通过增加节点数量，我们看到了性能有所提升。在第 20 个 epoch，该模型甚至产生了一个逗号。继续训练可能会进一步提升模型性能。

7.4.11 项目源码

LargeCorpusTextGeneration 的完整源码如清单 7-3 所示。

清单 7-3 LargeCorpusTextGeneration 完整源码

源码清单
链　接：https://pan.baidu.com/s/1NV0rimQ_8kRz22xfFHN-Cw
提取码：1218

```python
import sys
import requests
import numpy as np
import tensorflow as tf
from tensorflow.keras import Sequential
from tensorflow.keras.callbacks
            import ModelCheckpoint
from tensorflow.keras.layers
            import Dense, Activation,
                    Dropout, LSTM

from google.colab import drive
drive.mount('/content/drive')

cd '/content/drive/My Drive/TextGenerationDemo'

r = requests.get("https://cs.stanford.edu/people/karpathy/char-rnn/warpeace_input.txt")

raw_txt = r.text

chars = sorted(list(set(raw_txt)))
```

```python
print("Corpus: {}".format(len(raw_txt)))
print("Categories: {}".format(len(chars)))

ix_to_char = {ix:char for ix,
              char in enumerate(chars)}
char_to_ix = {char:ix for ix,
              char in enumerate(chars)}

maxlen = 10
x_data = []
y_data = []
for i in range(0, len(raw_txt) - maxlen, 1):
    in_seq = raw_txt[i: i + maxlen]
    out_seq = raw_txt[i + maxlen]
    x_data.append([char_to_ix[char]
                   for char in in_seq])
    y_data.append([char_to_ix[out_seq]])
nb_chars = len(x_data)
print('Number of sequences:',
      int(len(x_data)/maxlen))

# scale and transform data
x = np.reshape(x_data , (nb_chars , maxlen , 1))
n_vocab = len(chars)
x = x/float(n_vocab)

x.shape

y = tf.keras.utils.to_categorical(y_data)

print("The shape of x_training data : " ,x.shape)
print("The shape of y_training data : " ,y.shape)

Model = tf.keras.Sequential([
                tf.keras.layers.LSTM(800 ,
                input_shape = (len(x[1]) , 1) ,
                    return_sequences = True),
                tf.keras.layers.Dropout(0.2),
                    tf.keras.layers.LSTM(800,
                    return_sequences = True),
                tf.keras.layers.Dropout(0.2),
                tf.keras.layers.LSTM(800),
                tf.keras.layers.Dropout(0.2),
                tf.keras.layers.Dense(len(y[1]),
                    activation = 'softmax')
])

Model.compile(loss = '
              categorical_crossentropy' ,
              optimizer = 'adam')
```

```python
filepath = "model_weights_saved.hdf5"
checkpoint = ModelCheckpoint(filepath,
            monitor = 'loss', verbose = 1,
            save_best_only = True, mode = 'min')
model_callbacks = [checkpoint]

epoch_number = 0
filename = 'predictions.txt'
file = open(filename , 'w')
file.truncate()
file.close()
class CustomCallback(tf.keras.callbacks.Callback):

    def on_epoch_end(self , epoch , logs = None):
        global epoch_number
        epoch_number = epoch_number + 1

        filename = 'predictions.txt'
        file = open(filename , 'a')
        seed = "looking fo"

        pattern = []
        for i in seed:
            value = char_to_ix[i]
            pattern.append(value)
        file.seek(0)
        file.write("\n\n Epoch number :
                    {}\n\n".format(epoch_number))
        for i in range(100):
            X = np.reshape(pattern ,
                            (1, len(pattern) , 1))
            X = X/float(n_vocab)
            int_prediction = Model.predict(X ,
                                verbose = 0)
            index = np.argmax(int_prediction)
            prediction = ix_to_char[index]
            #sys.stdout.write(prediction)
            file.write(prediction)
            pattern.append(index)
            pattern = pattern[1:len(pattern)]
        file.close()

Model.fit(x, y , batch_size = 2000, epochs = 10 ,
            callbacks = [CustomCallback() ,
            model_callbacks])

try:
    Model.load_weights(filepath)
except Exception as error:
    print("Error loading in model :
```

```
            {}".format(error))
Model.fit(x, y , batch_size = 200, epochs = 25 ,
            callbacks = [CustomCallback() ,
            model_callbacks])
```

7.5 进一步工作

Andrej Karpathy 在文本生成方面所做的工作值得一提，发表在他著名的博客 *The Unreasonable Effectiveness of Recurrent Neural Networks* (http://karpathy.github.io/2015/05/21/rnn-effectiveness/) 中。在他的实验中，使用了莎士比亚文学、维基百科文章、LaTeX（代数几何语言文本），甚至是 Linux 源码。经过充分的训练后，网络模型产生的结果令人难以置信。该模型能够生成奇妙的数学方程，大多数情况下这些数学方程的句法都是正确的。该模型还能够生成几乎可以编译的计算机源代码。这个项目被认为是使用 LSTM 生成高质量文本的一个很好的证明。

从本章讨论的两个示例中，我们可以发现训练 LSTM 以生成高质量文本时需要大量的资源和时间，同时还需要进行一些实验才能获得最佳结果。以下是读者在微调文本生成应用程序时可以考虑的一些技巧。

① 为了节省训练时间，可以删除不需要的字符来减少词汇量。
② 添加更多的 LSTM 和 dropout 层，每层设置更多 LSTM 单元。
③ 尝试调整 batch_size 的大小、优化器和序列长度等参数，看看哪个效果最好。
④ 尝试更大的 epoch 数量。
⑤ 使用大型文本语料库。

总结

在本章中，我们学习了一种新的神经网络架构：RNN。LSTM 是 RNN 的一个特例。传统的 DNN 不具备记忆能力，而 LSTM 具有长短期记忆功能，因此，对于文本生成这样的问题，LSTM 做得非常出色。在本章中，我们学习了使用基于 LSTM 的神经网络生成婴儿名字，我们还学习了如何通过著名作家的小说生成高质量的文本段落。

在下一章，我们将介绍另一种用于机器翻译的语言模型，如从英语翻译为法语或者从西班牙语翻译为日语。

第 8 章
语言翻译

以前,当我们在国际机场转机时,由于语言不通,很难读懂机场指示牌。现在,我们只需将手机指向这些标志,手机中的应用程序就会提供各种语言的翻译。这些翻译是如何完成的?这背后涉及的技术不止一种,核心是一个机器学习模型,使用大量预定义单词训练单词到单词的翻译模型。显然,这种单词到单词或序列到序列的简单机器翻译在机场和路标等固定场景下可以非常准确,但可能无法在自然语言句子中产生理想的翻译结果。例如,"How are you today"这样的语句,不能简单地通过单个单词的翻译拼接成完整的语句翻译结果。为了实现此类较为复杂的翻译任务,Google 在 2016 年设计了基于统计的语言翻译方法,提出神经机器翻译模型(neural machine translation, NMT)。本章将介绍如何开发机器翻译模型。

8.1 sequence-to-sequence 模型

在第 7 章中,讲解了循环神经网络和长短期记忆人工神经网络。本章将根据这些网络模型创建编码器、解码器结构和注意力机制实现机器翻译模型。下面,详细解释什么是编码器、解码器和注意力机制。编码器、解码器模型实现了序列到序列(sequehce-to-sequence, seqzseq)的建模,可用于多种应用场景,包括情感分析、机器翻译、语音机器人、目标识别、文本生成等。例如,对于"How are you today?"这样的问题,语音机器人会给出答案:"I am fine, thank you! How are you doing today?"。这当然不是逐字对应的翻译。此类翻译需要对数量庞大的词库进行模型训练,并使用神经网络模型完成单词映射。这些模型与目前在本书中研究的传统机器学习模型有很大区别。

在本章中,我们将重点讲解机器翻译任务,实现将英语语句翻译成西班牙语,流程图如图 8-1 所示。

实际应用中,英语短语"How are you?"对应于西班牙语"¿Cómo estás?"。可见机器翻译模型的输出序列大小不必与其输入相同。为了理解机器翻译的完整实现过程,首先讲解编码器和解码器结构。

图 8-1 使用 seq2seq 模型将英语翻译为西班牙语

8.1.1 编码器、解码器

编码器和解码器是序列到序列的编码模型，本章的编码器和解码器都使用了 LSTM。之前讲到，LSTM 能够记住长序列，并且不会受到训练中梯度消失现象的影响，因此，LSTM 是语言翻译的理想组件。编码器和解码器由 LSTM 架构的两种不同组合实现。首先，介绍编码器架构。

1. 编码器

编码器结构如图 8-2 所示。

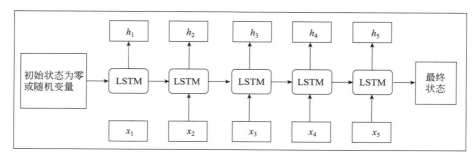

图 8-2 编码器结构

首先，将整个输入句子拆分成单词，并在不同时刻将单词输入到编码器中。以图 8-1 的示例语句 "Peter is a good boy" 为例，编码器将这个语句分 5 次输入 LSTM，如下所示。

X_1 = "Peter", X_2 = "is", X_3 = "a", X_4 = "good", X_5 = "boy"

LSTM 计算每个单词的隐藏状态值 h_i。隐藏状态与下一个单词在下一个时刻被送到解码器。通过这种方法，网络能够捕获输入序列的上下文信息。编码器的初始状态通常为零向量，编码器的最终状态用作解码器的输入，也被称为编码器的跳跃思维向量 (thought vector)。

2. 解码器

解码器结构如图 8-3 所示。

将解码器中 LSTM 单元的隐藏状态从前一个单元送到下一个单元。解码器被训练后，能够在给定前一个单词的情况下预测下一个单词。在目标序列提供给解码器之前，标志 <start> 和 <end> 被添加到序列的开头和结尾。

解码器测试时目标序列是未知的，因此，输入序列始终以 <start> 标记开始，第一个单词传递给解码器时，立即开始预测目标序列，遇到 <end> 标记表示语句结束。接下来，讨论预测过程中的解码步骤。

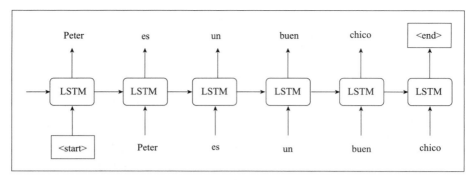

图 8-3　解码器结构

3. 模型预测

预测流程如下。

① 解码器预测过程以 <start> 标记开始，进行循环计算。
② 在预测过程中，解码器的 LSTM 被多次调用，每个时间步长或迭代中输出一个单词。
③ 解码器的初始状态等于编码器的最终状态。
④ 每个时间步长后解码器状态被保留并成为下一个时间步长或迭代期间的初始状态。
⑤ 每次迭代的预测输出作为下一次迭代的输入。
⑥ 解码循环在 <end> 标记处中断。

示例文本的完整预测过程如图 8-4 所示。

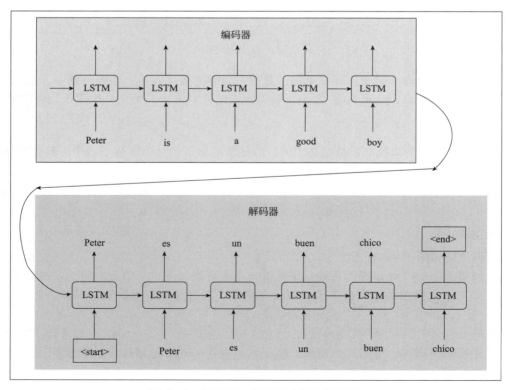

图 8-4　编码器、解码器中的预测过程

以上就是 seq2seq 模型的工作原理。接下来，讨论 seq2seq 模型的缺点。

8.1.2 Seq2seq 模型的缺点

seq2seq 模型由编码器、解码器架构组成，其中编码器处理输入序列并将信息编码为上下文向量。上下文向量通常有固定的长度，可以充分表达输入序列。通常使用上下文向量初始化解码器，但是这样做无法帮助解码器记住长序列，导致处理整个序列时会忘记序列的较早部分。为了解决这个问题，提出了注意力机制。

8.2 注意力模型

人类在看到或听到某事时，会有意识地关注图片或语言的一部分，如图 8-5 所示。

当我们看到这张图片时，注意力会转移到一匹马和一位女士身上。因此，如果必须为图片生成标题，有人会说"女士骑马"。同样，如果有人提出一个问题："你最喜欢哪项运动？"，我们会立即关注"运动"和"你"这两个词。但是，之前介绍的编码器、解码器模型没有关注输入序列中的某些关键字。因此，注意力机制的设计思路是在生成目标的同时增加输入序列中特定单词的重要性。

注意力模型不是从编码器的最后一个隐藏状态构建单个上下文向量，而是为整个输入序列创建一个上下文向量，如图 8-6 所示。

图 8-5 示例图片

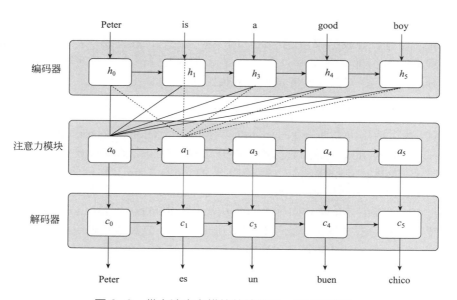

图 8-6 带有注意力模块的编码器、解码器架构

在图 8-6 中，输入序列是"Peter is a good boy"，这个序列被分成单独的词。解码器按时间步长从编码器输出中获取输入。如果使用简单的 seq2seq 建模，翻译质量不会令人满意。因此，在编码器和解码器之间引入了一个注意力模块。现在，每个时刻解码器输入为前一时刻解码器状态与输入序列注意力编码的加和。例如，生成单词"buen"时，解码器会更加关注 Peter、good、boy 等单词。注意力网络为输入的单词分配不同权重，用于生成解码器预测下一个词时需要的注意力上下文特征。

给定时间步长 t_s，网络的注意力权重 (α) 可以用如下数学公式计算。

$$\alpha_{t_s} = \frac{\exp\left(\text{score}\left(h_t, \overline{h}_s\right)\right)}{\sum_{s'=1}^{s} \exp\left(\text{score}\left(h_t, \overline{h}_{s'}\right)\right)}$$

那么上下文向量可以表示为：

$$c_t = \sum_s \alpha_{t_s} \overline{h}_s$$

其中，c_t 表示 t 时刻的上下文特征，α 和 h 分别表示注意力权重和隐藏状态。最终，注意力向量是关于 c_t 和 h_t 的函数，计算过程如下。

$$a_t = f(c_t, h_t) = \tanh\left(W_c[h_t, h_t]\right)$$

a_t 代表解码器的输入，该输入结合了之前时刻的解码状态。深入分析 a_t，能够更好地理解注意力网络，详细解读在下一小节介绍。

8.3 英语翻译为西班牙语

本节将使用本书目前为止讲解过的机器学习技术创建英语到西班牙语的翻译模型。构建带有注意力模块的编码器、解码器模型。为了训练机器翻译模型，需要两种语言之间的句子映射。已有公开数据库创建了这样的映射，并可以在网站"www.manythings.org/anki/"下载公开数据库，数据库包含了世界上许多流行语言之间的映射。数据库使用 TAB 符号分隔双语句子对，文件的每一行格式如下。

English + TAB + The Other Language + TAB + Attribution

在本项目中，使用由 Jeffrey Pennington 等人创建的 Global Vectors for Word Representations(GloVe) 公开数据库。在开始这个项目之前下载数据库到本地 (http://nlp.stanford.edu/data/glove.6B.zip)。解压缩下载的 glove *.zip 文件，此文件包含四个不同维度的词向量，维度分别为 50、100、200、300。glove 文件将在后文中详细介绍，包含名为 glove.6B.50d.txt、glove.6B.100d.txt、glove.6B.200d.txt、glove.6B.300d.txt 的文本文件。本章实验使用的数据量较大，使用了维度为 200 的数据。将所有这些文件复制到 Google 硬盘中，以便 Colab 项目可以访问。

接下来，可以创建机器翻译项目了。

8.3.1 创建项目

打开新的 Colab 文档并将其重命名为 NMT。使用如下代码导入所需的库。

```python
import tensorflow as tf

from tensorflow.keras.models import Model
from tensorflow.keras.layers import Input,
            Dense, LSTM, Embedding, Bidirectional,
            RepeatVector, Concatenate, Activation,
            Dot, Lambda
from tensorflow.keras.preprocessing.text
            import Tokenizer
from tensorflow.keras.preprocessing.sequence
            import pad_sequences
from keras import preprocessing, utils
import numpy as np
import matplotlib.pyplot as plt
```

8.3.2 下载数据集

使用如下语句从本书的存储库下载英语到西班牙语的翻译数据集。

```
!pip install wget
import wget
url = 'https://raw.githubusercontent.com/Apress/artificialneural-networks-with-tensorflow-2/main/ch08/spa.txt'
wget.download(url,'spa.txt')
```

8.3.3 创建数据集

使用以下语句读取翻译数据库中的文件内容。

```
# reading data
with open('/content/spa.txt',encoding='utf-8',errors='ignore')
as file:
    text=file.read().split('\n')
```

使用以下语句打印文件内容的前五行。

```
text[:5]
```

输出结果如图 8-7 所示。

```
['Go.\tVe.\tCC-BY 2.0 (France) Attribution: tatoeba.org #2877272 (CM) & #4986655 (cueyayotl)',
 'Go.\tVete.\tCC-BY 2.0 (France) Attribution: tatoeba.org #2877272 (CM) & #4986656 (cueyayotl)',
 'Go.\tVaya.\tCC-BY 2.0 (France) Attribution: tatoeba.org #2877272 (CM) & #4986657 (cueyayotl)',
 'Go.\tVáyase.\tCC-BY 2.0 (France) Attribution: tatoeba.org #2877272 (CM) & #6586271 (arh)',
 'Hi.\tHola.\tCC-BY 2.0 (France) Attribution: tatoeba.org #538123 (CM) & #431975 (Leono)']
```

图 8-7 翻译文件的内容

此文件中，原始语句和目标语言语句由 TAB 符号分隔。

使用以下循环语句检查 100～110 范围内的数据记录。

```
for t in text[100:110]:
    print(t)
```

输出结果如图 8-8 所示。

```
Be good.      Sed buenas.    CC-BY 2.0 (France) Attribution: tatoeba.org #7932304 (Seael) & #7932308 (Seael)
Be good.      Sed buenos.    CC-BY 2.0 (France) Attribution: tatoeba.org #7932304 (Seael) & #7932309 (Seael)
Be good.      Sea buena.     CC-BY 2.0 (France) Attribution: tatoeba.org #7932304 (Seael) & #7932310 (Seael)
Be good.      Sea bueno.     CC-BY 2.0 (France) Attribution: tatoeba.org #7932304 (Seael) & #7932311 (Seael)
Be good.      Sean buenas.   CC-BY 2.0 (France) Attribution: tatoeba.org #7932304 (Seael) & #7932313 (Seael)
Be good.      Sean buenos.   CC-BY 2.0 (France) Attribution: tatoeba.org #7932304 (Seael) & #7932315 (Seael)
Be kind.      Sean gentiles. CC-BY 2.0 (France) Attribution: tatoeba.org #1916315 (CK) & #2092229 (hayastan)
Be nice.      Sé agradable.  CC-BY 2.0 (France) Attribution: tatoeba.org #1916314 (CK) & #5769224 (arh)
Beat it.      Pírate. CC-BY 2.0 (France) Attribution: tatoeba.org #37902 (CM) & #5769215 (arh)
Call me.      Llamame.       CC-BY 2.0 (France) Attribution: tatoeba.org #1553532 (CK) & #1555788 (hayastan)
```

图 8-8　翻译文件的内容列表

第一组词或句子是输入数据集，TAB 后面的第二组语句为目标语言数据集，其余内容本章未使用。在图 8-8 中，英文文本"Be nice"翻译后的结果为"Sé agradable"。

> **注意**
>
> 相同的输入句子可以有多个翻译结果，所有翻译结果都传达相同的含义。

现在声明两个变量用于存储输入数据和目标数据，如下所示。

```
input_texts=[] #encoder input
target_texts=[] # decoder input
```

文档中共有 122937 个语句、2751187 个单词和 18127427 个字符，最后一句为空白。如果使用完整的数据集，则训练模型需要很长时间；使用较小的数据集，则会限制翻译词汇量。为了进行算法讲解，本章选择前 10000 个单词、句子进行训练。首先，根据 TAB 符号拆分语句，将输入与目标分开。使用以下代码将数据库中前 10000 个条目组合为两个数组。

```
NUM_SAMPLES = 10000
for line in text[:NUM_SAMPLES]:
  english, spanish = line.split('\t')[:2]
  target_text = spanish.lower()
  input_texts.append(english.lower())
  target_texts.append(target_text)
```

打印两个数组中的前五个条目以检查数据内容：

```
print(input_texts[:5],target_texts[:5])
```

输出如下。

```
['go.', 'go.', 'go.', 'go.', 'hi.'] ['ve.', 'vete.', 'vaya.', 'váyase.', 'hola.']
```

如上所示，文本中存在一个"."字符，需要在数据中去除。

8.3.4 数据预处理

现在进行一些文本处理，使下载后的文本符合机器翻译模型的输入要求。

1. 清除标点符号

首先，从目标数据和输入数据中删除所有标点符号。标点符号可以通过调用 string 类的 punctuation 方法列出，使用如下语句实现。

```
import string
print('Characters to be removed in preprocessing', string.punctuation)
```

上述语句的输出如下。

```
Characters to be removed in preprocessing !"#$%&'()*+,-./:;<=>?@[\]^_`{|}~
```

为了去除标点符号，定义了以下函数。

```
def remove_punctuation(s):
    out=s.translate(str.maketrans("","",string.punctuation))
    return out
```

调用 remove_punctuation 函数从输入数据和目标数据中删除标点符号，如下所示。

```
input_texts = [remove_punctuation(s)
                for s in input_texts]
target_texts = [remove_punctuation(s)
                for s in target_texts]
```

显示两个数组中的前五个项目，以此检查数据是否符合要求，使用如下语句实现。

```
input_texts[:5],target_texts[:5]
```

输出结果如下。

```
(['go', 'go', 'go', 'go', 'hi'], ['ve', 'vete', 'vaya', 'váyase', 'hola'])
```

2. 添加开始、结束标记

对于目标文本，需要添加开始和结束标签。在讨论编码器、解码器架构时，已经介绍了 <start> 和 <end> 标记的作用。使用以下语句添加标记。

```
# adding start and end tags
target_texts=['<start> ' + s + ' <end>'
            for s in target_texts]
```

使用以下语句检查是否成功调解标记。

```
target_texts[1]
```

输出结果如下。

```
'<start> vete <end>'
```

3. 编码输入数据集

输入数据是文本格式，神经网络不理解文本数据，因此需要对每个单词进行编码。首先，标记输入文本并创建词汇表。word_index 字典提供了此类映射，如 ('hello':133)。代码如下。

```
tokenizer_in=Tokenizer()
#tokenizing the input texts
tokenizer_in.fit_on_texts(input_texts)
#vocab size of input
input_vocab_size=len(tokenizer_in.word_index) + 1
```

使用如下语句检查输入词汇的数量。

```
input_vocab_size
```

数据量为 2332。这意味着本章使用的输入字典有 2332 个单词，使用这些单词进行模型训练。如果充分使用数据库中的 122937 个单词或句子，将获得由 13731 个单词或句子构成的词汇表。使用以下代码检查标记化字典中的几个项目。

```
# Listing few items
input_tokens = tokenizer_in.index_word
for k,v in sorted(input_tokens.items())[2000:2010]:
    print (k,v)
```

输出结果如下。

```
2001 visa
2002 trains
2003 poems
2004 forgetful
2005 insane
2006 flew
2007 harvard
2008 obvious
2009 lecture
2010 divorce
```

如上所示，每个单词或句子都与唯一的编号对应。

4. 编码输出数据集

与输入数据集一样，使用以下代码标记输出数据集。

```
#tokenizing output that is spanish translation
tokenizer_out=Tokenizer(filters='')
tokenizer_out.fit_on_texts(target_texts)

#vocab size of output
output_vocab_size=len(tokenizer_out.word_index) + 1
output_vocab_size
```

由于诸如 <start> 和 <end> 之类的特殊标记中使用了尖括号，不希望处理输出数据集时去除此类符号，因此，对于输出数据集，将目标符号设置为单引号以过滤掉输出标记。

通过数据处理后，打印了 4964 个输出，表明输出词汇表中有 4964 个西班牙语单词或句子。使用如下语句从字典中检查一些项目。

```
# Listing few items
output_tokens = tokenizer_out.index_word
for k,v in sorted(output_tokens.items())[2000:2010]:
    print (k,v)
```

输出结果如下。

```
2001 suyos
2002 ley
2003 palabras
2004 ausente
2005 delgaducho
2006 sucio
2007 adoptado
2008 violento
2009 roncando
2010 podríamos
```

使用以上方法构建输入数据集和输出数据集，用这两个数据集训练机器翻译模型的编码器、解码器。训练完成后，用户将能够使用此模型对数据集中出现的标准短语进行翻译。如果用户输入一个不在标准数据集中的短语，该短语将被拆分为单词，训练模型对每个单词进行翻译。每次翻译后，模型会在注意力模块的帮助下尝试猜测下一个单词。这里使用的注意力模块非常重要。解码器接收两个输入，包括编码器的先前状态和来自注意力模块的上下文向量。然后，解码器根据这两个输入以预测单词。

5. 创建输入序列

调用 texts_to_sequences 方法将输入文本转换为序列，这是数据输入训练模型之前的必要步骤。代码如下。

```
#converting tokenized sentence into sequences
tokenized_input = tokenizer_in.texts_to_sequences
                    ( input_texts )
```

字典中的单词没有固定的长度，然而，出于训练的要求，需要有固定长度的序列。因此，需要确定编码后输入单词的最大长度，并用零填充所有序列到此长度。使用如下代码段实现。

```
#max length of the input
maxlen_input = max( [ len(x)
                for x in tokenized_input ] )

#padding sequence to a maximum fixed length
padded_input = preprocessing.sequence.pad_sequences
        ( tokenized_input , maxlen=maxlen_input ,
```

```
                padding='post' )
```

使用如下语句检查部分序列以查看填充的效果。

```
padded_input[2000:2010]
```

输出结果如下。

```
array([[ 1, 613, 195, 0, 0],
       [ 1,  54, 109, 0, 0],
       [ 1,  54, 109, 0, 0],
       [ 1,  54, 109, 0, 0],
       [ 1,  54, 182, 0, 0],
       [ 1,  54,  14, 0, 0],
       [ 1,  54,  98, 0, 0],
       [ 1,  54,  98, 0, 0],
       [ 1,  54,  10, 0, 0],
       [ 1,  54,  10, 0, 0]], dtype=int32)
```

以上每个序列的末尾均为 0。以输入语句 "I voted yes" 为例，此语句被编码并填充为序列 [1,613,195,0,0]，序列结尾处的 0 即为填充值。

所以，所有输入序列的恒定长度为 5。如果使用整个词汇表，单词的最大长度将会不同，因此每个输入序列的大小将是最大值。

使用如下语句将输入数据转换为一个 NumPy 数组。

```
encoder_input_data = np.array( padded_input )
print( encoder_input_data.shape)
```

上述语句打印变量值为：

```
(10000, 5)
```

因此，有 10000 个固定宽度为 5 的输入序列。

6. 创建目标序列

与输入文本一样，使用如下代码将编码后的目标单词转换为序列。

```
#converting tokenized text into sequences
tokenized_output = tokenizer_out.texts_to_sequences
                        (target_texts)
```

使用以下代码检查输出词汇表的大小。

```
output_vocab_size=len(tokenizer_out.word_index) + 1
output_vocab_size
```

结果显示词汇表中有 4964 个西班牙语单词。

本章使用 Teacher Forcing 网络训练方法，以便模型训练过程更快地收敛。Teacher Forcing 方法向解码器提示下一个单词，使其学习更快。代码如下。

```
# teacher forcing
for i in range(len(tokenized_output)) :
    tokenized_output[i] = tokenized_output[i][1:]
```

> **注意**
> Teacher Forcing 方法仅在训练模型时使用,而不是在测试模式中使用。

确定输出的最大长度并填充所有输出标记。代码如下。

```
maxlen_output = max( [ len(x) 
                       for x in tokenized_output ] )
padded_output = preprocessing.sequence.pad_sequences
                       ( tokenized_output ,
                         maxlen=maxlen_output ,
                         padding='post' )
```

将输出数据转换为 NumPy 数组,使其符合机器学习的要求。代码如下。

```
# converting to numpy
decoder_input_data = np.array( padded_output )
```

使用如下语句打印解码器输入数据。

```
decoder_input_data[2000:2010]
```

输出结果如下。

```
array([[  11, 1147,  26, 108,   2,   0,   0,   0,   0],
       [  51,   17, 135,   2,   0,   0,   0,   0,   0],
       [  11,   51,  17, 135,   2,   0,   0,   0,   0],
       [  11,   51,  13, 224,   2,   0,   0,   0,   0],
       [  51,   59,   2,   0,   0,   0,   0,   0,   0],
       [  51,   20,   2,   0,   0,   0,   0,   0,   0],
       [  28,   51,   2,   0,   0,   0,   0,   0,   0],
       [  37,   51,   2,   0,   0,   0,   0,   0,   0],
       [  51,   88,   2,   0,   0,   0,   0,   0,   0],
       [  51,   19,   2,   0,   0,   0,   0,   0,   0]],
      dtype=int32)
```

上述每个元素使用 0 填充,使每个元素的长度相等。对于本章使用的数据集,将序列长度固定为 9。类似地,如果使用了完整的数据集,这个固定长度会有不同的值。使用 one-hot 编码来创建解码器目标输出。代码如下。

```
#decoder target output
decoder_target_one_hot=np.zeros((len(input_texts),
                                 maxlen_output,
                                 output_vocab_size),
                                 dtype='float32')
for i,d in enumerate(padded_output):
    for t,word in enumerate(d):
        decoder_target_one_hot[i,t,word]=1
```

使用如下语句打印变量值来检查目标输出。

decoder_target_one_hot[0]

输出结果如下。

```
array([[0., 0., 0., ..., 0., 0., 0.],
       [0., 0., 1., ..., 0., 0., 0.],
       [1., 0., 0., ..., 0., 0., 0.],
       ...,
       [1., 0., 0., ..., 0., 0., 0.],
       [1., 0., 0., ..., 0., 0., 0.],
       [1., 0., 0., ..., 0., 0., 0.]], dtype=float32)
```

8.3.5 GloVe 词嵌入

在开始项目之前，创建单词的向量数据集，该数据集给出了词之间的共现概率。换句话说，对于任何给定的词，向量数据集会给出所有其他词出现在该给定词旁边的概率是多少。例如，单词"cream"出现在单词"ice"旁边的概率可能最高。向量数据集依赖庞大的词汇表创建。了解共现概率有助于将某种形式的含义编码到输入语句中。本章将基于向量数据集为输入的英语词汇构建一个字典。首先，为输入序列提供一个嵌入层，该层将输入词转换为词向量。这些词向量将作为输入传递给编码器。

1. 词向量索引

GloVe 词嵌入提供了四种维度的映射：50、100、200 和 300。本项目使用一个 200 维的数据库。使用的数据维度越高，翻译效果就越好，但要以处理时间和资源为代价。

此处显示了维度为 50 的文件示例。

```
more 0.87943 -0.11176 0.4338 -0.42919 0.41989 0.2183 -0.3674
-0.60889 -0.41072 0.4899 -0.4006 -0.50159 0.24187 -0.1564
0.67703 -0.021355 0.33676 0.35209 -0.24232 -1.0745 -0.13775
0.29949 0.44603 -0.14464 0.16625 -1.3699 -0.38233 -0.011387
0.38127 0.038097 4.3657 0.44172 0.34043 -0.35538 0.30073
-0.09223 -0.33221 0.37709 -0.29665 -0.30311 -0.49652 0.34285
0.77089 0.60848 0.15698 0.029356 -0.42687 0.37183 -0.71368 0.30175
' -0.039369 1.2036 0.35401 -0.55999 -0.52078 -0.66988 -0.75417
-0.6534 -0.23246 0.58686 -0.40797 1.2057 -1.11 0.51235 0.1246
0.05306 0.61041 -1.1295 -0.11834 0.26311 -0.72112 -0.079739
0.75497 -0.023356 -0.56079 -2.1037 -1.8793 -0.179 -0.14498
-0.63742 3.181 0.93412 -0.6183 0.58116 0.58956 -0.19806 0.42181
-0.85674 0.33207 0.020538 -0.60141 0.50403 -0.083316 0.20239
0.443 -0.060769 -0.42807 -0.084135 0.49164 0.085654
```

以上为单词 more 的词索引。我们将矩阵文件中的每一行拆分为单词词符，使用第一个单词作为字典变量中的键，其余值是该词与其他词的共现概率，并作为字典变量中的值。创建字典的代码如下：

```
#creating dictionary of words corresponding to vectors
print('Indexing word vectors.')
```

```
embeddings_index = {}

# Use this open command in case of downloading using wget above
#f = open('glove.6B.200d.txt', encoding='utf-8')

# we can choose any dimensions 50 100 200 300
f = open('drive/My Drive/tfbookdata/glove.6B.200d.txt',
            encoding='utf-8')

for line in f:
    values = line.split()
    word = values[0]
    coefs = np.asarray(values[1:], dtype='float32')
    embeddings_index[word] = coefs
f.close()

print('Found %s word vectors.' %
            len(embeddings_index))
```

运行代码时，显示400000，表明在完整数据集中有400000个嵌入可用。使用如下语句查看结果。

```
embeddings_index["any"]
```

矩阵的部分输出如图8-9所示。

```
array([ 6.3113e-01,  4.3183e-01,  2.3103e-01, -6.4909e-01,  2.3744e-01,
        4.4619e-01, -8.6148e-01,  2.9341e-01,  8.0033e-02,  8.5633e-03,
        1.0165e-01,  6.2783e-01,  2.5047e-01,  5.2425e-02,  6.3045e-01,
       -4.0008e-02,  2.5212e-01,  6.2147e-01,  6.6967e-02, -7.9787e-02,
       -2.1607e-02,  3.4236e+00, -4.7925e-02,  2.4620e-01, -2.8834e-02,
        4.0330e-02, -2.7858e-01, -2.7939e-01,  3.5606e-01, -5.7373e-01,
       -9.9960e-02, -3.2374e-01,  1.7812e-01,  2.0671e-02,  2.3637e-01,
       -1.7074e-01, -4.9345e-01, -4.0289e-01, -3.4184e-01, -1.9405e-01,
```

图8-9 关键字 "any" 的嵌入矩阵

注意

对于 "any" 这样的词，有200个可用的映射，200为本项目使用的数据维度大小。

2. 创建子集

现在，使用简单的for循环提取英语输入词汇的条目，如下所示。

```
#embedding matrix
num_words = len(tokenizer_in.word_index)+1
word2idx_input = tokenizer_in.word_index
```

```
embedding_matrix = np.zeros((num_words, 200))
for word,i in word2idx_input.items():
    if i<num_words:
        embedding_vector = embeddings_index.get(word)
    if embedding_vector is not None:
        embedding_matrix[i] = embedding_vector
embedding_matrix.shape
```

矩阵创建完成后，其形状会打印在终端上，如下。

(2332, 200)

> **注意**
>
> 输入数据库中有 2332 个有标注的英文单词。对于每个单词，有 200 个共现概率。为什么是 200？因为本项目选择了一个 200 维的映射文件。

3. 定义嵌入层

声明两个变量在创建 LSTM 和嵌入层时使用，如下所示。

```
#lstm hidden dimensions
LATENT_DIM=256
#embeding layer dimensions
EMBEDDING_DIM=200
```

声明嵌入层，代码如下。

```
embedding_layer=Embedding(input_vocab_size,
                          EMBEDDING_DIM,
                          weights=[embedding_matrix],
                          input_length=maxlen_input)
```

输入词汇大小是 2332，维度是 200。将 weights 参数设置为之前创建的嵌入矩阵，该矩阵提供词到向量的映射。将 input_length 设置为填充序列的最大长度，此处为 5。现在，开始定义模型。首先，实例化一些方法和层。

8.3.6 定义编码器

现在定义编码器。对于编码器，首先定义输入层。代码如下。

```
#encoder
encoder_input = Input(shape=(maxlen_input,),
                      name = 'encoderinput')
```

该层的输入大小为 5，由 maxlen_input 参数值决定。使用以下语句创建一个嵌入层。

```
embedding_layer:
```

```
encoder_input = embedding_layer(encoder_input)
```

接下来，定义 LSTM 层。代码如下。

```
encoder = Bidirectional(LSTM(LATENT_DIM,
                             return_sequences=True,
                             dropout = 0.3),
                        name = 'encoder_bidirection')
```

本章使用双向 LSTM。LATENT_DIM 设置隐藏层数为 200。每层使用概率为 30% 的 dropout。最后，将输入数据传递给编码器。代码如下。

```
encoder_outputs = encoder(encoder_input)
```

上述语句得到编码器输出，通过调用 shape 方法来检查输出变量的形状。代码如下。

```
encoder_outputs.shape
```

输出结果如下。

```
TensorShape([None, 5, 512])
```

由于 LSTM 是双向的，维度为 512，即潜在维数 (LATENT_DIM) 为 256×2。潜在维度是 LSTM 层中的单元数。有了 LSTM 后，现在继续定义解码器。

8.3.7 定义解码器

首先为解码器模型定义输入层：

```
decoder_inputs = Input(shape=(maxlen_output,),
                       name='decoder_input')
```

其中，maxlen_output 为 9。解码器中嵌入层使用以下语句定义。

```
decoder_embedding = Embedding(output_vocab_size,
                              EMBEDDING_DIM,
                              name='decoder_embedding')
```

其中，output_vocab_size 为 4964，EMBEDDING_DIM 指定输出维度为 200。最后，创建一个解码器输入：

```
decoder_input = decoder_embedding(decoder_inputs)
```

调用 shape 方法打印此张量的形状：

```
decoder_input.shape
```

输出结果如下。

```
TensorShape([None, 9, 200])
```

> **注意**
> 输出词汇表的序列长度为 9。

定义解码层

现在定义解码层，需要一个 LSTM 层和一个使用 softmax 函数激活的线性连接层。LSTM 层定义如下。

```
decoder_lstm = LSTM(LATENT_DIM,
                    return_state = True,
                    name = 'decoder_lstm')
```

其中，LATENT_DIM 参数的值为 256，这是隐藏层的数量。
线性连接层定义如下。

```
decoder_dense = Dense(output_vocab_size,
                      activation='softmax',
                      name='decoder_dense')
```

output_vocab_size 参数设置线性连接层维度为 4964。该层将预测每个单词出现的概率。如果使用数据量更大的数据集，这一层的维度会非常高，导致解码时间增加。
接下来，定义注意力层。

8.3.8 注意力网络

首先，定义注意力网络模型。

1. 定义 Softmax

任何给定时间步长 t_s，注意力权重 (α) 可以用数学表示如下。

$$\alpha_{t_s} = \frac{\exp\left(\text{score}\left(h_t, \overline{h}_s\right)\right)}{\sum_{s'=1}^{s} \exp\left(\text{score}\left(h_t, \overline{h}_{s'}\right)\right)}$$

以下代码段实现了 softmax_attention 方法用于 α 的计算。

```
# Computing alphas
import tensorflow.keras.backend as k
def softmax_attention(x):
    assert(k.ndim(x)>2)

    e=k.exp(x-k.max(x,axis=1,keepdims=True))
    s=k.sum(e,axis=1,keepdims=True)
    return e/s
```

输入数据尺寸为 $N\times T\times D$，其中 N 是样本数，T 是序列长度，D 是向量维度。随着计算过程的深入，希望确保时间维度上的所有输出总和为 1，因此使用 softmax 函数。

2. 注意力层

接下来，定义几个层以构建注意力模型。将注意力层声明为全局变量，以便被解码器重复调用。创建一个 attention_repeat 变量，作为编码器和解码器模块之间的桥梁。代码如下。

```
attention_repeat = RepeatVector(maxlen_input)
```

其中，maxlen_input 为 5。下面，定义连接层如下。

```
attention_concat = Concatenate(axis=-1)
```

由一个具有十个节点的线性连接层和 tanh 激活层组成：

```
dense1_layer = Dense(10,activation='tanh')
```

使用单个节点创建另一个线性连接层：

```
dense2_layer = Dense(1,activation = softmax_attention)
```

最后，使用点乘得到最终输出：

```
dot_layer = Dot(axes=1)
```

3. 上下文注意力

现在编写一个名为 context_attention 的注意力函数，解码器在每个时间步长调用 context_attention 函数。代码如下。

```
def context_attention(h, st_1):
```

此函数有两个参数：h 是编码器当前的隐藏状态，st_1 是解码器之前的隐藏状态。因此，h 将在每次迭代中取不同的值，如 h_1，h_2，$\cdots$，h_t；同样，s 也将在每个时间步长采用不同的值，形状为 LATENT_DIM×2，在本章例子中为 512。因此，LATENT_DIM 定义为 256。本项目使用了双向 LSTM，因此 LATENT_DIM×2。

现在将注意力层复制 $s_{t-1}t_x$ 次，形状将为 (t_x, LATENT_DIM)：

```
st_1=attention_repeat(st_1)
```

将 h_t 与 s_{t-1} 连接起来：

```
x=attention_concat([h,st_1])
```

x 的形状为 (t_x, LATENT_DIM + LATENT_DIM × 2)。

使用第一个线性连接层学习 α 值：

```
x=dense1_layer(x)
```

将第二个线性连接层与自定义的 softmax 函数一起使用：

```
alphas=dense2_layer(x)
```

最终取所有 α 和 h 的点积：

```
context = dot_layer([alphas,h])
```

最终的上下文注意力权重返回给调用者：

```
return context
```

context_attention 的完整函数代码如清单 8-1 所示。

清单 8-1 计算注意力上下文的函数

```
# computing attention context
def context_attention(h, st_1)
    st_1=attention_repeat(st_1)
    x=attention_concat([h,st_1])
    x=dense1_layer(x)
    alphas=dense2_layer(x)
    context = dot_layer([alphas,h])
    return context
```

源码清单
链　接：https://pan.baidu.com/s/1NV0rimQ_8kRz22xfFHN-Cw
提取码：1218

4. 记录输出结果

现在，记录各个时刻点的输出。首先，需要创建初始状态提供给解码器：

```
initial_s = Input(shape=(LATENT_DIM,), name='s0')
initial_c = Input(shape=(LATENT_DIM,), name='c0')
```

编写一个 Concatenate 函数连接输出结果：

```
context_last_word_concat_layer = Concatenate(axis=2)
```

遍历所有时刻记录输出。将初始状态复制，稍后在 for 循环中将使用的两个局部变量如下。

```
s = initial_s
c = initial_c
```

声明一个用于记录输出结果的输出数组：

```
outputs = []
```

声明一个 for 循环来遍历所有时刻：

```
for t in range(maxlen_output): #ty times
```

词汇表的 maxlen_output 是 9 个，所以 for 循环将收集 9 个输出。对于初始状态，调用 context_attention 函数创建上下文。代码如下。

```
#get the context using attention mechanism
context=context_attention(encoder_outputs,s)
```

其中，第一个参数是编码器的输出，第二个参数是解码器的初始状态。
每个时刻都需要一个不同的层：

```
selector=Lambda(lambda x: x[:,t:t+1])
x_t=selector(decoder_input)
```

调用 context_last_word_concat_layer 函数为解码器创建一个输入：

```
decoder_lstm_input=context_last_word_concat_layer
                  ([context,x_t])
```

组合 [context,last word] 与 s 和 c 一起传递到 LSTM 中，以获得新的 s、c 和输出：

```
out,s,c=decoder_lstm(decoder_lstm_input, initial_state=[s,c])
```

调用最后的线性连接层得到下一个单词的预测：

```
decoder_outputs = decoder_dense(out)
```

将输出附加到列表中：

```
outputs.append(decoder_outputs)
```

记录输出结果的完整 for 循环如清单 8-2 所示。

清单 8-2 每个时刻解码器的输出

源码清单
链　接：https://pan.baidu.com/s/1NV0rimQ_8kRz22xfFHN-Cw
提取码：1218

```
s = initial_s
c = initial_c
outputs = []

#collect output in a list at first
for t in range(maxlen_output): #ty times
    #get the context using attention mechanism
    context=context_attention(encoder_outputs,s)

#we need a different layer for each time step
selector=Lambda(lambda x: x[:,t:t+1])
x_t=selector(decoder_input)

#combine
decoder_lstm_input=context_last_word_concat_layer
                  ([context,x_t])

#pass the combined [context,last word] into lstm
#along with [s,c]
#get the new[s,c] and output
out,s,c=decoder_lstm(decoder_lstm_input,
                     initial_state=[s,c])

#final dense layer to get next word prediction
decoder_outputs = decoder_dense(out)
outputs.append(decoder_outputs)
```

输出为一个长度为 ty 的列表。使用以下命令打印输出。

```
outputs
```

输出结果如下。

```
[<tf.Tensor 'decoder_dense_9/Identity:0' shape=(None, 4964)
dtype=float32>,
 <tf.Tensor 'decoder_dense_10/Identity:0' shape=(None, 4964)
dtype=float32>,
 <tf.Tensor 'decoder_dense_11/Identity:0' shape=(None, 4964)
dtype=float32>,
 <tf.Tensor 'decoder_dense_12/Identity:0' shape=(None, 4964)
dtype=float32>,
 <tf.Tensor 'decoder_dense_13/Identity:0' shape=(None, 4964)
dtype=float32>,

<tf.Tensor 'decoder_dense_14/Identity:0' shape=(None, 4964)
dtype=float32>,
<tf.Tensor 'decoder_dense_15/Identity:0' shape=(None, 4964)
dtype=float32>,
<tf.Tensor 'decoder_dense_16/Identity:0' shape=(None, 4964)
dtype=float32>,
<tf.Tensor 'decoder_dense_17/Identity:0' shape=(None, 4964)
dtype=float32>]
```

每个元素的形式都为 (batch size, output vocab size)。如果简单地将所有输出堆叠成一个张量,形状将是 $T \times N \times D$。希望第一个维度为 N,因此,定义了一个函数来堆叠和转置矩阵,如下所示。

```
def stack(x):
    x=k.stack(x)
    x=k.permute_dimensions(x,pattern=(1,0,2))
    return x
```

参数 x 是一个长度为 T 的列表;每个元素都是一个 batch_size × output_vocab_size 张量。

返回值是一个 batch_size × T × output_vocab_size 的张量。

调用如下函数生成最终输出。

```
stacker=Lambda(stack)
outputs=stacker(outputs)
outputs
```

输出结果如下。

```
<tf.Tensor 'lambda_18/Identity:0' shape=(None, 9, 4964)
dtype=float32>
```

> 转置矩阵后形状的改变的方式。

接下来，定义本章模型。

8.3.9 定义模型

本章模型包括一个编码器、一个解码器和作为输入的编码器的初始状态，输出为之前生成的最终输出。模型使用以下语句定义。

```
model=Model(inputs=[encoder_inputs,
                    decoder_inputs,
                    initial_s,
                    initial_c],
            outputs=outputs)
```

调用模型的 summary 方法来获取模型结构。summary 方法的输出很长，前几行的输出如图 8-10 所示。

```
model.summary()
Model: "model"
_____
Layer (type)                 Output Shape         Param #    Connected to
=================================================================
enocderinput (InputLayer)    [(None, 5)]          0
_____
embedding_1 (Embedding)      (None, 5, 200)       466400     enocderinput[0][0]
_____
s0 (InputLayer)              [(None, 256)]        0
_____
enocder_bidirection (Bidirectio (None, 5, 512)    935936     embedding_1[1][0]
_____
repeat_vector (RepeatVector) (None, 5, 256)       0          s0[0][0]
                                                             decoder_lstm[9][1]
                                                             decoder_lstm[10][1]
                                                             decoder_lstm[11][1]
                                                             decoder_lstm[12][1]
                                                             decoder_lstm[13][1]
                                                             decoder_lstm[14][1]
                                                             decoder_lstm[15][1]
                                                             decoder_lstm[16][1]
_____
concatenate (Concatenate)    (None, 5, 768)       0          enocder_bidirection[0][0]
                                                             repeat_vector[9][0]
                                                             enocder_bidirection[0][0]
                                                             repeat_vector[10][0]
```

图 8-10 模型嵌入编码器、解码器

模型的可训练参数数量如下所示。

```
Total params: 4,670,841
Trainable params: 4,670,841
Non-trainable params: 0
```

通过网络参数量可以轻松判断用于训练的语言模型的复杂性。如果使用完整的数据集，参数的数量会非常大。使用 plot_model 函数打印模型的可视化表示：

```
tf.keras.utils.plot_model (model)
```

输出结果如图 8-11 所示。

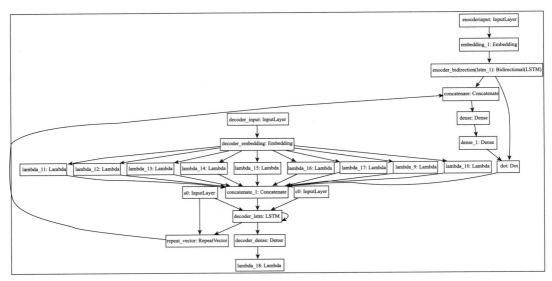

图 8-11　完整模型的可视化结果

从图 8-11 中，能够看到语言翻译的完整流程。最后，使用分类交叉熵和 Adam 优化器编译模型。代码如下。

```
model.compile(optimizer='adam',
              loss='categorical_crossentropy',
              metrics=['accuracy'])
```

8.3.10　模型训练

将零状态作为输入来创建初始状态：

```
initial_s_training=np.zeros((NUM_SAMPLES,LATENT_DIM))
initial_c_training =np.zeros(shape=(NUM_SAMPLES,LATENT_DIM))
```

使用以下语句训练模型。

```
R = model.fit([encoder_input_data,
               decoder_input_data,
               initial_s_training,
               initial_c_training],
              decoder_target_one_hot,
              batch_size=100,
              epochs=100,
              validation_split=0.3)
```

本章模型训练了 100 个 epoch，每个 epoch 在 GPU 上的计算时间大约需要 5s。

8.3.11　预测

现在准备使用真实数据测试本章模型。为此，定义一个用于对用户输入文本进行编码的编码器和一个进行翻译的解码器。

1. 编码

创建一个编码器模型来对输入文本进行编码：

encoder_model = Model(encoder_inputs, encoder_outputs)

模型概要如图 8-12 所示。

```
Model: "model_1"
_____
Layer (type)                 Output Shape              Param #
=================================================================
enocderinput (InputLayer)    [(None, 5)]               0
_____
embedding_1 (Embedding)      (None, 5, 200)            466400
_____
enocder_bidirection (Bidirec (None, 5, 512)            935936
=================================================================
Total params: 1,402,336
Trainable params: 1,402,336
Non-trainable params: 0
```

图 8-12　编码器模型概要

通过生成此模型的可视化图，可以将模型各个结构与图 8-12 中的模型概要相对应。代码如下。

tf.keras.utils.plot_model (encoder_model)

模型可视化图如图 8-13 所示。

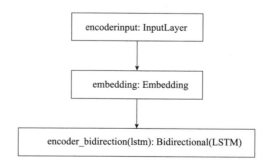

图 8-13　编码器模型可视化

本章在翻译输入文本时使用此模型。

2. 解码

为解码器模型创建一些变量：

encoder_outputs_as_input = Input(shape=
                (maxlen_input, LATENT_DIM* 2,))

> **注意**
>
> 长度参数为 LATENT_DIM × 2,因为使用了双向 LSTM。

另一个变量如下。

```
decoder_input_embedding = Input(shape=(1,))
```

调用 decoder_embedding 为解码器创建输入向量:

```
decoder_input_ = decoder_embedding
                    (decoder_input_embedding)
```

调用 one_step_attention 函数来计算每个时间步长的上下文注意力:

```
context = context_attention
          (encoder_outputs_as_input, initial_s)
```

调用 context_last_word_concat_layer 方法计算解码器输入:

```
decoder_lstm_input = context_last_word_concat_layer(
    [context, decoder_input_])
```

调用解码器并记录其输出:

```
out, s, c = decoder_lstm(decoder_lstm_input,
                    initial_state=[initial_s, initial_c])
decoder_outputs = decoder_dense(out)
```

最后,定义解码器模型如下。

```
decoder_model = Model(
  inputs=[
    decoder_input_embedding,
    encoder_outputs_as_input,
    initial_s,
    initial_c
  ],
  outputs=[decoder_outputs, s, c]
)
decoder_model.summary()
```

解码器概要如图 8-14 所示。

可训练参数的数量高达 300 万以上。

使用如下语句生成模型的可视化表示。

```
tf.keras.utils.plot_model (decoder_model)
```

```
Model: "model_2"
_____
Layer (type)                    Output Shape         Param #     Connected to
==================================================================================================
s0 (InputLayer)                 [(None, 256)]        0
_____
input_1 (InputLayer)            [(None, 5, 512)]     0
_____
repeat_vector (RepeatVector)    (None, 5, 256)       0           s0[0][0]
_____
concatenate (Concatenate)       (None, 5, 768)       0           input_1[0][0]
                                                                 repeat_vector[18][0]
_____
dense (Dense)                   (None, 5, 10)        7690        concatenate[18][0]
_____
dense_1 (Dense)                 (None, 5, 1)         11          dense[18][0]
_____
input_2 (InputLayer)            [(None, 1)]          0
_____
dot (Dot)                       (None, 1, 512)       0           dense_1[18][0]
                                                                 input_1[0][0]
_____
decoder_embedding (Embedding)   multiple             992800      input_2[0][0]
_____
concatenate_1 (Concatenate)     (None, 1, 712)       0           dot[18][0]
                                                                 decoder_embedding[1][0]
_____
c0 (InputLayer)                 [(None, 256)]        0
_____
decoder_lstm (LSTM)             [(None, 256), (None, 992256      concatenate_1[18][0]
                                                                 s0[0][0]
                                                                 c0[0][0]
_____
decoder_dense (Dense)           (None, 4964)         1275748     decoder_lstm[18][0]
==================================================================================================
Total params: 3,268,505
Trainable params: 3,268,505
Non-trainable params: 0
```

图 8-14 解码器概要

输出结果如图 8-15 所示。

接下来，准备尝试进行机器翻译。

3. 机器翻译

当用模型处理输入文本并将输出预测编码为整数时，需要创建从整数到单词索引的反向映射。代码如下。

```
word2index_input=tokenizer_in.word_index
word2index_output=tokenizer_out.word_index
#reverse mapping integer to words for english
idx2word_eng = {v:k for k, v in
                word2index_input.items()}
#reverse mapping integer to words for spanish
idx2word_trans = {v:k for k, v in
                  word2index_output.items()}
```

编写函数解码由若干单词组成的句子。函数定义如下。

```
def decode_sequence(input_seq):
```

在函数体中，将输入编码为状态向量：

```
enc_out = encoder_model.predict(input_seq)
```

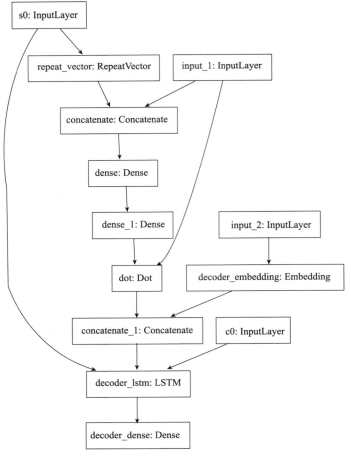

图 8-15 解码器的可视化结果

生成长度为 1 的空目标序列：

`target_seq = np.zeros((1, 1))`

使用 <start> 标记填充目标序列的第一个字符。

`target_seq[0, 0] = word2index_output['<start>']`

> **注意**
>
> 标记器将所有单词小写。

使用 <end> 标记跳出循环：

`End_statement = word2index_output['<end>']`

随着训练状态在每个循环中更新,需要在每次迭代时初始化训练状态。代码如下。

```
s = np.zeros((1,LATENT_DIM))
c = np.zeros((1,LATENT_DIM))
```

为了存储翻译后的序列,声明一个输出数组。代码如下。

```
output_sentence = []
```

定义一个 for 循环记录输出结果:

```
for _ in range(maxlen_output):
```

在每次迭代中,调用解码器的 predict 方法:

```
out, s, c = decoder_model.predict([target_seq,
                    enc_out, s, c])
```

根据解码器输出的预测概率选取下一个单词:

```
index = np.argmax(out.flatten())
```

如果到达序列的末尾,终止循环:

```
if End_statement == index:
    break
```

将预测词的索引转换为字符格式并将其添加到输出序列中:

```
word = ''
if index > 0:
    word = idx2word_trans[index]
    output_sentence.append(word)
```

更新解码器输入,生成输出单词:

```
target_seq[0, 0] = index
```

最后,将记录的输出序列返回给调用者:

```
return ' '.join(output_sentence)
```

完整的函数实现过程在清单 8-3 中给出,仅供参考。

清单 8-3 解码函数

源码清单
链　接:https://pan.baidu.com/s/1NV0rimQ_8kRz22xfFHN-Cw
提取码:1218

```
def decode_sequence(input_seq):
    # Encode the input as state vectors.
    enc_out = encoder_model.predict(input_seq)

    # Generate empty target sequence of length 1.
    target_seq = np.zeros((1, 1))

    # Populate the first character of target sequence with the
      start character.
    # NOTE: tokenizer lower-cases all words
```

```python
    target_seq[0, 0] = word2index_output['<start>']

    # if we get this we break
    End_statement = word2index_output['<end>']

    # [s, c] will be updated in each loop iteration
    s = np.zeros((1,LATENT_DIM))
    c = np.zeros((1,LATENT_DIM))

    # Create the translation
    output_sentence = []
    for _ in range(maxlen_output):
      out, s, c = decoder_model.predict([target_seq,
                            enc_out, s, c])
    # Get next word
    index = np.argmax(out.flatten())

    # End sentence
    if End_statement == index:
      break

    word = ''
    if index > 0:
      word = idx2word_trans[index]
      output_sentence.append(word)

    # Update the decoder input
    # which is just the word just generated
    target_seq[0, 0] = index

  return ' '.join(output_sentence)
```

接下来，看一些翻译示例。首先，定义了一个循环用于接收用户输入。用户可以输入英文语句，然后得到西班牙文的翻译版本。循环代码如下。

```
count=0
while (count<5):

  input_text=[str(input('input the sentence : '))]
  seq=tokenizer_in.texts_to_sequences(input_text)

  input_seq=pad_sequences(seq,maxlen=maxlen_input,
                        padding='post')
  translation=decode_sequence(input_seq)

  print('Predicted translation:', translation)
  count+=1
```

机器翻译模型所做的翻译示例如下。

```
input the sentence : Hello
```

```
Predicted translation: hola
input the sentence : How are you?
Predicted translation: ¿qué están
input the sentence : What are you doing?
Predicted translation: ¿cómo están usted
input the sentence : see you later
Predicted translation: te luego
input the sentence : bye
Predicted translation: ¡ataque
```

并非所有翻译都是正确的，这是由于本项目没有使用足够的数据实例，因此模型准确率较低且损失值偏高。通过增加词汇量、调整嵌入层维度、改变编码器和解码器的潜在维度、尝试不同数量的隐藏层及拥有更深的模型来微调模型，可以提高翻译质量。

8.3.12 项目源码

清单 8-4 给出了本章实验的完整源代码，仅供参考。

清单 8-4 NMT 项目的完整源代码

源码清单
链　接：https://pan.baidu.com/s/1NV0rimQ_8kRz22xfFHN-Cw
提取码：1218

```python
import tensorflow as tf

from tensorflow.keras.models import Model
from tensorflow.keras.layers import Input, \
            Dense, LSTM, Embedding, Bidirectional, \
            RepeatVector, Concatenate, Activation, \
            Dot, Lambda
from tensorflow.keras.preprocessing.text \
            import Tokenizer
from tensorflow.keras.preprocessing.sequence \
            import pad_sequences
from keras import preprocessing, utils
import numpy as np
import matplotlib.pyplot as plt

!pip install wget
import wget
url = 'https://raw.githubusercontent.com/Apress/artificialneural-
networks-with-tensorflow-2/main/ch08/spa.txt'
wget.download(url,'spa.txt')

# reading data
with open('/content/spa.txt',encoding='utf-8',errors='ignore') \
as file:
  text=file.read().split('\n')

len(text)

# input and target data is separated by tab '\t'
text[:5]
```

```python
for t in text[100:110]:
    print(t)

input_texts=[] #encoder input
target_texts=[] # decoder input

# we will select subset of the whole data
NUM_SAMPLES = 10000
for line in text[:NUM_SAMPLES]:
    english, spanish = line.split('\t')[:2]
    target_text = spanish.lower()
    input_texts.append(english.lower())
    target_texts.append(target_text)

print(input_texts[:5],target_texts[:5])

import string
print('Characters to be removed in preprocessing', string.punctuation)

#removing punctuation from target and input
def remove_punctuation(s):
    out=s.translate(str.maketrans("","",
                    string.punctuation))

    return out

input_texts = [remove_punctuation(s)
                for s in input_texts]
target_texts = [remove_punctuation(s)
                for s in target_texts]
input_texts[:5],target_texts[:5]

# adding start and end tags
target_texts=['<start> ' + s + ' <end>'
        for s in target_texts]

target_texts[1]

tokenizer_in=Tokenizer()
#tokenizing the input texts
tokenizer_in.fit_on_texts(input_texts)
#vocab size of input
input_vocab_size=len(tokenizer_in.word_index) + 1

input_vocab_size

# Listing few items
input_tokens = tokenizer_in.index_word
for k,v in sorted(input_tokens.items())[2000:2010]:
```

```python
        print (k,v)

#tokenizing output that is spanish translation
tokenizer_out=Tokenizer(filters='')
tokenizer_out.fit_on_texts(target_texts)
#vocab size of output
output_vocab_size=len(tokenizer_out.word_index) + 1
output_vocab_size

# Listing few items
output_tokens = tokenizer_out.index_word
for k,v in sorted(output_tokens.items())[2000:2010]:
    print (k,v)

#converting tokenized sentence into sequences
tokenized_input = tokenizer_in.texts_to_sequences
                    ( input_texts )
#max length of the input
maxlen_input = max( [ len(x)
                    for x in tokenized_input ] )
#padding sequence to the maximum length
padded_input = preprocessing.sequence.pad_sequences
        ( tokenized_input , maxlen=maxlen_input ,
            padding='post' )

padded_input[2000:2010]

encoder_input_data = np.array( padded_input )
print( encoder_input_data.shape)

#converting tokenized text into sequences
tokenized_output = tokenizer_out.texts_to_sequences
                    (target_texts)

output_vocab_size=len(tokenizer_out.word_index) + 1
output_vocab_size

# teacher forcing
for i in range(len(tokenized_output)) :
    tokenized_output[i] = tokenized_output[i][1:]

# padding
maxlen_output = max( [ len(x)
                    for x in tokenized_output ] )
padded_output = preprocessing.sequence.pad_sequences
                ( tokenized_output ,
                    maxlen=maxlen_output ,
                    padding='post' )

# converting to numpy
```

```python
decoder_input_data = np.array( padded_output )
decoder_input_data[2000:2010]
#decoder target output
decoder_target_one_hot=np.zeros((len(input_texts),
                                 maxlen_output,
                                 output_vocab_size),
                                dtype='float32')
for i,d in enumerate(padded_output):
    for t,word in enumerate(d):
        decoder_target_one_hot[i,t,word]=1

decoder_target_one_hot[0]

from google.colab import drive
drive.mount('/content/drive', force_remount=True)

#creating dictionary of words corresponding to vectors
print('Indexing word vectors.')

embeddings_index = {}

# Use this open command in case of downloading using wget above
#f = open('glove.6B.200d.txt', encoding='utf-8')

# we can choose any dimensions 50 100 200 300
f = open('drive/My Drive/tfbookdata/glove.6B.200d.txt',
         encoding='utf-8')

for line in f:
    values = line.split()
    word = values[0]
    coefs = np.asarray(values[1:], dtype='float32')
    embeddings_index[word] = coefs
f.close()
print('Found %s word vectors.' %
      len(embeddings_index))

embeddings_index["any"]

#embedding matrix
num_words = len(tokenizer_in.word_index)+1
word2idx_input = tokenizer_in.word_index
embedding_matrix = np.zeros((num_words, 200))
for word,i in word2idx_input.items():
    if i<num_words:
        embedding_vector = embeddings_index.get(word)
    if embedding_vector is not None:
        embedding_matrix[i] = embedding_vector
embedding_matrix.shape
```

```python
#lstm hidden dimensions
LATENT_DIM=256
#embeding layer dimensions
EMBEDDING_DIM=200

embedding_layer=Embedding(input_vocab_size,
                          EMBEDDING_DIM,
                          weights=[embedding_matrix],
                          input_length=maxlen_input)

#encoder
encoder_inputs = Input(shape=(maxlen_input,),
                       name = 'encoderinput')
encoder_input = embedding_layer(encoder_inputs)
encoder = Bidirectional(LSTM(LATENT_DIM,
                             return_sequences=True,
                             dropout = 0.3),
                        name = 'encoder_bidirection')
encoder_outputs = encoder(encoder_input)

encoder_outputs.shape

decoder_inputs = Input(shape=(maxlen_output,),
                       name='decoder_input')
decoder_embedding = Embedding(output_vocab_size,
                              EMBEDDING_DIM,
                              name='decoder_embedding')
decoder_input=decoder_embedding(decoder_inputs)

decoder_input.shape

#decoder lstm
decoder_lstm = LSTM(LATENT_DIM,
                    return_state = True,
                    name = 'decoder_lstm')
#decoder dense with softmax for predicting each word
decoder_dense = Dense(output_vocab_size,
                      activation='softmax',
                      name='decoder_dense')

# Computing alphas
import tensorflow.keras.backend as k
def softmax_attention(x):
    assert(k.ndim(x)>2)
    
    e=k.exp(x-k.max(x,axis=1,keepdims=True))
    s=k.sum(e,axis=1,keepdims=True)
    return e/s

# nerual network layers for our repeated use
```

```python
attention_repeat = RepeatVector(maxlen_input)
attention_concat = Concatenate(axis=-1)
dense1_layer = Dense(10,activation='tanh')
dense2_layer = Dense(1,activation = softmax_attention)
dot_layer = Dot(axes=1)

# computing attention context
def context_attention(h, st_1):
    st_1=attention_repeat(st_1)
    x=attention_concat([h,st_1])
    x=dense1_layer(x)
    alphas=dense2_layer(x)
    context = dot_layer([alphas,h])
    return context

#initial states to be fed
initial_s = Input(shape=(LATENT_DIM,), name='s0')
initial_c = Input(shape=(LATENT_DIM,), name='c0')
context_last_word_concat_layer = Concatenate(axis=2)

s = initial_s
c = initial_c
outputs = []

#collect output in a list at first
for t in range(maxlen_output): #ty times
    #get the context using attention mechanism
    context=context_attention(encoder_outputs,s)
    
    #we need a different layer for each time step
    selector=Lambda(lambda x: x[:,t:t+1])
    x_t=selector(decoder_input)
    
    #combine
    decoder_lstm_input=context_last_word_concat_layer
                        ([context,x_t])
    #pass the combined [context,last word] into lstm
    #along with [s,c]
    #get the new[s,c] and output
    out,s,c=decoder_lstm(decoder_lstm_input,
                        initial_state=[s,c])
    
    #final dense layer to get next word prediction
    decoder_outputs = decoder_dense(out)
    outputs.append(decoder_outputs)

def stack(x):
    x=k.stack(x)
    x=k.permute_dimensions(x,pattern=(1,0,2))
    return x
```

```
stacker=Lambda(stack)
outputs=stacker(outputs)
outputs

model=Model(inputs=[encoder_inputs,
                    decoder_inputs,
                    initial_s,
                    initial_c],
            outputs=outputs)

model.summary()

tf.keras.utils.plot_model (model)

model.compile(optimizer='adam',
              loss='categorical_crossentropy',
              metrics=['accuracy'])

initial_s_training=np.zeros((NUM_SAMPLES,LATENT_DIM))
#initial s c
initial_c_training =np.zeros(shape=(NUM_SAMPLES,LATENT_DIM))
r = model.fit([encoder_input_data,
               decoder_input_data,
               initial_s_training,
               initial_c_training],
              decoder_target_one_hot,
              batch_size=100,
              epochs=100,
              validation_split=0.3)

encoder_model = Model(encoder_inputs,
                      encoder_outputs)

encoder_model.summary()
tf.keras.utils.plot_model (encoder_model)

#input will have the length double of latent dimension due to
bidirectional
encoder_outputs_as_input = Input(shape=
                    (maxlen_input, LATENT_DIM* 2,))
#we are going to predict one word at a time with input of one
word
decoder_input_embedding = Input(shape=(1,))
decoder_input_ = decoder_embedding
                    (decoder_input_embedding)

#calculating context
context = context_attention
            (encoder_outputs_as_input, initial_s)
```

```python
decoder_lstm_input = context_last_word_concat_layer(
    [context, decoder_input_])
out, s, c = decoder_lstm(decoder_lstm_input,
                    initial_state=[initial_s,initial_c])
decoder_outputs = decoder_dense(out)

decoder_model = Model(
  inputs=[
    decoder_input_embedding,
    encoder_outputs_as_input,
    initial_s,
    initial_c
  ],
  outputs=[decoder_outputs, s, c]
)
decoder_model.summary()

#tf.keras.utils.plot_model (decoder_model,to_file='decoder_
model.jpg')
tf.keras.utils.plot_model (decoder_model)

word2index_input=tokenizer_in.word_index
word2index_output=tokenizer_out.word_index
#reverse mapping integer to words for english
idx2word_eng = {v:k for k, v in
                    word2index_input.items()}
#reverse mapping integer to words for spanish
idx2word_trans = {v:k for k, v in
                    word2index_output.items()}

def decode_sequence(input_seq):
  # Encode the input as state vectors.
  enc_out = encoder_model.predict(input_seq)
  # Generate empty target sequence of length 1.
  target_seq = np.zeros((1, 1))

  # Populate the first character of target sequence with the
    start character.
  # NOTE: tokenizer lower-cases all words
  target_seq[0, 0] = word2index_output['<start>']

  # if we get this we break
  End_statement = word2index_output['<end>']

  # [s, c] will be updated in each loop iteration
  s = np.zeros((1,LATENT_DIM))
  c = np.zeros((1,LATENT_DIM))

  # Create the translation
  output_sentence = []
```

```
    for _ in range(maxlen_output):
      out, s, c = decoder_model.predict([target_seq,
                            enc_out, s, c])

      # Get next word
      index = np.argmax(out.flatten())

      # End sentence
      if End_statement == index:
        break
      word = ''
      if index > 0:
        word = idx2word_trans[index]
        output_sentence.append(word)
      # Update the decoder input
      # which is just the word just generated
      target_seq[0, 0] = index

    return ' '.join(output_sentence)

count=0
while (count<5):

  input_text=[str(input('input the sentence : '))]
  seq=tokenizer_in.texts_to_sequences(input_text)

  input_seq=pad_sequences(seq,maxlen=maxlen_input,
                              padding='post')
  translation=decode_sequence(input_seq)

  print('Predicted translation:', translation)
  count+=1
```

总结

在本章中，我们学习了用于机器翻译的复杂深度学习应用程序，也称为 NMT。NMT 用于将文本从一种语言翻译成另一种语言。本章我们构建了从英语到西班牙语的 NMT 模型。为了提高翻译质量，使用了词向量词典。NMT 模型本质上是一个 seq2seq 模型，由一个编码器和一个解码器组成。简单的编码器、解码器模型无法对较长的句子提供高质量的翻译，因此，在机器翻译模型中添加了注意力网络以提高翻译质量。

在下一章中，我们将学习另一种用于语言理解的深度学习模型。

第 9 章

自然语言理解

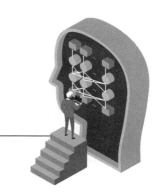

在第 8 章，我们学习了使用 seq2seq 模型和注意力模型进行文本翻译，本章将介绍一种更加复杂的自然语言处理技术，以及一种用于自然语言处理的新模型，即 Transformer 模型。Transformer 模型缓解了上一章算法对 LSTM 的依赖，并且 Transformer 模型取得了比基于 LSTM 的 seq2seq 模型更好的结果。现在让我们了解一下什么是 Transformer 模型。

9.1 Transformer 简介

第 8 章在机器翻译任务中证明了注意力模块的重要性，注意力模块为句子中提供重要语义信息的词分配更高的权重，然后将注意力权重与待翻译文本一起提供给解码器。在了解句子中哪些部分更加重要的前提下，解码器能够提供更好的翻译结果。

Transformer 模型的示意图如图 9-1 所示。

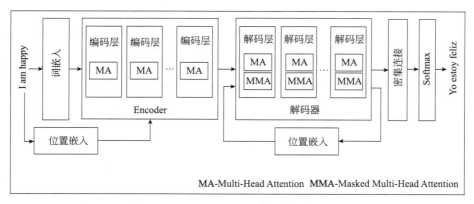

图 9-1　Transformer 模型的一般架构

如图 9-1 所示，Transformer 模型主要由一个编码器和一个解码器组成，此结构与第 8 章使用的模型结构类似，主要区别在于编码器和解码器都不包含 LSTM 模型。相反，Transformer 模型的特征编码和解码阶段使用了由 Keras 设计的多层结构，如图 9-1 中编码器和解码器部分所示。上述多层结构的层数可由用户自行配置。编码器包含一个特殊的注意力模块，称为多头注意力 (multi-head attention, MA) 机制。同样地，解码器也包含此类注意力模块，此外，解码器中包含另一种名为 masked multi-Head attention (MMA) 的注意力模块。Transformer 中的特征提取过程与传统的特征提取方式不同，特征传递不是按从编码器到解码器的串联方式进行的，而是将输入语句分为若干单词，以并联方式输入特征提取网络。上述数据输入方式被称为多头 (multi-head) 输入，此方式有利于分布式训练和推理。综上所述，Transformer 模型优于第 8 章中讨论的基于注意力机制的编码器、解码器模型。

除了上述结构外，图 9-1 中的 Transformer 结构使用了额外的嵌入 (embedding) 模块。通常在自然语言处理任务中使用词嵌入 (word embedding) 模块。在 Transformer 中，在输入的语句中加入了一个额外的嵌入模块，称为位置嵌入 (positional embedding)，如图 9-1 所示。位置嵌入用于指定特定单词在输入语句中的相对位置，其作用与 LSTM 结构中的编码器、解码器相同，因此 Transformer 模型不需要使用 LSTM 机制。对于开发人员来说，理解 Transformer 模块的各部分代码，可以更好地了解 Transformer 的内在原理与工作机制。本章详细地展示了 Transformer 的工作代码，并通过自上而下的方法，解释每个模块。

现在，让我们开始研究 Transformer 模块吧。

9.2 Transformer 详解

首先，创建一个新的 Colab 项目并将其重命名为 NLP-transformer，NLP 为自然语言处理的简称 (natural language processing, NLP)。接下来导入本章项目所需的库，如下代码段所示。

```
import tensorflow as tf
from tensorflow.keras.models import Model
from tensorflow.keras.layers
            import Input,Dense,LSTM,Embedding,
                Bidirectional,RepeatVector,
                Concatenate,Activation,Dot,Lambda
from tensorflow.keras.preprocessing.text
            import Tokenizer
from tensorflow.keras.preprocessing.sequence
            import pad_sequences
from keras import preprocessing,utils
import numpy as np
import matplotlib.pyplot as plt
import tensorflow_datasets as tfds
import os
import re
import numpy as np
import string
```

9.2.1 下载原始数据

使用如下语句从本书的官方网站下载原始数据。

```
!pip install wget
import wget
url = 'https://raw.githubusercontent.com/Apress/artificialneural-
networks-with-tensorflow-2/main/ch08/spa.txt'
wget.download(url,'spa.txt')
```

本章使用的数据集与第 8 章使用的数据集相同，因此，本章省略了对数据集结构的详细介绍。

9.2.2 创建数据集

本章创建数据集的代码与第 8 章介绍的开发代码相同，提供如下代码段供读者参考。

```
# reading data
with open('/content/spa.txt',encoding='utf-8',
          errors='ignore') as file:
  text=file.read().split('\n')

input_texts=[] #encoder input
target_texts=[] # decoder input

# we will select subset of the whole data
NUM_SAMPLES = 10000
for line in text[:NUM_SAMPLES]:
  english, spanish = line.split('\t')[:2]

  target_text = spanish.lower()
  input_texts.append(english.lower())
  target_texts.append(target_text)
```

与之前介绍的情况相同，本章仅使用整个语料库的一个子集进行算法研究，约 10000 个样本。

9.2.3 数据预处理

首先，使用以下代码从原始数据中删除标点符号：

```
regex = re.compile('[%s]' %re.escape(string.punctuation))
for s in input_texts:
  regex.sub('', s)
for s in target_texts:
  regex.sub('', s)
```

删除原始数据集中的标点符号后，需要进行语料分词和填充工作。

9.2.4 构建语料库

本章使用 SubWordTextEncoder 类创建分词器，分词器将原始文本与文本编码 (ID) 关联，进行字典构建。如果存在特殊文本不在现有字典中，分词器将其分解为若干子词进行处理。

首先，为本章项目的输入数据集构建一个分词器，使用如下代码段实现。

```
tokenizer_input = tfds.features.text.
                  SubwordTextEncoder.build_from_corpus(
    input_texts, target_vocab_size=2**13)
```

其中，第二个参数 target_vocab_size 可以指定一个任意大的值。

下面，为了测试分词器的有效性，尝试向分词器输入字符串。例如，使用以下代码段对字符串进行编码。

```
# example showing how this tokenizer works
tokenized_string1=tokenizer_input.encode
                  ('hello i am good')
tokenized_string1
```

上述代码输出如下。

```
[2269, 1, 41, 89]
```

将上面的输出以表格形式打印，以显示输入文本与文本 ID 的映射视图，使用如下语句实现。

```
for token in tokenized_string1:
  print ('{} ----> {}'.format(token,
                   tokenizer_input.decode([token])))
```

上述代码输出如下。

```
2269 ----> hello
1 ----> i
41 ----> am
89 ----> good
```

观察以上结果得出，每个单词都在字典中与唯一的 ID 关联。接下来，尝试输入字典中没有的新语句，如下所示。

```
# if the word is not in dictionary
tokenized_string2=tokenizer_input.encode
                  ('how is the moon')

for token in tokenized_string2:
  print ('{} ----> {}'.format(token,
                   tokenizer_input.decode([token])))
```

上述代码输出如下。

```
64 ----> how
4 ----> is
21 ----> the
2827 ----> m
2829 ----> o
75 ----> on
```

241

单词 moon 没有在字典中出现过，以 moon 为例，观察分词器对未知单词的处理过程，分词器将其拆分成 m、o、on 分别处理。

上述代码构建了以英语为对象的语料库。类似地，以同样的方式处理翻译目标语言，即西班牙语文本，如下所示。

```
tokenizer_out=tfds.features.text.SubwordTextEncoder.
                                build_from_corpus(
    target_texts, target_vocab_size=2**13)
```

接下来，为输入语句和输出语句添加起始和结束标记，使用如下代码段实现。

```
START_TOKEN_in=[tokenizer_input.vocab_size]
#input start token
END_TOKEN_in=[tokenizer_input.vocab_size+1]
#input end token
START_TOKEN_out=[tokenizer_out.vocab_size]
#output start token
END_TOKEN_out=[tokenizer_out.vocab_size+1]
#output end token/
```

使用如下语句查看文本的 ID。

```
START_TOKEN_in,
        END_TOKEN_in,START_TOKEN_out,END_TOKEN_out
```

上述代码的输出如下。

([2974], [2975], [5737], [5738])

接下来，编写一个用于填充输入数据集和输出数据集的函数，将最大长度设为 10，用于构建相同长度的输入和输出语句，如下所示。

```
MAX_LENGTH = 10

# Tokenize, filter and pad sentences
def tokenize_and_padding(inputs, outputs):
  tokenized_inputs, tokenized_outputs = [], []

  for (input_sentence, output_sentence)
                    in zip(inputs, outputs):

    # tokenize sentence
    input_sentence = START_TOKEN_in +
            tokenizer_input.encode(input_sentence) +
                        END_TOKEN_in
    output_sentence = START_TOKEN_out +
            tokenizer_out.encode(output_sentence) +
                        END_TOKEN_out
    # check tokenized sentence max length
    #if len(input_sentence) <= MAX_LENGTH and
              len(output_sentence) <= MAX_LENGTH:
    tokenized_inputs.append(input_sentence)
    tokenized_outputs.append(output_sentence )
```

```python
    # pad tokenized sentences
    tokenized_inputs = 
        tf.keras.preprocessing.sequence.pad_sequences(
            tokenized_inputs, maxlen=MAX_LENGTH,
                                    padding='post')
    tokenized_outputs = 
        tf.keras.preprocessing.sequence.pad_sequences(
            tokenized_outputs, maxlen=MAX_LENGTH,
                                    padding='post')

    return tokenized_inputs, tokenized_outputs

english, spanish = tokenize_and_padding
                        (input_texts,target_texts)
```

打印上述函数输出结果中的一个元素以检查此函数的结果，使用如下语句。

```
english[1],spanish[1]
```

输出结果如下：

```
(array([2974, 50, 2764, 2975, 0, 0, 0, 0, 0, 0],
        dtype=int32),
 array([5737, 1017, 5527, 5738, 0, 0, 0, 0, 0, 0],
        dtype=int32))
```

上面的填充结果显示，每个语句的长度固定为 10。其中，英语语句为"go."，单词"go"的编码为 50，"."的编码为 2764。此外，英语语句中 2974 为起始标记，2975 为结束标记。

9.2.5 准备训练集数据

接下来，准备用于训练的数据集。为了创建数据输入管道，使用 tf.data.Dataset 函数提供的 cache 和 prefetch 等功能加快数据输入进程。如以下代码段所示。

```
BATCH_SIZE = 32
BUFFER_SIZE = 10000

# decoder inputs use the previous target as input
# remove START_TOKEN from targets
dataset = tf.data.Dataset.from_tensor_slices((
    {
        'inputs': english,
        'decoder_inputs': spanish[:, :-1]
    },
    {
        'outputs':spanish[:, 1:]
    },
))

dataset = dataset.cache()
```

```
dataset = dataset.shuffle(BUFFER_SIZE)
dataset = dataset.batch(BATCH_SIZE)
dataset = dataset.prefetch
            (tf.data.experimental.AUTOTUNE)
```

其中，prefetch 功能在预处理数据和模型执行数据之间并行提取数据，即当模型执行第 n 个训练时，输入管道读取第 $n+1$ 个步骤的数据。上述过程可以加快训练进程。

9.2.6　Transformer 模型

接下来，采取自上而下的方式解释 Transformer 模型的实现过程。Transformer 模型由多个组件构成，构建完整模型依赖于创建各部分组件。为了更好地理解模型的构建方式，首先概述模型的主要组件，然后深入了解每个组件及其详细的实现细节。本章的项目代码主要基于 TensorFlow 作者发布的 Transformer Chatbot 实现，并获取了 Apache 2.0 版的许可。当然，Transformer Chatbot 代码编写和结构十分合理，不需要进行改进，因此，直接使用已有代码进行算法讲解。

Transformer 模型的整体视图如图 9-2 所示。

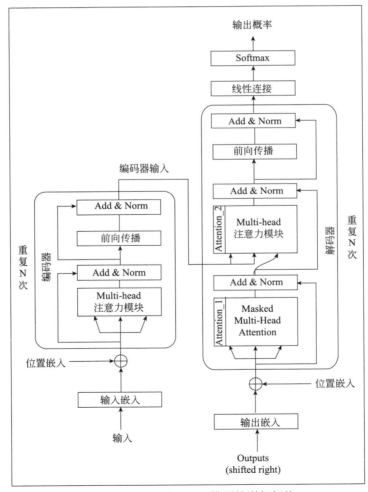

图 9-2　Transformer 模型的详细架构

如图 9-2 所示，Transformer 模型主要由一个编码器、一个解码器和若干多头注意力（机制）模块组成。其中，多头注意力（机制）模块是整个架构的核心模块。接下来，本节将对多头注意力（机制）、编码器、解码器分别进行详细的讲解。最后，介绍如何训练和使用 Transformer 模型进行运算。实验结果表明，Transformer 比之前的文本翻译模型（seq2seq 模型）有更好的自然语言理解能力，能够将英语翻译为西班牙语。

9.2.7 多头注意力（机制）

多头注意力（机制）是 Transformer 的核心，以网络层的形式出现在 Transformer 中。Transformer 结构多次使用多头注意力（机制），因此，为多头注意力（机制）编写了一个类，该类派生自 Keras Layer (tf.keras.layers.Layer) 类。多头注意力（机制）示意图如图 9-3 所示。

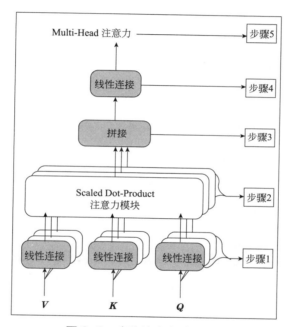

图 9-3　多头注意力（机制）

图 9-3 中的多头注意力（机制）有三部分输入，即 Q (query)、K (key)、V (value)。这三个向量被分成多个输入头，并通过线性连接层传递给下一步。通过这种方式，Transformer 模型可以同时处理输入句子中的多个单词，促进分布式训练。

对图 9-3 中的每个输入向量使用按比缩放的点积 (scaled dot-product) 注意力模块，并根据是否在编码器或解码器中使用此多头注意力（机制），对多头注意力（机制）进行编码，后续介绍两种常用的编码方式。接下来，Scaled Dot-Product 注意力模块的输出使用 tf.transpose 函数和 tf.reshape 函数连接起来，并通过一个线性连接层 (Linear) 产生最终的注意力向量。

尽管多头注意力（机制）原理图看起来比较复杂，但它的内部结构并不复杂，由五个主要组件实现，编号为步骤 1 到步骤 5。在 Transformer 代码中通过定义 MultiHeadAttention 类实现。下面给出 MultiHeadAttention 类的完整实现，并解释步骤 1

到步骤 5 在类定义中的实现代码。MultiHeadAttention 类的实现过程如清单 9-1 所示。

清单 9-1 MultiHeadAttention 类定义

源码清单
链　接：https://pan.baidu.com/s/1NV0rimQ_8kRz22xfFHN-Cw
提取码：1218

```python
class MultiHeadAttention(tf.keras.layers.Layer):
    def __init__(self, d_model, num_heads,
                 name="multi_head_attention"):
        super(MultiHeadAttention, self)
                    .__init__(name=name)
        self.num_heads = num_heads
        self.d_model = d_model
        self.depth = d_model // self.num_heads
        
        self.query_dense = 
                tf.keras.layers.Dense(units=d_model)
        self.key_dense = 
                tf.keras.layers.Dense(units=d_model)
        self.value_dense = 
                tf.keras.layers.Dense(units=d_model)
        
        self.dense = tf.keras.layers.Dense(units=d_model)
    
    def split_heads(self, inputs, batch_size):
        inputs = tf.reshape(
            inputs, shape=(batch_size, -1,
                self.num_heads, self.depth))
        return tf.transpose(inputs, perm=[0, 2, 1, 3])
    
    def call(self, inputs):
        query, key, value, mask = inputs['query'], inputs['key'],
                                  inputs['value'], inputs['mask']
        batch_size = tf.shape(query)[0]
        
        # linear layers
        query = self.query_dense(query)
        key = self.key_dense(key)
        value = self.value_dense(value)
        
        # split heads
        query = self.split_heads(query, batch_size)
        key = self.split_heads(key, batch_size)
        value = self.split_heads(value, batch_size)
        # scaled dot-product attention
        scaled_attention = scaled_dot_product_attention
                        (query, key, value, mask)
        scaled_attention = tf.transpose
                        (scaled_attention,
                         perm=[0, 2, 1, 3])
        
        # concatenation of heads
```

```
concat_attention = tf.reshape(scaled_attention,
                              (batch_size, -1,
                               self.d_model))

# final linear layer
outputs = self.dense(concat_attention)

return outputs
```

MultiHeadAttention 类派生自 Keras Layer 类, 如下所示。

```
class MultiHeadAttention(tf.keras.layers.Layer):
```

在初始化类构造函数时,传递的参数较少,这些参数存储在类变量中以备后用,如下所示。

```
def __init__(self, d_model, num_heads,
             name="multi_head_attention"):
    super(MultiHeadAttention, self)
                .__init__(name=name)
    self.num_heads = num_heads
    self.d_model = d_model
    self.depth = d_model // self.num_heads
```

Transformer 的网络输入通过一个线性连接组合,网络输入为 ***V***、***K*** 和 ***Q***,如图 9-3 所示。其中,***Q*** 为包含全部查询向量的矩阵,***K*** 和 ***V*** 为序列中所有单词的向量表示。上述三个输入层的定义如下。

```
self.query_dense = tf.keras.layers.Dense(units=d_model)
self.key_dense = tf.keras.layers.Dense(units=d_model)
self.value_dense = tf.keras.layers.Dense(units=d_model)
```

此外,MultiHeadAttention 类还声明了一个 Dense 层,用于得到注意力模型的输出,如下所示。

```
self.dense = tf.keras.layers.Dense(units=d_model)
```

上述过程完成了 MultiHeadAttention 类构造函数的定义。

接下来,定义一个 split_heads 函数,对输入句子中的不同单词进行并行处理,有利于分布式训练。split_heads 函数对数据进行整形、转置,并将处理后的输入数据返回给调用函数。代码如下。

```
def split_heads(self, inputs, batch_size):
    inputs = tf.reshape(
        inputs, shape=(batch_size, -1,
            self.num_heads, self.depth))
    return tf.transpose(inputs, perm=[0, 2, 1, 3])
```

接下来介绍函数调用,通过函数调用构建了整个 Transformer 网络,首先将输入数据分为四部分,代码如下。

```
def call(self, inputs):
```

```
query, key, value, mask = inputs['query'], inputs['key'],
                          inputs['value'], inputs['mask']
batch_size = tf.shape(query)[0]
```

下面，为三组输入数据 (*Q*, *K*, *V*) 创建线性连接层，即图 9-3 中的步骤 1，如下所示。

```
# linear layers
query = self.query_dense(query)
key = self.key_dense(key)
value = self.value_dense(value)
```

使用 split_heads 函数拆分输入数据，即图 9-3 中的步骤 2，如下所示。

```
# split heads
query = self.split_heads(query, batch_size)
key = self.split_heads(key, batch_size)
value = self.split_heads(value, batch_size)
```

接下来，计算 Scaled Dot-Product 注意力模块输出，即图 9-3 中的步骤 3，如下所示。

```
# scaled dot-product attention
scaled_attention = scaled_dot_product_attention
                    (query, key, value, mask)
scaled_attention = tf.transpose
                    (scaled_attention,
                     perm=[0, 2, 1, 3])
```

上段代码使用了 scaled_dot_product_attention 函数。目前为止，尚未定义 scaled_dot_product_attention 函数，此函数将在后续介绍中进行定义。接下来，使用如下语句连接各组输入数据的计算结果，即图 9-3 中的步骤 4。

```
# concatenation of heads
concat_attention = tf.reshape(scaled_attention,
                              (batch_size, -1,
                               self.d_model))
```

最后，定义线性连接输出层，即图 9-3 中的步骤 5，如下所示。

```
outputs = self.dense(concat_attention)
```

完成上述计算后将输出返回给调用函数。
下面定义之前使用的 scaled_dot_product_attention 函数。

9.2.8　Scaled Dot-Product 注意力模块

Transformer 使用的 Scaled Dot-Product 模块接收三组输入数据 (*Q*, *K*, *V*)，使用如下方程计算。

$$\text{Attention}(Q, K, V) = \text{softmax}_k \left(\frac{QK^T}{\sqrt{d_k}} \right) V$$

首先，计算 QK^T，使用如下语句实现。

```
QxK_transpose = tf.matmul
                (query, key, transpose_b=True)
```

接下来，计算 QK^T 除以 $\sqrt{d_k}$ 的结果，使用如下语句实现。

```
depth = tf.cast(tf.shape(key)[-1], tf.float32)
logits = QxK_transpose / tf.math.sqrt(depth)
```

对以上结果通过 softmax 激活，使用如下语句实现。

```
attention_weights = tf.nn.softmax(logits, axis=-1)
```

最后，通过计算注意力权重和输入向量 V 之间的矩阵乘法得到输出结果，使用如下语句实现。

```
output = tf.matmul(attention_weights, value)
```

将计算输出返回给调用函数，如下所示。

```
return output
```

scaled_dot_product_attention 函数的实现代码如清单 9-2 所示。

清单 9-2 scaled_dot_product_attention 函数

源码清单
链　接：https://pan.baidu.com/s/1NV0rimQ_8kRz22xfFHN-Cw
提取码：1218

```
def scaled_dot_product_attention
                (query, key, value, mask):
  QxK_transpose = tf.matmul
                (query, key, transpose_b=True)

  depth = tf.cast(tf.shape(key)[-1], tf.float32)
  logits = QxK_transpose / tf.math.sqrt(depth)

  if mask is not None:
    logits += (mask * -1e9)

  # softmax is normalized on the last axis (seq_len_k)
    attention_weights = tf.nn.softmax(logits, axis=-1)

    output = tf.matmul(attention_weights, value)

    return output
```

定义了 Transformer 的主要组件多头注意力（机制）模块后，接下来定义编码器和解码器。

9.2.9 编码器结构

编码器结构如图 9-4 所示。

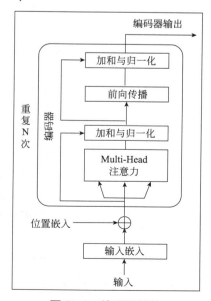

图 9-4　编码器结构

如图 9-4 所示，编码器包含一个由 Keras 定义的编码层，编码器在 Transformer 中重复 N 次。进一步讲，编码器由若干层组成，其中包括之前定义的多头注意力（机制）、加和与归一化层、前向传播层。编码器接收由输入嵌入模块和位置嵌入模块编码后的输入数据。位置嵌入模块定义单词在句子中的相对位置，给出单词的位置编码。在定义编码层之前，定义两个辅助函数，用于位置编码和屏蔽预处理数据中的填充标记。

1. 定义 create_padding_mask 函数

输入数据预处理时，网络可能会对输入语句进行填充处理，即补 0，并将补 0 后的语句作为网络输入。例如，上文提到的第一个英文语句为"go."，其编码如下。

```
(array([2974, 50, 2764, 2975, 0, 0, 0, 0, 0, 0],
       dtype=int32),
```

编码序列包含开始和结束标签，对于网络识别有重要作用。但是，不希望模型将所有的零值作为有效输入，因此，需要编写一个函数用来屏蔽语料编码后添加的额外填充值。该函数定义如下。

```
def create_padding_mask(x):
  mask = tf.cast(tf.math.equal(x, 0), tf.float32)
  # (batch_size, 1, 1, sequence length)
  return mask[:, tf.newaxis, tf.newaxis, :]
```

使用有效序列测试函数功能，如下所示。

```
x=tf.constant([[2974, 50, 2764, 2975, 0, 0,
                0, 0, 0, 0]])
create_padding_mask(x)
```

函数输出如下。

```
<tf.Tensor: shape=(1, 1, 1, 10), dtype=float32,
numpy=array([[[[0., 0., 0., 0., 1., 1., 1., 1., 1., 1.]]]],
dtype=float32)>
```

从此结果可以清楚地看到，原始语料中所有填充的零值都被赋值为 1，其余有效编码被赋值为 0。

除此以外，编写一个 create_look_ahead_mask 函数，用于定义语料库中的无用单词。例如，要预测第三个词，将只使用第一个词和第二个词。类似地，要预测第四个词，只会使用第一个、第二个和第三个词，以此类推。create_look_ahead_mask 函数定义如下。

```
def create_look_ahead_mask(x):
  seq_len = tf.shape(x)[1]
  look_ahead_mask = 1 - tf.linalg.band_part
             (tf.ones((seq_len, seq_len)), -1, 0)
  padding_mask = create_padding_mask(x)
  return tf.maximum(look_ahead_mask, padding_mask)
```

上述辅助函数作为 tf.keras.layers.Lambda 层。接下来，实现位置嵌入模块。

2. PositionalEncoding 类

与上一章介绍的 seq2seq 模型一样，Transformer 模型没有使用 LSTM 机制，不带有记忆功能。因此，为了提供给定句子中单词相对位置的信息，向 Transformer 网络提供了称为位置编码的附加输入，将两个向量，即位置编码和位置嵌入，一起输入到 Transformer 网络中。位置嵌入将具有相似含义的单词聚集在一起，但不指定单词在句子中的相对位置。因此，将位置编码结果和位置嵌入结果相加可以根据单词含义的相似性和它们在句子中的位置使单词更接近。

计算位置编码的公式如下。

$$\mathrm{PE}_{(pos,2i)} = \sin\left(pos/10000^{2i/d_{model}}\right)$$

$$\mathrm{PE}_{(pos,2i+1)} = \cos\left(pos/10000^{2i/d_{model}}\right)$$

此公式的证明超出了本书的范围，详见 https://arxiv.org/pdf/1706.03762.pdf。公式的实现过程发展，如下面给出的类定义所示。

```python
class PositionalEncoding(tf.keras.layers.Layer):

    def __init__(self, position, d_model):
        super(PositionalEncoding, self).__init__()
        self.pos_encoding = self.positional_encoding
                            (position, d_model)

    def get_angles(self, position, i, d_model):
        angles = 1 / tf.pow(10000, (2 * (i // 2)) /
                    tf.cast(d_model, tf.float32))
        return position * angles

    def positional_encoding(self, position, d_model):
        angle_rads = self.get_angles(
            position=tf.range(position, dtype=tf.float32)
                        [:, tf.newaxis],
            i=tf.range(d_model, dtype=tf.float32)
                        [tf.newaxis, :],
            d_model=d_model)
        # apply sine to even index in the array
        sines = tf.math.sin(angle_rads[:, 0::2])
        # apply cosine to odd index in the array
        cosines = tf.math.cos(angle_rads[:, 1::2])

        pos_encoding = tf.concat([sines, cosines],
                            axis=-1)
        pos_encoding = pos_encoding[tf.newaxis, ...]
        return tf.cast(pos_encoding, tf.float32)

    def call(self, inputs):
        return inputs + self.pos_encoding
                    [:, :tf.shape(inputs)[1], :]
```

定义了辅助函数后，现在继续定义编码层，编码层在编码器中多次使用。

3. 编码层

如图 9-4 中的编码器示意图所示，重复 N 次的编码层由以下部分组成。

① 多头注意力（机制）。

② 两个 Dense 层及 dropout 层。

上述两部分中均含有归一化层，用于处理深度神经网络中的梯度消失问题。

以下代码段定义了编码层，可以很容易理解编码层网络结构。

```python
def encoder_layer(units, d_model, num_heads, dropout,
                  name="encoder_layer"):
    inputs = tf.keras.Input(shape=(None, d_model),
                            name="inputs")
    padding_mask = tf.keras.Input(shape=(1, 1, None),
                                  name="padding_mask")

    # multi-head attention with padding mask
    attention = MultiHeadAttention(
        d_model, num_heads, name="attention")({
            'query': inputs,
            'key': inputs,
            'value': inputs,
            'mask': padding_mask
        })
    attention = tf.keras.layers.Dropout
                (rate=dropout)(attention)
    attention = tf.keras.layers.LayerNormalization(
        epsilon=1e-6)(inputs + attention)

    # two dense layers followed by a dropout
    outputs = tf.keras.layers.Dense(units=units,
                    activation='relu')(attention)
    outputs = tf.keras.layers.Dense(units=d_model)
                    (outputs)
    outputs = tf.keras.layers.Dropout(rate=dropout)
                    (outputs)
    outputs = tf.keras.layers.LayerNormalization(
        epsilon=1e-6)(attention + outputs)

    return tf.keras.Model(
        inputs=[inputs, padding_mask],
            outputs=outputs, name=name)
```

基于上述结构，现在准备定义编码器。

9.2.10 编码器

编码器由以下组件组成：输入嵌入模块 (input embedding)、位置嵌入模块 (position embedding)、编码层。

编码层的重复次数可以根据需要定义，本项目中重复次数为 2 次。第一个编码层的输入是输入嵌入和位置编码的总和。第二个编码层的输出为编码器的最终输出，也是解码器的输入。编码器定义如下。

```python
def encoder(vocab_size,
            num_layers,
            units,
            d_model,
            num_heads,
            dropout,
            name="encoder"):

    inputs = tf.keras.Input(shape=(None,),
                            name="inputs")

    # create padding mask
    padding_mask = tf.keras.Input(shape=(1, 1, None),
                                  name="padding_mask")

    # create combination of word embedding + positional encoding
    embeddings = tf.keras.layers.Embedding
                 (vocab_size, d_model)(inputs)
    embeddings *= tf.math.sqrt(tf.cast
                  (d_model, tf.float32))
    embeddings = PositionalEncoding
                 (vocab_size, d_model)(embeddings)

    outputs = tf.keras.layers.Dropout(rate=dropout)
                                     (embeddings)

    # repeat the Encoder Layer two times
    for i in range(num_layers):
      outputs = encoder_layer(
          units=units,
          d_model=d_model,
          num_heads=num_heads,
          dropout=dropout,
          name="encoder_layer_{}".format(i),
      )([outputs, padding_mask])

    return tf.keras.Model(
        inputs=[inputs, padding_mask], outputs=outputs,
                                      name=name)
```

> **注意**
>
> 在两个嵌入层之间共享了权重矩阵，将这些权重与变量 d_model 的平方根相乘后输入到 PositionalEncoding 中。

为了可视化编码器架构，使用以下代码生成编码器模型图。

```
sample_encoder = encoder(
    vocab_size=8192,
    num_layers=5,
    units=512,
    d_model=128,
    num_heads=4,
    dropout=0.3,
    name="sample_encoder")

tf.keras.utils.plot_model(
    sample_encoder, to_file='encoder.png')
```

> **注意**
>
> 此处编码层重复了 5 次，只是为了代码演示。 在实际项目中，只将编码层重复了 2 次。上述代码输出的编码器网络图如图 9-5 所示。
>
> 注意网络输入的实现过程，网络输入在送到第一个编码层之前与位置编码相结合。 编码层本身重复 5 次，每层的输入都应用了填充掩码。编码器的输出将成为解码器的输入。为了清楚起见，在网络图中我们将编码层重复了 5 次。在项目代码中，编码层只重复了 2 次以节省训练时间。

接下来，继续定义解码器。

9.2.11 解码器结构

解码器结构如图 9-6 所示。

解码器的输入是编码器的输出。与之前介绍的使用重复编码层的编码器一样，解码器也由重复的解码层组成。解码器第一层接收来自编码器的输入，最后一层通过线性连接层及 Softmax 层，得到目标单词的输出概率。解码器的其他输入是输出词嵌入和位置编码的组合。接下来，描述可重复的解码层构造。如图 9-6 所示，解码层包含两个多头注意力层。

① 多头注意力（机制）（如图 9-6 中 Attention_2 所示）。
② Masked 多头注意力（如图 9-6 中 Attention_1 所示）。

上述第一个多头注意力，即 Attention_2，接收来自编码器的输出最为输入，输出为 K 向量和 V 向量。此注意力的另一个输入为 Masked 多头注意力层的输出。上述第二个 Masked 多头注意力，即 Attention_1，使用 Masked 前缀的原因是它的输入为前一个解码器输出和位置编码的组合。Attention_2 的输出经过两个线性连接层及 dropout 层。上述为解码层的结构，解码层在解码器中重复 N 次，使用如下代码段实现。

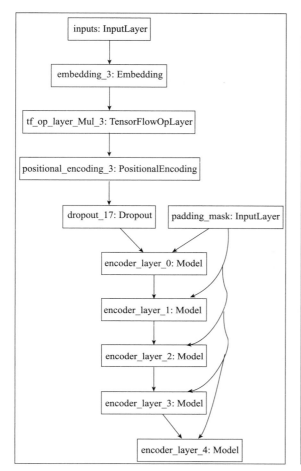

图 9-5 编码器网络图

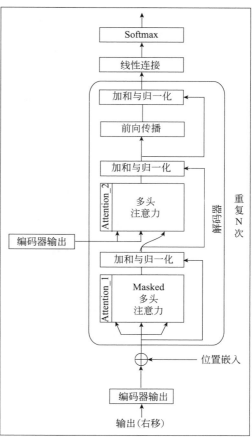

图 9-6 解码器结构 1

```
def decoder_layer(units, d_model, num_heads,
        dropout, name="decoder_layer"):
    inputs = tf.keras.Input(shape=
                (None, d_model), name="inputs")
    enc_outputs = tf.keras.Input(shape=(None, d_model),
                name="encoder_outputs")
    look_ahead_mask = tf.keras.Input(
        shape=(1, None, None), name="look_ahead_mask")
    padding_mask = tf.keras.Input(shape=(1, 1, None),
                name='padding_mask')

    attention1 = MultiHeadAttention(
        d_model, num_heads, name="attention_1")(inputs={
            'query': inputs,
            'key': inputs,
            'value': inputs,
            'mask': look_ahead_mask
    })
```

```python
    attention1 = tf.keras.layers.LayerNormalization(
        epsilon=1e-6)(attention1 + inputs)

    attention2 = MultiHeadAttention(
        d_model, num_heads, name="attention_2")(inputs={
            'query': attention1,
            'key': enc_outputs,
            'value': enc_outputs,
            'mask': padding_mask
        })
    attention2 = tf.keras.layers.Dropout(
                        (rate=dropout)(attention2)
    attention2 = tf.keras.layers.LayerNormalization(
        epsilon=1e-6)(attention2 + attention1)

    outputs = tf.keras.layers.Dense(units=units,
                activation='relu')(attention2)
    outputs = tf.keras.layers.Dense(units=d_model)
                    (outputs)
    outputs = tf.keras.layers.Dropout(rate=dropout)
                    (outputs)
    outputs = tf.keras.layers.LayerNormalization(
        epsilon=1e-6)(outputs + attention2)

    return tf.keras.Model(
        inputs=[inputs, enc_outputs, look_ahead_mask,
                        padding_mask],

        outputs=outputs,
        name=name)
```

上述代码中，Attention_1 为 Masked 多头注意力（机制），为列表中的第 2 项，Attention_2 为列表中的第 1 项。Attention_2 的输出经过两个密集连接模块，然后通过 dropout 层。

可以通过执行以下代码来可视化解码器的网络架构。

```python
sample_decoder_layer = decoder_layer(
    units=512,
    d_model=128,
    num_heads=4,
    dropout=0.3,
    name="sample_decoder_layer")

tf.keras.utils.plot_model(
    sample_decoder_layer,
        to_file='decoder_layer.png')
```

解码器的结构如图 9-7 所示。

重点关注上图中两个多头注意力（机制）的位置，以及它们的输入和输出。接下来，将定义解码器模块。

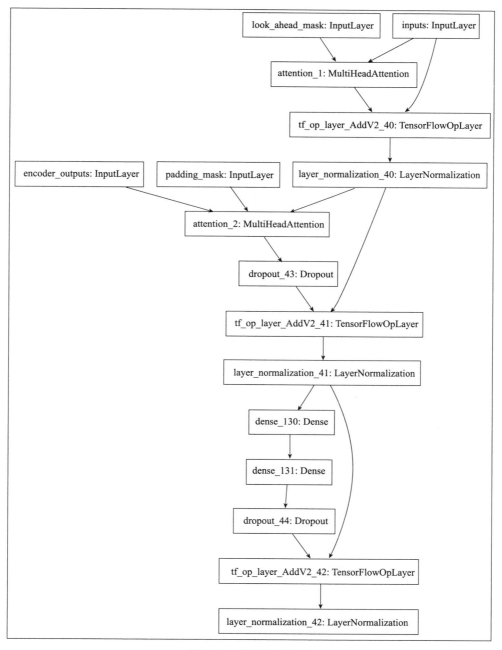

图 9-7　解码器结构 2

9.2.12　定义解码器

与编码器的情况一样,解码器网络是通过将解码层重复 N 次完成构建的。之前讲解了解码器的各个输入,解码器的输出经过线性连接层和 softmax 分类层得到。

使用如下代码段定义解码器。

```python
def decoder(vocab_size,
            num_layers,
            units,
            d_model,
            num_heads,
            dropout,
            name='decoder'):
  inputs = tf.keras.Input(shape=(None,),
              name='inputs')
  enc_outputs = tf.keras.Input(shape=(None, d_model),
              name='encoder_outputs')
  look_ahead_mask = tf.keras.Input(
      shape=(1, None, None), name='look_ahead_mask')
  padding_mask = tf.keras.Input(shape=(1, 1, None),
              name='padding_mask')

  embeddings = tf.keras.layers.Embedding
              (vocab_size, d_model)(inputs)
  embeddings *= tf.math.sqrt(tf.cast
              (d_model, tf.float32))
  embeddings = PositionalEncoding
              (vocab_size, d_model)(embeddings)

  outputs = tf.keras.layers.Dropout(rate=dropout)
              (embeddings)

  for i in range(num_layers):
    outputs = decoder_layer(
        units=units,
        d_model=d_model,
        num_heads=num_heads,
        dropout=dropout,
        name='decoder_layer_{}'.format(i),
    )(inputs=[outputs, enc_outputs, look_ahead_mask,
              padding_mask])

  return tf.keras.Model(
      inputs=[inputs, enc_outputs, look_ahead_mask,
              padding_mask],
      outputs=outputs,
      name=name)
```

使用以下代码实例化解码器并输出结构。

```python
sample_decoder = decoder(
    vocab_size=8192,
    num_layers=2,
    units=512,
    d_model=128,
    num_heads=4,
    dropout=0.3,
```

```
        name="sample_decoder")

tf.keras.utils.plot_model(
    sample_decoder, to_file='decoder.png')
```

上述代码生成的解码器结构如图 9-8 所示。

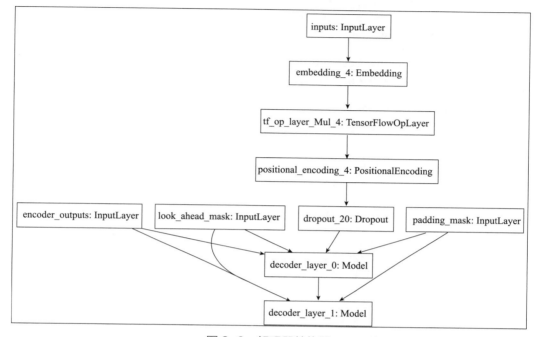

图 9-8　解码器结构图

接下来，构建最终的 Transformer 模型。

9.2.13　Transformer 模型

Transformer 模型由编码器、解码器和最终的线性连接层组成。解码器的输出是线性连接层的输入。使用如下代码段定义 Transformer 模型。

```
def transformer(input_vocab_size,
                target_vocab_size,
                num_layers,
                units,
                d_model,
                num_heads,
                dropout,
                name="transformer"):
    inputs = tf.keras.Input(shape=(None,),
                name="inputs")
    dec_inputs = tf.keras.Input(shape=(None,),
                name="decoder_inputs")
    enc_padding_mask = tf.keras.layers.Lambda(
```

```
        create_padding_mask, output_shape=(1, 1, None),
        name='enc_padding_mask')(inputs)
    # mask the future tokens for decoder inputs at the 1st
attention block
    look_ahead_mask = tf.keras.layers.Lambda(
        create_look_ahead_mask,
        output_shape=(1, None, None),
        name='look_ahead_mask')(dec_inputs)
    # mask the encoder outputs for the 2nd attention block
    dec_padding_mask = tf.keras.layers.Lambda(
        create_padding_mask, output_shape=(1, 1, None),
        name='dec_padding_mask')(inputs)

    enc_outputs = encoder(
        vocab_size=input_vocab_size,
        num_layers=num_layers,
        units=units,
        d_model=d_model,
        num_heads=num_heads,
        dropout=dropout,
    )(inputs=[inputs, enc_padding_mask])

    dec_outputs = decoder(
        vocab_size=target_vocab_size,
        num_layers=num_layers,
        units=units,
        d_model=d_model,
        num_heads=num_heads,
        dropout=dropout,
    )(inputs=[dec_inputs, enc_outputs, look_ahead_mask,
              dec_padding_mask])

    outputs = tf.keras.layers.Dense(units=target_vocab_size,
            name="outputs")(dec_outputs)

    return tf.keras.Model(inputs=[inputs, dec_inputs],
            outputs=outputs, name=name)
```

运行以下代码查看 Transformer 的网络架构。

```
sample_transformer = transformer(
    input_vocab_size = 100,
    target_vocab_size = 100,
    num_layers=4,
    units=512,
    d_model=128,
    num_heads=4,
    dropout=0.3,
    name="sample_transformer")

tf.keras.utils.plot_model(
    sample_transformer, to_file='transformer.png')
```

上述代码生成的 Transformer 网络架构如图 9-9 所示。

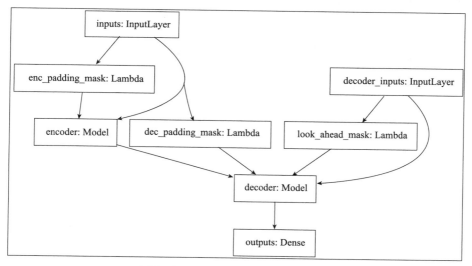

图 9-9　Transformer 网络架构图

从图 9-9 可以看出，Transformer 的网络结构简单、易懂，主要包含一个编码器和一个解码器。现在，在定义了 Transformer 模型后，需要进行模型创建和编译工作。

9.2.14　创建训练模型

要创建用于训练的模型，需要实例化一个 Transformer 类，使用如下代码实现。

```
D_MODEL = 256
model = transformer(
    tokenizer_input.vocab_size+2,
    tokenizer_out.vocab_size+2,
    num_layers = 2,
    units = 512,

    d_model = D_MODEL,
    num_heads = 8,
    dropout = 0.1)
```

如前所述，本章将编码层数设置为 2，而参考论文 (https://arxiv.org/abs/1706.03762) 将此值设置为 6。与此参数类似，其他参数被赋予较低的值以进行更快的训练。为了获得更准确的结果，可以尝试使用更大的 num_layers、units 和 d_model 值。

为了编译 Transformer 模型，需要定义损失函数和优化器。

9.2.15　损失函数

使用稀疏分类熵函数测量模型输出的预测值 y 和真实值之间的差异。损失函数使用以下代码实现。

```
def loss(y_true, y_pred):
    y_true = tf.reshape(y_true, shape=(-1, 10 - 1))
```

```
loss = tf.keras.losses.SparseCategoricalCrossentropy(
    from_logits=True, reduction='none')
             (y_true, y_pred)

mask = tf.cast(tf.not_equal(y_true, 0), tf.float32)
loss = tf.multiply(loss, mask)

return tf.reduce_mean(loss)
```

9.2.16 优化器

使用 Adam 优化器来训练模型。参考文献 (https://arxiv.org/pdf/1706.03762.pdf) 建议对优化器使用自定义的学习率。自定义学习率由以下等式计算。

$$\text{lrate} = d_{model}^{-0.5} \min\left(\text{step\_num}^{-0.5}, \text{step\_num} * \text{warmup\_steps}^{-1.5}\right)$$

根据以上方程，学习率最初会随着训练次数（由 warmup_steps 决定）的增加而线性增加，随后学习率的变化与训练次数（step_num）的平方根成正比。本项目中的 warmup_steps 为 4000。

为了实现学习率变化公式，创建了一个 CustomSchedule 类，如下所示。

```
class CustomSchedule
(tf.keras.optimizers.schedules.LearningRateSchedule):

  def __init__(self, d_model, warmup_steps=4000):
    super(CustomSchedule, self).__init__()
    self.d_model = d_model
    self.d_model = tf.cast(self.d_model, tf.float32)
    self.warmup_steps = warmup_steps

  def __call__(self, step):
    arg1 = tf.math.rsqrt(step)
    arg2 = step * (self.warmup_steps**-1.5)

    return tf.math.rsqrt(self.d_model) *
              tf.math.minimum(arg1, arg2)
```

9.2.17 编译

使用如下优化器和损失函数编译模型。

```
model.compile(optimizer=optimizer, loss=loss)
```

9.2.18 训练

最后，调用 fit 方法开始训练模型:

```
EPOCHS = 20
model.fit(dataset, epochs=EPOCHS)
```

It took me approximately 70 seconds per epoch to train the model.

9.2.19 预测

本项目将给定的英语句子翻译成德语,通过创建名为 translate 的函数实现。首先,使用之前介绍的预处理方法对输入语句进行编码,并向其中添加开始和结束标记。由变量 MAX_LENGTH 定义语句的最长长度为 10。因此,创建了一个循环一迭代的方式输入 10 个单词,注意力机制给出每个单词在全句中的重要程度,逐词进行语句翻译。translate 函数代码如下所示。

```
def translate (input_sentence):

    input_sentence = START_TOKEN_in +
                    tokenizer_input.encode
                    (input_sentence) + END_TOKEN_in
    encoder_input = tf.expand_dims(input_sentence, 0)
    decoder_input = [tokenizer_out.vocab_size]
    output = tf.expand_dims(decoder_input, 0)

    for i in range(MAX_LENGTH):
      predictions = model(inputs=[encoder_input,
                 output], training=False)

      # select the last word
      predictions = predictions[:, -1:, :]
      predicted_id = tf.cast(tf.argmax(predictions,
                 axis=-1), tf.int32)

      # terminate on END_TOKEN
      if tf.equal(predicted_id, END_TOKEN_out[0]):
        break

      # concatenated the predicted_id to the output
      output = tf.concat([output, predicted_id],
                 axis=-1)

    return tf.squeeze(output, axis=0)
```

9.2.20 测试

接下来,使用真实语句测试 Transformer 模型。可以设置一组输入句子并定义一个循环测试代码执行翻译,将翻译结果在控制台上打印输出。循环测试代码如下所示。

```
test_sentences = ['i am sorry', 'how are you']
for s in test_sentences:
  prediction = translate(s)

  predicted_sentence = tokenizer_out.decode(
      [i for i in prediction if i <
                  qtokenizer_out.vocab_size])
  print('Input: {}'.format(s))
  print('Output: {}'.format(predicted_sentence))
```

You will see the following output:

```
Input: i am sorry
Output: lo siento.
Input: how are you
Output: cómo estás.
```

9.2.21 项目源码

清单 9-3 给出了本项目的完整源代码，仅供参考。

清单 9-3 NLP_TRANSFORMER

源码清单
链　接：https://pan.baidu.com/s/1NV0rimQ_8kRz22xfFHN-Cw
提取码：1218

```python
import tensorflow as tf

from tensorflow.keras.models import Model
from tensorflow.keras.layers
            import Input,Dense,LSTM,Embedding,
                Bidirectional,RepeatVector,
                Concatenate,Activation,Dot,Lambda
from tensorflow.keras.preprocessing.text
            import Tokenizer
from tensorflow.keras.preprocessing.sequence
            import pad_sequences
from keras import preprocessing,utils
import numpy as np
import matplotlib.pyplot as plt
import tensorflow_datasets as tfds
import os
import re
import numpy as np
import string

!pip install wget
import wget
url = 'https://raw.githubusercontent.com/Apress/artificialneural-networks-with-tensorflow-2/main/ch08/spa.txt'
wget.download(url,'spa.txt')

# reading data
with open('/content/spa.txt',encoding='utf-8',
                errors='ignore') as file:
  text=file.read().split('\n')

input_texts=[] #encoder input
target_texts=[] # decoder input

# we will select subset of the whole data
NUM_SAMPLES = 10000
for line in text[:NUM_SAMPLES]:
  english, spanish = line.split('\t')[:2]
```

```python
    target_text = spanish.lower()
    input_texts.append(english.lower())
    target_texts.append(target_text)

regex = re.compile('[%s]' %
                re.escape(string.punctuation))
for s in input_texts:
  regex.sub('', s)
for s in target_texts:
  regex.sub('', s)
input_texts[1],target_texts[1]

tokenizer_input = tfds.features.text.
            SubwordTextEncoder.build_from_corpus(
    input_texts, target_vocab_size=2**13)

# example showing how this tokenizer works
tokenized_string1=tokenizer_input.encode
                ('hello i am good')

tokenized_string1

for token in tokenized_string1:
  print ('{} ----> {}'.format(token,
                tokenizer_input.decode([token])))

# if the word is not in dictionary
tokenized_string2=tokenizer_input.encode
                    ('how is the moon')
for token in tokenized_string2:
  print ('{} ----> {}'.format(token,
                tokenizer_input.decode([token])))

# tokenize Spanish text
tokenizer_out=tfds.features.text.SubwordTextEncoder.
                        build_from_corpus(
    target_texts, target_vocab_size=2**13)

START_TOKEN_in=[tokenizer_input.vocab_size]
#input start token
END_TOKEN_in=[tokenizer_input.vocab_size+1]
#input end token
START_TOKEN_out=[tokenizer_out.vocab_size]
#output start token
END_TOKEN_out=[tokenizer_out.vocab_size+1]
#output end token/

START_TOKEN_in,
        END_TOKEN_in,START_TOKEN_out,END_TOKEN_out
```

```python
MAX_LENGTH = 10

# Tokenize, filter and pad sentences
def tokenize_and_padding(inputs, outputs):
  tokenized_inputs, tokenized_outputs = [], []

  for (input_sentence, output_sentence) 
                     in zip(inputs, outputs):

    # tokenize sentence
    input_sentence = START_TOKEN_in + 
          tokenizer_input.encode(input_sentence) + 
                    END_TOKEN_in
    output_sentence = START_TOKEN_out + 
           tokenizer_out.encode(output_sentence) + 
                    END_TOKEN_out
    # check tokenized sentence max length
    #if len(input_sentence) <= MAX_LENGTH and 
               len(output_sentence) <= MAX_LENGTH:
    tokenized_inputs.append(input_sentence)
    tokenized_outputs.append(output_sentence )

  # pad tokenized sentences
  tokenized_inputs = 
     tf.keras.preprocessing.sequence.pad_sequences(
     tokenized_inputs, maxlen=MAX_LENGTH, 
                               padding='post')
  tokenized_outputs = 
     tf.keras.preprocessing.sequence.pad_sequences(
     tokenized_outputs, maxlen=MAX_LENGTH, 
                               padding='post')
  return tokenized_inputs, tokenized_outputs

english, spanish = tokenize_and_padding
                      (input_texts,target_texts)

english[1],spanish[1]

BATCH_SIZE = 32
BUFFER_SIZE = 10000

# decoder inputs use the previous target as input
# remove START_TOKEN from targets
dataset = tf.data.Dataset.from_tensor_slices((
    {
        'inputs': english,
        'decoder_inputs': spanish[:, :-1]
    },
    {
        'outputs':spanish[:, 1:]
```

```python
        },
    ))

dataset = dataset.cache()
dataset = dataset.shuffle(BUFFER_SIZE)
dataset = dataset.batch(BATCH_SIZE)
dataset = dataset.prefetch
            (tf.data.experimental.AUTOTUNE)

class MultiHeadAttention(tf.keras.layers.Layer):

    def __init__(self, d_model, num_heads,
                 name="multi_head_attention"):
        super(MultiHeadAttention, self)
                    .__init__(name=name)
        self.num_heads = num_heads
        self.d_model = d_model
        self.depth = d_model // self.num_heads

        self.query_dense =
                tf.keras.layers.Dense(units=d_model)
        self.key_dense =
                tf.keras.layers.Dense(units=d_model)
        self.value_dense =
                tf.keras.layers.Dense(units=d_model)

        self.dense = tf.keras.layers.Dense(units=d_model)

    def split_heads(self, inputs, batch_size):
        inputs = tf.reshape(
            inputs, shape=(batch_size, -1,
                self.num_heads, self.depth))
        return tf.transpose(inputs, perm=[0, 2, 1, 3])

    def call(self, inputs):
        query, key, value, mask = inputs['query'], inputs['key'],
                                  inputs['value'], inputs['mask']
        batch_size = tf.shape(query)[0]

        # linear layers
        query = self.query_dense(query)
        key = self.key_dense(key)
        value = self.value_dense(value)

        # split heads
        query = self.split_heads(query, batch_size)
        key = self.split_heads(key, batch_size)
        value = self.split_heads(value, batch_size)

        # scaled dot-product attention
```

```python
        scaled_attention = scaled_dot_product_attention
                            (query, key, value, mask)
        scaled_attention = tf.transpose
                            (scaled_attention,
                             perm=[0, 2, 1, 3])

        # concatenation of heads
        concat_attention = tf.reshape(scaled_attention,
                                      (batch_size, -1,
                                       self.d_model))

        # final linear layer
        outputs = self.dense(concat_attention)

        return outputs

def scaled_dot_product_attention
                    (query, key, value, mask):
    QxK_transpose = tf.matmul
                    (query, key, transpose_b=True)

    depth = tf.cast(tf.shape(key)[-1], tf.float32)
    logits = QxK_transpose / tf.math.sqrt(depth)

    if mask is not None:
        logits += (mask * -1e9)

    # softmax is normalized on the last axis (seq_len_k)
    attention_weights = tf.nn.softmax(logits, axis=-1)

    output = tf.matmul(attention_weights, value)

    return output

def create_padding_mask(x):
    mask = tf.cast(tf.math.equal(x, 0), tf.float32)
    # (batch_size, 1, 1, sequence length)
    return mask[:, tf.newaxis, tf.newaxis, :]

# function testing
x=tf.constant([[2974, 50, 2764, 2975, 0, 0,
                0, 0, 0, 0]])
create_padding_mask(x)

def create_look_ahead_mask(x):
    seq_len = tf.shape(x)[1]
    look_ahead_mask = 1 - tf.linalg.band_part
                (tf.ones((seq_len, seq_len)), -1, 0)
    padding_mask = create_padding_mask(x)
    return tf.maximum(look_ahead_mask, padding_mask)
```

```python
class PositionalEncoding(tf.keras.layers.Layer):

    def __init__(self, position, d_model):
        super(PositionalEncoding, self).__init__()
        self.pos_encoding = self.positional_encoding(
                            (position, d_model)

    def get_angles(self, position, i, d_model):
        angles = 1 / tf.pow(10000, (2 * (i // 2)) /
                    tf.cast(d_model, tf.float32))
        return position * angles

    def positional_encoding(self, position, d_model):
        angle_rads = self.get_angles(
            position=tf.range(position, dtype=tf.float32)
                    [:, tf.newaxis],
            i=tf.range(d_model, dtype=tf.float32)
                    [tf.newaxis, :],
            d_model=d_model)
        # apply sin to even index in the array
        sines = tf.math.sin(angle_rads[:, 0::2])
        # apply cos to odd index in the array
        cosines = tf.math.cos(angle_rads[:, 1::2])

        pos_encoding = tf.concat([sines, cosines],
                            axis=-1)
        pos_encoding = pos_encoding[tf.newaxis, ...]
        return tf.cast(pos_encoding, tf.float32)

    def call(self, inputs):
        return inputs + self.pos_encoding
                    [:, :tf.shape(inputs)[1], :]

def encoder_layer(units, d_model, num_heads, dropout,
                    name="encoder_layer"):
    inputs = tf.keras.Input(shape=(None, d_model),
                    name="inputs")
    padding_mask = tf.keras.Input(shape=(1, 1, None),
                    name="padding_mask")

    # multi-head attention with padding mask
    attention = MultiHeadAttention(
        d_model, num_heads, name="attention")({
            'query': inputs,
            'key': inputs,
            'value': inputs,
            'mask': padding_mask
        })
    attention = tf.keras.layers.Dropout
```

```python
                    (rate=dropout)(attention)
    attention = tf.keras.layers.LayerNormalization(
        epsilon=1e-6)(inputs + attention)
    # two dense layers followed by a dropout
    outputs = tf.keras.layers.Dense(units=units,
                    activation='relu')(attention)
    outputs = tf.keras.layers.Dense(units=d_model)
                    (outputs)
    outputs = tf.keras.layers.Dropout(rate=dropout)
                    (outputs)
    outputs = tf.keras.layers.LayerNormalization(
        epsilon=1e-6)(attention + outputs)

    return tf.keras.Model(
        inputs=[inputs, padding_mask],
            outputs=outputs, name=name)

def encoder(vocab_size,
            num_layers,
            units,
            d_model,
            num_heads,
            dropout,
            name="encoder"):

    inputs = tf.keras.Input(shape=(None,),
                        name="inputs")

    # create padding mask
    padding_mask = tf.keras.Input(shape=(1, 1, None),
                        name="padding_mask")

    # create combination of word embedding + positional encoding
    embeddings = tf.keras.layers.Embedding
                    (vocab_size, d_model)(inputs)
    embeddings *= tf.math.sqrt(tf.cast
                    (d_model, tf.float32))
    embeddings = PositionalEncoding
                    (vocab_size, d_model)(embeddings)

    outputs = tf.keras.layers.Dropout(rate=dropout)
                                        (embeddings)
    # repeat the Encoder Layer two times
    for i in range(num_layers):
        outputs = encoder_layer(
            units=units,
            d_model=d_model,
            num_heads=num_heads,
            dropout=dropout,
            name="encoder_layer_{}".format(i),
```

```python
    )([outputs, padding_mask])

    return tf.keras.Model(
        inputs=[inputs, padding_mask], outputs=outputs,
                                        name=name)

sample_encoder = encoder(
    vocab_size=8192,
    num_layers=5,
    units=512,
    d_model=128,
    num_heads=4,
    dropout=0.3,
    name="sample_encoder")

tf.keras.utils.plot_model(
    sample_encoder, to_file='encoder.png')

def decoder_layer(units, d_model, num_heads,
        dropout, name="decoder_layer"):
inputs = tf.keras.Input(shape=
            (None, d_model), name="inputs")
enc_outputs = tf.keras.Input(shape=(None, d_model),
            name="encoder_outputs")
look_ahead_mask = tf.keras.Input(
    shape=(1, None, None), name="look_ahead_mask")
padding_mask = tf.keras.Input(shape=(1, 1, None),
            name='padding_mask')

attention1 = MultiHeadAttention(
    d_model, num_heads, name="attention_1")(inputs={
        'query': inputs,
        'key': inputs,
        'value': inputs,
        'mask': look_ahead_mask
    })
attention1 = tf.keras.layers.LayerNormalization(
    epsilon=1e-6)(attention1 + inputs)

attention2 = MultiHeadAttention(
    d_model, num_heads, name="attention_2")(inputs={
        'query': attention1,
        'key': enc_outputs,
        'value': enc_outputs,
        'mask': padding_mask
    })
attention2 = tf.keras.layers.Dropout
                    (rate=dropout)(attention2)
attention2 = tf.keras.layers.LayerNormalization(
    epsilon=1e-6)(attention2 + attention1)
```

```python
    outputs = tf.keras.layers.Dense(units=units,
                activation='relu')(attention2)
    outputs = tf.keras.layers.Dense(units=d_model)
                (outputs)
    outputs = tf.keras.layers.Dropout(rate=dropout)
                (outputs)
    outputs = tf.keras.layers.LayerNormalization(
        epsilon=1e-6)(outputs + attention2)

    return tf.keras.Model(
        inputs=[inputs, enc_outputs, look_ahead_mask,
                        padding_mask],

        outputs=outputs,
        name=name)

sample_decoder_layer = decoder_layer(
    units=512,
    d_model=128,
    num_heads=4,
    dropout=0.3,
    name="sample_decoder_layer")

tf.keras.utils.plot_model(
    sample_decoder_layer,
        to_file='decoder_layer.png')

def decoder(vocab_size,
            num_layers,
            units,
            d_model,
            num_heads,
            dropout,
            name='decoder'):
    inputs = tf.keras.Input(shape=(None,),
                name='inputs')
    enc_outputs = tf.keras.Input(shape=(None, d_model),
                name='encoder_outputs')
    look_ahead_mask = tf.keras.Input(
        shape=(1, None, None), name='look_ahead_mask')
    padding_mask = tf.keras.Input(shape=(1, 1, None),
                name='padding_mask')

    embeddings = tf.keras.layers.Embedding
                (vocab_size, d_model)(inputs)
    embeddings *= tf.math.sqrt(tf.cast
                (d_model, tf.float32))
    embeddings = PositionalEncoding
                (vocab_size, d_model)(embeddings)
```

```python
    outputs = tf.keras.layers.Dropout(rate=dropout)
            (embeddings)

    for i in range(num_layers):
        outputs = decoder_layer(
            units=units,
            d_model=d_model,
            num_heads=num_heads,
            dropout=dropout,
            name='decoder_layer_{}'.format(i),
        )(inputs=[outputs, enc_outputs, look_ahead_mask,
                padding_mask])

    return tf.keras.Model(
        inputs=[inputs, enc_outputs, look_ahead_mask,
                padding_mask],
        outputs=outputs,
        name=name)
sample_decoder = decoder(
    vocab_size=8192,
    num_layers=2,
    units=512,
    d_model=128,
    num_heads=4,
    dropout=0.3,
    name="sample_decoder")

tf.keras.utils.plot_model(
    sample_decoder, to_file='decoder.png')

def transformer(input_vocab_size,
                target_vocab_size,
                num_layers,
                units,
                d_model,
                num_heads,
                dropout,
                name="transformer"):
    inputs = tf.keras.Input(shape=(None,),
                name="inputs")
    dec_inputs = tf.keras.Input(shape=(None,),
                name="decoder_inputs")

    enc_padding_mask = tf.keras.layers.Lambda(
        create_padding_mask, output_shape=(1, 1, None),
        name='enc_padding_mask')(inputs)
    # mask the future tokens for decoder inputs at the 1st
attention block
    look_ahead_mask = tf.keras.layers.Lambda(
        create_look_ahead_mask,
```

```python
        output_shape=(1, None, None),
        name='look_ahead_mask')(dec_inputs)
    # mask the encoder outputs for the 2nd attention block
    dec_padding_mask = tf.keras.layers.Lambda(
        create_padding_mask, output_shape=(1, 1, None),
        name='dec_padding_mask')(inputs)

    enc_outputs = encoder(
        vocab_size=input_vocab_size,
        num_layers=num_layers,
        units=units,
        d_model=d_model,
        num_heads=num_heads,
        dropout=dropout,
    )(inputs=[inputs, enc_padding_mask])

    dec_outputs = decoder(
        vocab_size=target_vocab_size,
        num_layers=num_layers,
        units=units,
        d_model=d_model,
        num_heads=num_heads,
        dropout=dropout,
    )(inputs=[dec_inputs, enc_outputs, look_ahead_mask,
              dec_padding_mask])

    outputs = tf.keras.layers.Dense(units=target_vocab_size,
              name="outputs")(dec_outputs)

    return tf.keras.Model(inputs=[inputs, dec_inputs],
              outputs=outputs, name=name)

sample_transformer = transformer(
    input_vocab_size = 100,
    target_vocab_size = 100,
    num_layers=4,
    units=512,
    d_model=128,
    num_heads=4,
    dropout=0.3,
    name="sample_transformer")

tf.keras.utils.plot_model(
    sample_transformer, to_file='transformer.png')

D_MODEL = 256
model = transformer(
    tokenizer_input.vocab_size+2,
    tokenizer_out.vocab_size+2,
    num_layers = 2,
```

```python
        units = 512,
        d_model = D_MODEL,
        num_heads = 8,
        dropout = 0.1)

    def loss(y_true, y_pred):
        y_true = tf.reshape(y_true, shape=(-1, 10 - 1))

        loss =
            tf.keras.losses.SparseCategoricalCrossentropy(
            from_logits=True, reduction='none')
                (y_true, y_pred)

        mask = tf.cast(tf.not_equal(y_true, 0), tf.float32)
        loss = tf.multiply(loss, mask)

        return tf.reduce_mean(loss)
    class CustomSchedule
    (tf.keras.optimizers.schedules.LearningRateSchedule):

        def __init__(self, d_model, warmup_steps=4000):
            super(CustomSchedule, self).__init__()
            self.d_model = d_model
            self.d_model = tf.cast(self.d_model, tf.float32)
            self.warmup_steps = warmup_steps

        def __call__(self, step):
            arg1 = tf.math.rsqrt(step)
            arg2 = step * (self.warmup_steps**-1.5)

            return tf.math.rsqrt(self.d_model) *
                    tf.math.minimum(arg1, arg2)

learning_rate = CustomSchedule(D_MODEL)

optimizer = tf.keras.optimizers.Adam(
    learning_rate, beta_1=0.9, beta_2=0.98,
            epsilon=1e-9)

model.compile(optimizer=optimizer, loss=loss)

EPOCHS = 20
model.fit(dataset, epochs=EPOCHS)

def translate (input_sentence):

    input_sentence = START_TOKEN_in +
                    tokenizer_input.encode
                    (input_sentence) + END_TOKEN_in
    encoder_input = tf.expand_dims(input_sentence, 0)
```

```python
    decoder_input = [tokenizer_out.vocab_size]
    output = tf.expand_dims(decoder_input, 0)

    for i in range(MAX_LENGTH):
        predictions = model(inputs=[encoder_input,
                    output], training=False)

        # select the last word
        predictions = predictions[:, -1:, :]
        predicted_id = tf.cast(tf.argmax(predictions,
                    axis=-1), tf.int32)

        # terminate on END_TOKEN
        if tf.equal(predicted_id, END_TOKEN_out[0]):
            break

        # concatenated the predicted_id to the output
        output = tf.concat([output, predicted_id],
                    axis=-1)

    return tf.squeeze(output, axis=0)

test_sentences = ['i am sorry', 'how are you']
for s in test_sentences:
    prediction = translate(s)

    predicted_sentence = tokenizer_out.decode(
        [i for i in prediction if i <
                    qtokenizer_out.vocab_size])

    print('Input: {}'.format(s))
    print('Output: {}'.format(predicted_sentence))
```

9.3 下一步是什么

近年来，一种名为 BERT 的语言表示模型被广泛使用。此模型由 Jacob Devlin 等人提出，发表于 2018 年 10 月题为 *BERT: Pre-training of Deep Bidirectional Transformers for Language Understanding* 的论文中 (https://arxiv.org/pdf/1810.04805.pdf)。BERT 代表使用 Transformers 的双向编码器。此模型的设计巧妙之处在于，可以使用一个预先训练好的 BERT 模型结合一个额外的层对它进行微调后应用于不同任务中，如问答、语言翻译等。感兴趣的读者可以参考 Devlin 等人的原始论文，了解更多信息。

总结

在本章中，介绍了另一种自然语言处理模型，即 Transformer 模型。与之前介绍的带有注意力机制的 seq2seq 模型相比，Transformer 模型更好地完成了自然语言理解任务。

Transformer 模型没有使用 LSTM 的记忆功能，相反地，Transformer 模型使用位置嵌入来了解句子中重要单词的相对位置。Transformer 使用了一种不同的注意力模型，称为多头注意力模型，将输入分成多个通道创建的多头输入，这有利于分布式训练和预测。总体说来，Transformer 模型比前文中讨论的其他模型要更加先进。

在下一章中，我们将讲解图像描述模型，学习设计一个能够为给定图像输出语义信息的深度神经网络。

第 10 章
图像描述

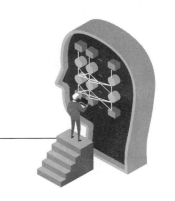

设想一下，度假时您拍摄了几张美丽的风景照，然后希望在这些照片上添加一些描述，并将照片发布到社交网络上。如果这时有一个可以自动制作照片描述的应用程序，这不是一件很方便的事情吗？在本章中，我们将学习如何创建和训练神经网络来为照片创建描述。

图像描述是一个非常具有挑战性的问题。为了解决这个问题，需要同时具备计算机视觉和自然语言处理的相关知识。计算机视觉可帮助神经网络理解图像的内容，完成图像目标检测。自然语言处理将图像内容转化为按正确顺序排列的单词，组成描述语句。

解决图像描述问题的通用框架如图 10-1 所示。

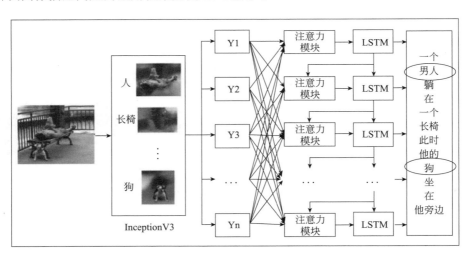

图 10-1 图像描述的通用架构

首先，使用针对图像分类任务设计的预训练模型处理原始图像，如 InceptionV3 网络或 VGG16 网络。具体做法是使用此类网络的卷积层提取图像特征，忽略网络的分类层。以图 10-1 中的图像为例，输入图像中一个人躺在长椅上，旁边坐着一只狗。当使用图像分类网络处理这张图像时，会提取出图像的各个部分，如人、狗、长椅等。检测到图像中的目标后，生成与图像中全部目标相关的说明，这一步不能简单地列出对象名称，而应该使用这些检测到的目标名称创建一个恰当且有意义的句子。这是图像描述应用程序中最具挑战性的部分。为了得到完整的语句描述，需要一个与前一章类似的注意力机制。下面回想一下，第 9 章使用了带有多头注意力模块的 Transformer 模型进行文本翻译。针对本章中的自然语言处理任务，将使用不同的注意力模块，即 Bahdanau 注意力模块，此注意力模块由 Dzmitry Bahdanau 提出 (https://arxiv.org/pdf/1409.0473.pdf)，通过编码器和解码器的线性组合，学习在给定的序列中关联、排列单词。最初，Bahdanau 注意力模块用于改进机器翻译任务中基于编码器、解码器的循环神经网络 (recurrent neural network, RNN)。

Bahdanau 注意力模块示意图如图 10-2 所示。

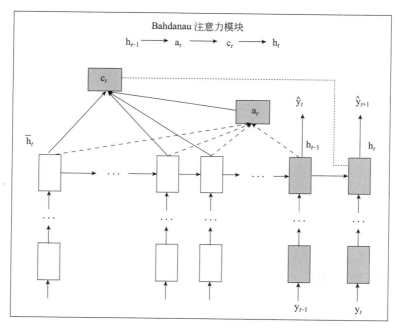

图 10-2　Bahdanau 注意力模块

图 10-2 中 Bahdanau 结构的计算过程如下。

$$a_t(s) = \text{align}(h_{t-1}, \underline{h}_s)$$

$$c_t = \sum a_t h_s$$

$$h_t = \text{RNN}(h_{t-1}^{l-1}, [c_t; h_{t-1}])$$

和上一章类似，本章将通过一个实际应用解释 Bahdanau 注意力模块。下面，让我们直接开始开发一个图像描述项目。

10.1 项目简介

与任何机器学习项目一样，构建图像描述网络模型最重要的是建立合适的训练数据和测试数据。Flickr 已公开了用于图像描述的多个数据集。其中，Flickr8k 数据集包含约 8000 张图像，Flickr30k 数据集包含约 30000 张图像。如果希望使用数据量更大的数据集，MS COCO 数据集包含约 180000 张图像。上述数据集中的图像均被正确标注，部分图像可能有多个标注。使用数据量更大的数据集训练神经网络，能够对未知图像提供更加准确的预测结果，但是网络训练过程也需要更多的计算资源。本项目以学习图像描述任务为目的，不追求过高的模型精度，因此 Flickr8k 数据集的数据量已经足够，此数据集中每个图像都带有五个相关标注。

10.2 创建项目

新建一个 Colab 项目并将其命名为 ImageCaptioning。使用如下语句导入附加依赖库。

```python
import os
import time
import pickle
import numpy as np
import tensorflow as tf
import matplotlib.pyplot as plt
from sklearn.utils import shuffle
from sklearn.model_selection import train_test_split

from tensorflow.keras.applications \
        import InceptionV3
from os import listdir
from tqdm import tqdm
from PIL import Image
```

10.3 下载数据

对于本项目，需要下载两种类型的数据集，包括原始图像及其标注。如之前提到的，Flickr 制作了公开数据集 (http://academictorrents.com/details/9dea07ba660a722ae1008c4c8afdd303b6f6e53b)，数据由图像和图像标注组成。使用以下代码下载数据。

```
!wget --no-check-certificate -r 'https://drive.google.com/uc?export=download&id=1c7yGTpizf5egVD9dc3Q2lrxS8wtOAV42' -O Flickr8k_text.zip
```

使用如下语句解压已下载的文件。

```
!mkdir captions images
!unzip 'Flickr8k_text.zip' -d '/content/captions'
```

完成上述操作后，磁盘上的 captions 文件夹包含多个文件。训练图像和测试图像保存在两个单独的文件夹中。Flickr_8k.trainImages.txt 文件记录了一长串扩展名为"jpg"的文件，为图像数据。Flickr8k.token.txt 文件中包含每个图像对应的标注，下面显示此文件的前几行。

```
1000268201_693b08cb0e.jpg#1    A girl going into a wooden
building .
1000268201_693b08cb0e.jpg#2    A little girl climbing into a
wooden playhouse .
1000268201_693b08cb0e.jpg#3    A little girl climbing the
stairs to her playhouse .
1000268201_693b08cb0e.jpg#4    A little girl in a pink dress
going into a wooden cabin .
1001773457_577c3a7d70.jpg#0    A black dog and a spotted dog
are fighting
```

使用以下代码下载并解压缩图像数据集。

```
!wget --no-check-certificate -r 'https://drive.google.com/uc?export=download&id=1126G_E2OpvULyvTmOKz_oMhOzv8CkiW1' -O Flickr8k_Dataset.zip
!unzip 'Flickr8k_Dataset.zip' -d '/content/images'
```

解压后图像存储在 images 文件夹中，图 10-3 显示了一个示例图像。

图 10-3　示例图像

与此图像对应的描述可以在之前下载的标注文件 Flickr8k.token.txt 中找到。图像标注如下所示。

```
1003163366_44323f5815.jpg#0    A man lays on a bench while his
dog sits by him .
1003163366_44323f5815.jpg#1    A man lays on the bench to which
a white dog is also tied .
1003163366_44323f5815.jpg#2    a man sleeping on a bench
outside with a white and black dog sitting next to him .
1003163366_44323f5815.jpg#3    A shirtless man lies on a park
bench with his dog .
1003163366_44323f5815.jpg#4    man laying on bench holding
leash of dog sitting on ground
```

使用如下代码检查数据集中的图像数量。

```
image_dir = '/content/images/Flicker8k_Dataset'
images = listdir(image_dir)
print("The number of jpg flies in Flicker8k: {}"
      .format(len(images)))
```

结果显示本项目使用的数据集包含 8091 张图片，满足本项目的基本要求。

10.4 解析 Token 文件

现在解析 token 文件以创建图像名称和相应的标注列表。10.3 节已经展示了 token 文件的结构，即 Flickr8k.token.txt。该文件包含一个文件名及一个标注文本。接下来创建两个列表，一个存储图像路径，另一个存储图像的标注。为了节省训练时间，本项目仅使用每张图像的一个标注。实际上，使用数据集中给定的全部图像标注，将得到更加准确的模型。

接下来，编写一个函数将 token 文件内容加载到内存中。

10.4.1 加载数据

以下函数将 token 文件的内容加载到内存中并返回一个字符串。

```
# load doc into memory
def load(filename):
    file = open(filename, 'r')
    text = file.read()
    file.close()
    return text

filename = '/content/captions/Flickr8k.token.txt'
doc = load(filename)
```

上述代码中，doc 变量保存了完整的文件内容。解析 doc 变量可以创建本项目需要的两个单独列表，即原始图像和图像标注。

创建一个迭代器用来遍历图像文件夹中的文件：

```
dirs = listdir('/content/images/Flicker8k_Dataset')
```

使用如下语句显示文件夹的部分内容。

```
dirs[:5]
```

输出结果如下。

```
['3583065748_7d149a865c.jpg',
 '3358621566_12bac2e9d2.jpg',
 '509778093_21236bb64d.jpg',
 '2094323311_27d58b1513.jpg',
 '3314180199_2121e80368.jpg']
```

10.4.2 创建列表

现在编写一个函数完成列表创建，使用如下代码实现。

```
def load_small(doc):
    PATH = '/content/images/Flicker8k_Dataset/'
    img_path = []
    img_id = []
    img_cap = []
    for line in doc.split('\n'):
        tokens = line.split()
        if len(line) < 2:
            continue
        image_id , image_desc = tokens[0] ,
                                tokens[1:]
        image_id = image_id.split('.')[0]
        image_id = image_id + '.jpg'
        image_desc = ' '.join(image_desc)
        if image_id not in img_id:
            if len(img_id) <= 8000:
                img_id.append(image_id)
                image_path = PATH + image_id
                image_desc = '<start> ' + image_desc
                                    + ' <end>'
                if image_id in dirs:
                    img_path.append(image_path)
                    img_cap.append(image_desc)
            else:
                continue
    return img_path , img_cap
```

此函数将图像数量限制为 8000，每张图像只能读取一个标注，并对每个标注添加了 <start> 和 <end> 标记。使用如下语句调用上面的函数。

```
all_image_path , all_image_captions = load_small(doc)
```

检查列表的大小和图像列表中的一些条目，使用如下语句实现。

```
print('Number of images: ', len(all_image_path))
all_image_path[:5]
```

输出结果如下。

```
Number of images: 8000
['/content/images/Flicker8k_Dataset/1000268201_693b08cb0e.jpg',
 '/content/images/Flicker8k_Dataset/1001773457_577c3a7d70.jpg',
 '/content/images/Flicker8k_Dataset/1002674143_1b742ab4b8.jpg',
 '/content/images/Flicker8k_Dataset/1003163366_44323f5815.jpg',
 '/content/images/Flicker8k_Dataset/1007129816_e794419615.jpg']
```

另外，检查标注列表内容，使用如下语句实现。

```
print('Number of captions: ',
              len(all_image_captions))
all_image_captions[:5]
```

输出结果如下。

```
Number of captions: 8000
['<start> A child in a pink dress is climbing up a set of
stairs in an entry way . <end>',
 '<start> A black dog and a spotted dog are fighting <end>',
 '<start> A little girl covered in paint sits in front of a
painted rainbow with her hands in a bowl . <end>',
 '<start> A man lays on a bench while his dog sits by him .
<end>',
 '<start> A man in an orange hat staring at something . <end>']
```

> **注意**
>
> 此列表包含 8000 个标注，即每个图像一个标注。

下面使用如下语句打乱训练数据的顺序。

```
train_captions, img_name_vector =
              shuffle(all_image_captions,
                      all_image_path,
                      random_state=1)
```

10.5　加载 InceptionV3 模型

本项目使用 InceptionV3 模型提取图像特征，使用以下语句加载模型。

```
image_model = InceptionV3(include_top=False,
                          weights='imagenet')
```

其中，weights 参数为"imagenet"，指代 ImageNet 数据集。ImageNet 是一个包含超过 1500 万张带标注的高分辨率图像的数据集，约有 22000 个类别。InceptionV3 模型在 120 万张图像上进行了训练，另外使用 5 万张图像用于验证和 10 万张图像用于测试。本项目使用这种训练方式并加载预训练的权重。

本项目应用中取消了 InceptionV3 中用于图像分类的顶层，使用余下部分进行特征提取。由于本章只对特征提取感兴趣，所以不需要顶层权重。接下，基于 image_model 使用 tf.keras 创建本项目使用的图像特征提取模型。首先，提取 InceptionV3 模型的最后一层：

```
new_input = image_model.input
hidden_layer = image_model.layers[-1].output
```

此处的特征提取模型通过提取 InceptionV3 模型的输入部分并使用 hidden_layer 作为输出层实现,隐藏层位于最后一个 softmax 层之前。使用如下语句构建特征提取模型。

```
image_features_extract_model = tf.keras.Model
                  (new_input, hidden_layer)
```

通过 InceptionV3 模型的最后一个卷积层后得到图像的特征图,输出特征图的尺寸为 8×8×2048。

10.6 准备数据集

InceptionV3 模型需要输入大小为 299×299 的图像,同时,必须将图像像素进行归一化到 [-1,1] 之间。

用于加载和调整图像大小的函数如下。

```
def load_image(image_path):
    img = tf.io.read_file(image_path)
    img = tf.image.decode_jpeg(img, channels=3)
    img = tf.image.resize(img, (299, 299))
    img = tf.keras.applications.inception_v3.preprocess_input
                            (img)
    return img, image_path
```

调用 from_tensor_slices 并使用 load_image 函数进行预处理和创建图像数据集,如下所示。

```
encode_train = sorted(set(img_name_vector))
image_dataset = tf.data.Dataset.from_tensor_slices
                    (encode_train)
image_dataset = image_dataset.map(load_image,
            num_parallel_calls=tf.data.experimental.
            AUTOTUNE).batch(16)
```

10.7 提取特征

对于数据集中的每个图像,调用之前创建的 image_features_extract_model 模型提取特征。数据重采样后,将其保存到 physical 文件中,使用以下 for 循环实现。

```
for img, path in tqdm(image_dataset):
  batch_features = image_features_extract_model(img)
  batch_features = tf.reshape(batch_features,
                    (batch_features.shape[0],
                    -1,batch_features.shape[3]))

  for bf, p in zip(batch_features, path):
    path_of_feature = p.numpy().decode("utf-8")
    np.save(path_of_feature, bf.numpy())
```

注意

虽然将特征保存到内存中会提高效率,但由于每个图像需要使用 8×8×2048 个浮点数表示,会占用大量资源。并且,由于 Colab 内存限制为 12GB,在 GPU 上运行前面的循环需要时间较长,约 2min。

10.8 创建词汇表

现在将创建一个包含所有单词的词汇表,使用如下代码实现。

```
tokenizer = tf.keras.preprocessing.text.Tokenizer
        (filters='!"#$%&()*+.,-/:;=?@[\]^_`{|}~ ')
tokenizer.fit_on_texts(train_captions)
max_size = len(tokenizer.word_index)
```

10.9 创建输入序列

使用以下代码创建图像标注词的输入序列。

```
train_seqs = tokenizer.texts_to_sequences (train_captions)
```

使用如下语句打印序列示例。

```
train_seqs[:5]
```

输出结果如下。

```
[[2, 1, 2339, 8, 155, 2340, 1198, 19, 2341, 1390, 24, 480, 554, 3],
 [2, 21, 1714, 7, 1199, 1715, 1, 108, 2342, 19, 5, 173, 3],
 [2, 1, 11, 4, 1, 28, 32, 506, 1, 507, 3],
 [2, 1, 101, 102, 12, 1, 26, 3],
 [2, 63, 34, 4, 1, 272, 3]]
```

如上所示,不同序列具有不同的长度。对于本项目的模型开发,需要全部序列具有相同的长度,因此,需要对序列进行填充。使用如下语句实现。

```
max_length = max(len(t) for t in train_seqs)

cap_vector = tf.keras.preprocessing.
                    sequence.pad_sequences
                            (train_seqs, padding='post')
```

使用如下语句打印 cap_vector 变量并检查其内容。

```
cap_vector[:5]
```

cap_vector 变量的内容如图 10-4 所示。

```
array([[    2,    1, 2339,    8,  155, 2340, 1198,   19, 2341, 1390,   24,
          480,  554,    3,    0,    0,    0,    0,    0,    0,    0,    0,
            0,    0,    0,    0,    0,    0,    0,    0,    0,    0,    0,
            0,    0],
       [    2,   21, 1714,    7, 1199, 1715,    1,  108, 2342,   19,    5,
          173,    3,    0,    0,    0,    0,    0,    0,    0,    0,    0,
            0,    0,    0,    0,    0,    0,    0,    0,    0,    0,    0,
            0,    0],
       [    2,    1,   11,    4,    1,   28,   32,  506,    1,  507,    3,
            0,    0,    0,    0,    0,    0,    0,    0,    0,    0,    0,
            0,    0,    0,    0,    0,    0,    0,    0,    0,    0,    0,
            0,    0],
       [    2,    1,  101,  102,   12,    1,   26,    3,    0,    0,    0,
            0,    0,    0,    0,    0,    0,    0,    0,    0,    0,    0,
            0,    0,    0,    0,    0,    0,    0,    0,    0,    0,    0,
            0,    0],
       [    2,   63,   34,    4,    1,  272,    3,    0,    0,    0,    0,
            0,    0,    0,    0,    0,    0,    0,    0,    0,    0,    0,
            0,    0,    0,    0,    0,    0,    0,    0,    0,    0,    0,
            0,    0]], dtype=int32)
```

图 10-4　填充后的序列示例

10.10　创建训练数据集

首先，声明用于创建数据集的变量，如下所示。

```
BATCH_SIZE = 64
BUFFER_SIZE = 1000
embedding_dim = 256
units = 512
vocab_size = max_size + 1
num_steps = len(img_name_vector) // BATCH_SIZE
```

以下函数将预先保存的每个图像特征向量加载到张量中。

```
def map_func(img_name, cap):
    img_tensor = np.load(img_name.decode
                    ('utf-8')+'.npy')
    return img_tensor, cap
```

使用以下函数创建数据集。

```
def create_dataset(img_name_train,caption_train):
    dataset = tf.data.Dataset.from_tensor_slices
            ((img_name_train, caption_train))

    # Use map to load the numpy files in parallel
    dataset = dataset.map(lambda item1, item2:
        tf.numpy_function(map_func, [item1, item2],
        [tf.float32, tf.int32]),num_parallel_calls=
        tf.data.experimental.AUTOTUNE)
```

```
# Shuffle and batch
dataset = dataset.shuffle(BUFFER_SIZE).batch(BATCH_SIZE)
.prefetch(buffer_size=tf.data.experimental.AUTOTUNE)
return dataset
```

上述代码中，map 函数加载 NumPy 文件，并打乱图像顺序创建输入数据流。上述函数的调用方法如下。

```
dataset = create_dataset(img_name_vector,cap_vector)
```

10.11 创建模型

使用 Bahdanau 注意力机制和门控循环单元 (gated recurrent unit, GRU) 创建序列到序列的映射模型。GRU 在 RNN 中提供门控机制，该机制与第 9 章中使用的带有遗忘门的 LSTM 类似，但参数比 LSTM 少。由于门控机制的训练参数较少，因此训练速度比 LSTM 快，使用的内存更少，执行速度也更快。然而，门控机制的缺点在于处理长序列时不如 LSTM 准确。

Bahdanau 注意力机制的工作原理如下。
① 为给定的输入图像编码，得到图像的隐藏状态。
② 计算对齐分数，对齐分数表示每个编码器隐藏状态和解码器隐藏状态之间的关联。
③ 使用 Softmax 激活对齐分数。
④ 计算上下文向量。
⑤ 解码输出。
⑥ 重复步骤 2 到步骤 5，直到遇到结束标记。
详细了解每一步的实现过程将更好地理解上述步骤，后续内容进行详细讨论。

10.12 创建编码器

编码器将提取的特征作为输入并将其传递给全连接层。编码器定义如下。

```
class Inception_Encoder(tf.keras.Model):
    def __init__(self, embedding_dim):
        super(Inception_Encoder, self).__init__()
        # shape after fc = (batch_size, 64, embedding_dim)
        self.fc = tf.keras.layers.Dense (embedding_dim)
    def call(self, x):
        x = self.fc(x)
        x = tf.nn.relu(x)
        return x
```

10.13 创建解码器

创建解码器是本项目最重要的部分。本章解码器嵌入了 Bahdanau 注意力机制。首先，简单介绍一下 Bahdanau 注意力机制。

10.13.1 Bahdanau 注意力机制

根据论文 *Show, Attend and Tell*(https://arxiv.org/pdf/1502.03044.pdf) 的研究，本项目从每幅图像中提取一组特征向量，将图像的相应部分使用 N 维向量表示。编码器通过全连接层传递这些特征向量。

Show, Attend and Tell 论文介绍了两种注意力机制。
① Bahdanau 注意力机制，一种确定性的"软"注意力。
② Luong 注意力机制，一种随机"硬"注意力。

本项目使用确定性的"软"注意力，即 Bahdanau 注意力机制。这种注意力机制计算每个 input_vector 的注意力权重和图像上下文向量，完成图像特征提取。简单来说，上下文向量是 t 时刻输入图像相关部分的动态表示。本项目定义了一种注意力机制，从输入向量计算上下文向量，这些向量表示在不同图像位置提取的特征。对于图像中的每个位置，称其为 loc，注意力机制生成一个正权重，在代码中表示为得分。得分可以表示为位置 loc 生成下一个正确单词的概率，或者表示为是否应该提高该 loc 的重要性。每个输入向量的注意力权重由注意力模型计算得到，为了计算这些注意力权重，使用多层感知器，其将解码器隐藏状态和当前输入向量在编码器中的隐藏状态作为输入。

10.13.2 解码器功能

为了预测目标单词，解码器使用以下内容作为输入。
① 上下文向量(注意力权重和解码器输出的加权相乘)。
② 上一时刻解码器的输出。
③ 之前解码器的隐藏状态。

10.13.3 解码器初始化

使用如下代码段声明 RNN_Decoder 类。

```
class RNN_Decoder(tf.keras.Model):
```

在类初始化时，创建了一个 GRU 层:

```
self.gru = tf.keras.layers.GRU(self.units,
                return_sequences=True,
                return_state=True,
                recurrent_initializer='glorot_uniform')
```

使用批量归一化来加速训练过程，并提高模型的性能。批量归一化层会自动归一化深度学习神经网络中某个层的输入。代码如下。

```
self.batchnormalization =
    tf.keras.layers.BatchNormalization
            (axis=-1,
             momentum=0.99,
             epsilon=0.001,
             center=True,
             scale=True,
             beta_initializer='zeros',
```

```
                gamma_initializer='ones',
                moving_mean_initializer='zeros',
                moving_variance_initializer='ones',
                beta_regularizer=None,
                gamma_regularizer=None,
                beta_constraint=None,
                gamma_constraint=None)
```

为了实现注意力机制，声明了 3 个线性连接层：

```
self.W1 = tf.keras.layers.Dense(units)
self.W2 = tf.keras.layers.Dense(units)
self.V = tf.keras.layers.Dense(1)
```

10.13.4 解码器调用方法

解码器需要三个输入，包括编码器输出、隐藏状态（初始化为 0）、解码器输入。

解码器调用方法如下。

```
def call(self, x, features, hidden):
```

其中，x 代表解码器输入，features 代表编码器输出，hidden 是解码器隐藏状态，并初始化为零。

首先，通过扩增解码器输出的维度，以改变先前解码器隐藏状态。

```
hidden_with_time_axis = tf.expand_dims(hidden, 1)
```

然后，计算注意力得分。

10.13.5 注意力得分

注意力得分使用以下公式计算。

$$\text{score}(h_t, \underline{h}_s) = v_a^T \tanh(W_1 h_t + W_2 \underline{h}_s)$$

计算 Bahdanau 注意力得分的伪代码如下：

```
score = FC(tanh(FC(EO) + FC(H)))
```

实现过程如下。

```
score = tf.nn.tanh(self.W1(features) +
        self.W2(hidden_with_time_axis))
```

接下来，计算注意力权重。

10.13.6 注意力权重

注意力权重的数学表示为：

$$\alpha_{ts} = \frac{\exp(\text{score}(h_t, \underline{h}_s))}{\sum_{s'=1}^{s} \exp(\text{score}(h_t, \underline{h}_{s'}))}$$

注意力权重的计算代码如下。

```
attention_weights = tf.nn.softmax(self.V(score), axis=1)
```

将 softmax 激活函数应用于注意力得分，用于获得注意力权重。softmax 激活函数的输出值表示概率，总和为 1，有助于表示每个输入序列在全部序列中的权重。输入序列的注意力权重越高，对预测目标词的影响就越大。

接下来，计算上下文向量。

10.13.7　上下文向量

上下文向量的数学表达为：

$$c_t = \sum_s a_{ts} \underline{h}_s$$

上述公式分两步实现。首先，计算每个输入的上下文向量，如下所示。

```
context_vector = attention_weights * features
```

> 上下文向量是编码器隐藏状态及其注意力权重的乘积。

接下来，对全部上下文向量求和。代码如下。

```
context_vector = tf.reduce_sum(context_vector, axis=1)
```

这一步对输出结果进行解码，即上下文向量与解码器输出连接，并将当前时刻与先前解码器隐藏状态一起输入解码器循环神经网络，以产生新的输出。基本上，通过前面语句中的上下文向量来做到这一点。

以上步骤就完成了 Bahdanau 注意力模型的实现。下面继续构建本章使用的解码器。

10.13.8　解码器实现

首先，通过一个嵌入层将标注的索引转换为向量，使用如下语句实现。

```
x = self.embedding(x)
```

接下来，将上下文向量与图像标注 (x) 进行关联，并将它们组合成一个向量，使用如下语句实现。

```
x = tf.concat([tf.expand_dims(context_vector, 1), x], axis=-1)
```

现在，将向量传递给 GRU 模块，使用如下语句实现。

```
output, state = self.gru(x)
```

将 GRU 层的输出传递给线性连接层，使用如下语句实现。

```
x = self.fc1(output)
```

这个阶段输出 x 的形状为 (batch_size, max_length, hidden_size)。接下来，将 x 的形状调整为 (batch_size × max_length, hidden_size)，使用如下语句实现。

```
x = tf.reshape(x, (-1, x.shape[2]))
```

添加 dropout 和批量归一化层，使用如下语句实现。

```
x = self.dropout(x)
x = self.batchnormalization(x)
```

此时网络输出的形状为 (64 × 512)，将其通过线性连接层后形状转换为 (64 × 8329)，8329 为本项目数据集包含的词汇量。

```
x = self.fc2(x)
```

最后，将计算出的网络输出值返回给调用者：

```
return x, state, attention_weights
```

为 RNN_Decoder 类中添加一个函数，用于重置解码器的初始状态，如下所示。

```
def reset_state(self, batch_size):
    return tf.zeros((batch_size, self.units))
```

清单 10-1 给出了解码器实现的完整代码，仅供参考。

清单 10-1 带有 Bahdanau 注意力模块的 Decoder 类

源码清单
链　接：https://pan.baidu.com/s/1NV0rimQ_8kRz22xfFHN-Cw
提取码：1218

```python
class RNN_Decoder(tf.keras.Model):
    def __init__(self, embedding_dim, units,
                 vocab_size):
        super(RNN_Decoder, self).__init__()
        self.units = units

        self.embedding = tf.keras.layers.Embedding
                        (vocab_size, embedding_dim)
        self.gru = tf.keras.layers.GRU(self.units,
                    return_sequences=True,
                    return_state=True,
                    recurrent_initializer=
                    'glorot_uniform')

        self.fc1 = tf.keras.layers.Dense(self.units)

        self.dropout = tf.keras.layers.Dropout
                    (0.5, noise_shape=None, seed=None)
        self.batchnormalization =
            tf.keras.layers.BatchNormalization
                (axis=-1,
                 momentum=0.99,
                 epsilon=0.001,
```

```python
                        center=True,
                        scale=True,
                        beta_initializer='zeros',
                        gamma_initializer='ones',
                        moving_mean_initializer='zeros',
                        moving_variance_initializer='ones',
                        beta_regularizer=None,
                        gamma_regularizer=None,
                        beta_constraint=None,
                        gamma_constraint=None)

    self.fc2 = tf.keras.layers.Dense(vocab_size)

    # Implementing Attention Mechanism
    self.W1 = tf.keras.layers.Dense(units)
    self.W2 = tf.keras.layers.Dense(units)
    self.V = tf.keras.layers.Dense(1)

def call(self, x, features, hidden):

    hidden_with_time_axis = tf.expand_dims(hidden, 1)

    # Attention Function
    # computing scores
    score = tf.nn.tanh(self.W1(features) +
                self.W2(hidden_with_time_axis))

    # Probability using Softmax
    attention_weights = tf.nn.softmax(self.V(score),
                                    axis=1)

    # Compute context vector
    context_vector = attention_weights * features
    context_vector = tf.reduce_sum(context_vector,
                                    axis=1)

    # passing the input caption index(integer) to
    # embedding layer to convert it to vector
    x = self.embedding(x)

    # Map the context vector with the input vector
    # (the vectors of caption) and then concatenate them
    x = tf.concat([tf.expand_dims(context_vector, 1),
                                    x], axis=-1)

    # Pass concatenated vector to the GRU
    output, state = self.gru(x)

    # shape == (batch_size, max_length, hidden_size)
    # Pass output of GRU layer through a Dense layer
```

```
        x = self.fc1(output)
        
        # x shape == (batch_size * max_length,
                                  hidden_size)
        x = tf.reshape(x, (-1, x.shape[2]))
        
        # Add Dropout and BatchNorm Layers
        x = self.dropout(x)
        x = self.batchnormalization(x)
        # output shape == (64 * 512)
        x = self.fc2(x)
        # shape : (64 * 8329(vocab))
        return x, state, attention_weights
    
    def reset_state(self, batch_size):
        return tf.zeros((batch_size, self.units))
```

10.14 编码器、解码器实例化

使用如下代码进行编码器和解码器实例化。

```
encoder = Inception_Encoder(embedding_dim)
decoder = RNN_Decoder(embedding_dim, units, vocab_size)
```

使用如下代码绘制编码器和解码器的模型图。

```
tf.keras.utils.plot_model (encoder)
tf.keras.utils.plot_model (decoder)
```

模型图如图 10-5 所示。

图 10-5 为简要的模型结构图,完整的数据计算在模型内部完成。

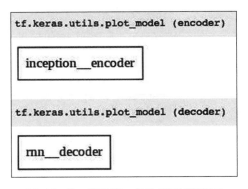

图 10-5 编码器、解码器的模型图

10.15 定义优化器和损失函数

使用 Adam 优化器和稀疏分类交叉熵作为损失函数,使用如下代码段实现。

```
optimizer = tf.keras.optimizers.Adam()
loss_object = tf.keras.losses.
              SparseCategoricalCrossentropy(
    from_logits=True, reduction='none')

def loss_function(real, pred):
  mask = tf.math.logical_not(tf.math.equal(real, 0))
  
  loss_ = loss_object(real, pred)
```

```
mask = tf.cast(mask, dtype=loss_.dtype)
loss_ *= mask

return tf.reduce_mean(loss_)
```

接下来使用一个例子解释如何计算网络损失。计算损失函数需要两个参数，即真实的标注向量和网络的预测值。首先，使用 tf.keras.losses.SparseCategoricalCrossentropy 模块中的 loss_object 方法计算这两个向量之间的损失。然后，使用 tf.cast 函数将编码向量转换为 float32 类型。tf.cast 函数将向量转换为不同的数据类型，以方便进一步操作。最后，将 loss_ 与 mask 相乘，将真实值（数据集中存在的标注）映射为损失函数。

下面用一个例子来说明上述操作。真实标注如下所示。

```
real(passed as a parameter) : tf.Tensor(
[   0    0    0    0    0    0    0    0    0    0    0    0    0
    0    4    0    0    0    0 1760    0  367    0    0    4    0    0
    0    0    0    0    0    0    0    9  453    0    0    0    0    0
    0    0    0    0    0    0  132    0    0    0    0    0    0
    0    0    0    4    0    0    0    5], shape=(64,), dtype=int32)
```

使用如下语句将前面的向量转化为实数，tf.math.equal 函数将向量转化为 bool 值，tf.math.equal 函数需要两个输入参数：x 和 y。如果 x 等于 y，则函数返回 true。

```
tf.math.equal(real , 0) tf.Tensor
```

上述语句输出如下。

```
[True  True  True  True  True  True  True  True  True  True  True  True
 True  True  True False  True  True  True  True False  True False  True
 True False  True  True  True  True  True  True  True  True  True False
False  True  True  True  True  True  True  True  True  True  True  True
False  True  True  True  True  True  True  True  True  True  True False
 True  True  True False], shape=(64,), dtype=bool)
```

tf.math.logical_not 函数接收一个布尔参数并对其执行逻辑非操作，如下所示。

```
mask = tf.math.logical_not(tf.math.equal(real, 0))
```

上述语句输出结果如下。

```
tf.Tensor(
[False False False False False False False False False False False False
 False False False  True False False False False  True False  True False
 False  True False False False False False False False False False  True
  True False False False False False False False False False False False
  True False False False False False False False False False False  True
 False False False  True], shape=(64,), dtype=bool)
```

接下来，使用如下语句输出损失值。

```
loss_ = loss_object(real, pred)
```

上述语句输出结果如下。

```
loss_ = loss_object(real, pred)
tf.Tensor(
[13.458616   11.725777   13.339547   13.877813    13.6512375  13.609352
 12.680449   13.963526   12.929108   12.504114    12.995626   13.473895
 13.966334   13.3766165  13.607654    0.10513641  13.231352   13.313489
 13.727711   14.456019   10.560667   13.632038     4.2983437  14.144966
 14.331357    0.28515333 13.97144    13.087602    15.597718   13.351999
 13.649492   12.489752   12.744471   12.558954    13.255367    1.8581532
  3.1811125  13.873036   12.329573   12.222642    13.126439   14.233135
 12.379726   11.951986   12.869691   13.468082    12.732171   12.240744
  3.8898373  12.682398   13.192276   12.453615    15.758832   14.152502
 13.160431   11.863881   12.530688   13.764532    13.640175    0.7283469
 14.0648575  12.560375   14.25197     0.53315634], shape=(64,),
                                                    dtype=float32)
```

使用如下语句对 mask 编码。

```
mask = tf.cast(mask, dtype=loss_.dtype)
```

输出结果如下。

```
tf.Tensor(
[0. 0. 0. 0. 0. 0. 0. 0. 0. 0. 0. 0. 0. 0. 0. 1. 0. 0. 0. 0. 1. 0. 1. 0.
 0. 1. 0. 0. 0. 0. 0. 0. 0. 0. 1. 1. 0. 0. 0. 0. 0. 0. 0. 0. 0. 0. 0. 0.
 1. 0. 0. 0. 0. 0. 0. 0. 0. 1. 0. 0. 0. 1.], shape=(64,), dtype=float32)
```

最后，将 loss_ 与 mask 相乘

```
loss_ *= mask
```

输出结果如下。

```
loss tf.Tensor(
[ 0.          0.          0.          0.          0.          0.
  0.          0.          0.          0.          0.          0.
  0.          0.          0.          0.10513641  0.          0.
  0.          0.         10.560667    0.          4.2983437   0.
  0.          0.28515333  0.          0.          0.          0.
  0.          0.          0.          0.          13.255367    1.8581532
  3.1811125   0.          0.          0.          0.          0.
  0.          0.          0.          0.          0.          0.
  3.8898373   0.          0.          0.          0.          0.
  0.          0.          0.          0.          0.          0.7283469
  0.          0.          0.          0.53315634], shape=(64,),
                                                   dtype=float32)
```

10.16 创建 checkpoints

创建一个单独的文件夹用于保存网络训练过程的 checkpoints（中间结果），本项目在文件夹中最多保存五个 checkpoints。使用如下代码段实现。

```
checkpoint_path = "./checkpoints/train"
ckpt = tf.train.Checkpoint(encoder=encoder, decoder=decoder,
                           optimizer = optimizer)
ckpt_manager = tf.train.CheckpointManager
         (ckpt, checkpoint_path, max_to_keep=5)
```

声明一个 start_epoch 变量,方便训练中断后从当前结果重新开始训练。代码如下。

```
start_epoch = 0
```

使用如下语句检查上次保存的状态。

```
if ckpt_manager.latest_checkpoint:
  start_epoch = int
    (ckpt_manager.latest_checkpoint.split('-')[-1])
  # restoring the latest checkpoint in checkpoint_path
  ckpt.restore(ckpt_manager.latest_checkpoint)
```

如果当前轮次的训练状态没有保存,可以通过调用以下代码恢复。

```
ckpt.restore(tf.train.latest_checkpoint
             (checkpoint_path))
```

10.17 训练函数

现在编写一个函数执行网络训练,并根据预定义的 epoch 数调用此函数,函数定义如下。

```
loss_plot = []

def train_step(img_tensor, target):
  loss = 0

  # initialize the hidden state for each batch
  # because the captions are not related from image to image
  hidden = decoder.reset_state (batch_size=target.shape[0])

  dec_input = tf.expand_dims([tokenizer.word_index
                    ['<start>']] * batch_size, 1)

  with tf.GradientTape() as tape:
      features = encoder(img_tensor)

      for i in range(1, target.shape[1]):
          # Pass the features through the decoder
          predictions, hidden, _ = decoder(dec_input,
                                    features, hidden)

          loss += loss_function(target[:, i],
                                predictions)
```

```
            # Use teacher forcing
            dec_input = tf.expand_dims(target[:, i], 1)

    total_loss = (loss / int(target.shape[1]))
    trainable_variables = encoder.trainable_variables +
                          decoder.trainable_variables
    gradients = tape.gradient
                    (loss, trainable_variables)
    optimizer.apply_gradients(zip(gradients,
                              trainable_variables))

    return loss, total_loss
```

该函数首先调用解码器模型中定义的 reset_state，为每轮数据输入初始化解码器的隐藏状态。

> 每轮输入数据的相应标注均不相同。在第一轮训练数据中添加 <start> 标记，随后使用 tape.Gradient 函数遍历输入数据并在每次迭代时更新梯度。

10.18 模型训练

根据所需的次数调用 train_step 函数训练模型，如下所示。

```
for epoch in range(start_epoch, 20):
    start = time.time()
    total_loss_train = 0
    for (batch, (img_tensor, target))
                in enumerate(dataset):
        batch_loss, t_loss = train_step (img_tensor, target)
        total_loss_train += t_loss
    if epoch % 5 == 0:
      ckpt_manager.save()
    print ('Epoch {} Train-Loss {:.4f}'.format
              (epoch + 1, (total_loss_train/num_steps)))
print ('Time taken for this epoch {}
         sec\n'.format(time.time() - start))
```

10.19 模型预测

为新图像生成标注，采取与训练期间图像计算相同的步骤，通过编写如下评估函数实现。

```python
def evaluate(image):
    hidden = decoder.reset_state(batch_size=1)

    temp_input = tf.expand_dims
                    (load_image(image)[0], 0)
    img_tensor_val = image_features_extract_model
                    (temp_input)
    img_tensor_val = tf.reshape(img_tensor_val,
                    (img_tensor_val.shape[0],
                     -1,
                     img_tensor_val.shape[3]))

    features = encoder(img_tensor_val)

    dec_input = tf.expand_dims([tokenizer.word_index
                                    ['<start>']], 0)
    result = []

    for i in range(max_length):
        predictions, hidden, attention_weights =
                    decoder(dec_input,
                    features, hidden)

        predicted_id = tf.random.categorical
                    (predictions, 1)[0][0].numpy()
        result.append(tokenizer.index_word[predicted_id])

        if tokenizer.index_word[predicted_id] ==
                                    '<end>':
            return result

        dec_input = tf.expand_dims([predicted_id], 0)

    return result
```

函数实现很简单，与训练图像预处理步骤相同。首先，创建输入图像张量。然后，调用编码器提取其特征。最后，调用 max_length 次解码器完成语义预测。

max_length 在之前代码中给出，用于固定序列长度。

接下来，编写 predict 函数，此函数需要图像 URL 和图像名称两个参数。下载的镜像文件存放在 /root/.keras 文件夹中，每个下载文件使用不同的文件名，以防止文件名相同导致图像内容覆盖。代码如下。

```
def predict(image_url , random_name):
    image_extension = image_url[-4:]
    image_path = tf.keras.utils.get_file
                ('image'+ random_name +
                 image_extension,
                 origin=image_url)
    result = evaluate(image_path)
    print ('Prediction Caption:', ' '.join(result))
    Image.open(image_path)
    return image_path
```

现在在测试图像上调用 predict 函数，如下所示。

```
image_url = 'https://tensorflow.org/images/surf.jpg'
path = predict(image_url , 'surfee')
Image.open(path)
```

测试结果如图 10-6 所示。

图 10-6　示例图像与预测标注

第二幅图像的测试代码如下，预测结果如图 10-7 所示。

```
image_url = 'https://farm4.staticflickr.com/3296/2765087292_535
6df67ce_z.jpg'
path = predict(image_url , 'baseball')
Image.open(path)
```

第三幅图像的测试代码如下，预测结果如图 10-8 所示。

```
image_url = 'https://farm8.staticflickr.com/7139/8156048469_084
7c7ce15_z.jpg'
path = predict(image_url , 'dog')
Image.open(path)
```

图 10-7　第二幅测试图像和预测标注

图 10-8　第三幅测试图像和预测标注

10.20　项目源码

清单 10-2 给出了本项目的完整源代码，仅供参考。

清单 10-2 图像描述

```
import os
import time
import pickle
import numpy as np
import tensorflow as tf
import matplotlib.pyplot as plt
from sklearn.utils import shuffle
from sklearn.model_selection import train_test_split
from tensorflow.keras.applications \
        import InceptionV3
from os import listdir
from tqdm import tqdm
from PIL import Image
```

源码清单

链　接：https://pan.baidu.com/s/1NV0rimQ_
　　　　8kRz22xfFHN-Cw

提取码：1218

```
!wget --no-check-certificate -r 'https://drive.google.com/uc?
export=download&id=1c7yGTpizf5egVD9dc3Q2lrxS8wtOAV42' -O
Flickr8k_text.zip

!mkdir captions images

!unzip 'Flickr8k_text.zip' -d '/content/captions'

!wget --no-check-certificate -r 'https://drive.google.com/uc?
export=download&id=1126G_E2OpvULyvTmOKz_oMhOzv8CkiW1' -O
Flickr8k_Dataset.zip

!unzip 'Flickr8k_Dataset.zip' -d '/content/images'
## The location of the Flickr8K_ photos
image_dir = '/content/images/Flicker8k_Dataset'
images = listdir(image_dir)
print("The number of jpg flies in Flicker8k: {}"
      .format(len(images)))

# load doc into memory
def load(filename):
  file = open(filename, 'r')
  text = file.read()
  file.close()
  return text

filename = '/content/captions/Flickr8k.token.txt'
doc = load(filename)

dirs = listdir('/content/images/Flicker8k_Dataset')

dirs[:5]

def load_small(doc):
    PATH = '/content/images/Flicker8k_Dataset/'
    img_path = []
    img_id = []
    img_cap = []
    for line in doc.split('\n'):
        tokens = line.split()
        if len(line) < 2:
            continue
        image_id , image_desc = tokens[0] ,
                                tokens[1:]
        image_id = image_id.split('.')[0]
        image_id = image_id + '.jpg'
        image_desc = ' '.join(image_desc)
        if image_id not in img_id:
            if len(img_id) <= 8000:
```

```python
                    img_id.append(image_id)
                    image_path = PATH + image_id
                    image_desc = '<start> ' + image_desc
                                        + ' <end>'
                    if image_id in dirs:
                        img_path.append(image_path)
                        img_cap.append(image_desc)
                else:
                    continue
    return img_path , img_cap

all_image_path , all_image_captions = load_small(doc)

print('Number of images: ', len(all_image_path))
all_image_path[:5]

print('Number of captions: ',
            len(all_image_captions))
all_image_captions[:5]

train_captions, img_name_vector = 
            shuffle(all_image_captions,
                all_image_path,
                random_state=1)

image_model = InceptionV3(include_top=False,
                    weights='imagenet')

new_input = image_model.input
hidden_layer = image_model.layers[-1].output
image_features_extract_model = tf.keras.Model
            (new_input, hidden_layer)

def load_image(image_path):
    img = tf.io.read_file(image_path)
    img = tf.image.decode_jpeg(img, channels=3)
    img = tf.image.resize(img, (299, 299))
    img = tf.keras.applications.inception_v3.preprocess_input
                        (img)
    return img, image_path

encode_train = sorted(set(img_name_vector))
image_dataset = tf.data.Dataset.from_tensor_slices
                (encode_train)
image_dataset = image_dataset.map(load_image,
            num_parallel_calls=tf.data.experimental.
            AUTOTUNE)
                .batch(16)

for img, path in tqdm(image_dataset):
```

```python
    batch_features = image_features_extract_model(img)
    batch_features = tf.reshape(batch_features,
                              (batch_features.shape[0],
                               -1,
                               batch_features.shape[3]))

    for bf, p in zip(batch_features, path):
      path_of_feature = p.numpy().decode("utf-8")
      np.save(path_of_feature, bf.numpy())

tokenizer = tf.keras.preprocessing.text.Tokenizer
        (filters='!"#$%&()*+.,-/:;=?@[\]^_`{|}~ ')
tokenizer.fit_on_texts(train_captions)
max_size = len(tokenizer.word_index)

train_seqs = tokenizer.texts_to_sequences
                    (train_captions)

train_seqs[:5]

max_length = max(len(t) for t in train_seqs)
cap_vector = tf.keras.preprocessing.
                        sequence.pad_sequences
                                (train_seqs,
                                 padding='post')

cap_vector[:5]

BATCH_SIZE = 64
BUFFER_SIZE = 1000
embedding_dim = 256
units = 512
vocab_size = max_size + 1
num_steps = len(img_name_vector) // BATCH_SIZE

# Loading previously extracted features
def map_func(img_name, cap):
  img_tensor = np.load(img_name.decode
                      ('utf-8')+'.npy')
  return img_tensor, cap

def create_dataset(img_name_train,caption_train):
  dataset = tf.data.Dataset.from_tensor_slices
            ((img_name_train, caption_train))

  # Use map to load the numpy files in parallel
  dataset = dataset.map(lambda item1, item2:
      tf.numpy_function(map_func, [item1, item2],
      [tf.float32, tf.int32]),num_parallel_calls=
      tf.data.experimental.AUTOTUNE)
```

```python
# Shuffle and batch
dataset = dataset.shuffle(BUFFER_SIZE).batch(BATCH_SIZE)
.prefetch(buffer_size=tf.data.experimental.AUTOTUNE)
return dataset
dataset = create_dataset(img_name_vector,cap_vector)

class Inception_Encoder(tf.keras.Model):
    def __init__(self, embedding_dim):
        super(Inception_Encoder, self).__init__()
        # shape after fc = (batch_size, 64,
                            embedding_dim)
        self.fc = tf.keras.layers.Dense
                        (embedding_dim)

    def call(self, x):
        x = self.fc(x)
        x = tf.nn.relu(x)
        return x

class RNN_Decoder(tf.keras.Model):
  def __init__(self, embedding_dim, units,
            vocab_size):
    super(RNN_Decoder, self).__init__()
    self.units = units

    self.embedding = tf.keras.layers.Embedding
                    (vocab_size, embedding_dim)
    self.gru = tf.keras.layers.GRU(self.units,
                    return_sequences=True,
                    return_state=True,
                    recurrent_initializer=
                    'glorot_uniform')

    self.fc1 = tf.keras.layers.Dense(self.units)

    self.dropout = tf.keras.layers.Dropout
                (0.5, noise_shape=None, seed=None)
    self.batchnormalization =
        tf.keras.layers.BatchNormalization
                (axis=-1,
                 momentum=0.99,
                 epsilon=0.001,
                 center=True,
                 scale=True,
                 beta_initializer='zeros',
                 gamma_initializer='ones',
                 moving_mean_initializer='zeros',
                moving_variance_initializer='ones',
                 beta_regularizer=None,
                 gamma_regularizer=None,
```

```python
                    beta_constraint=None,
                    gamma_constraint=None)

        self.fc2 = tf.keras.layers.Dense(vocab_size)

        # Implementing Attention Mechanism
        self.W1 = tf.keras.layers.Dense(units)
        self.W2 = tf.keras.layers.Dense(units)
        self.V = tf.keras.layers.Dense(1)

    def call(self, x, features, hidden):

        hidden_with_time_axis = tf.expand_dims(hidden, 1)

        # Attention Function
        # computing scores
        score = tf.nn.tanh(self.W1(features) +
                    self.W2(hidden_with_time_axis))

        # Probability using Softmax
        attention_weights = tf.nn.softmax(self.V(score),
                                            axis=1)
        # Compute context vector
        context_vector = attention_weights * features
        context_vector = tf.reduce_sum(context_vector,
                                        axis=1)

        # passing the input caption index(integer) to
        # embedding layer to convert it to vector
        x = self.embedding(x)

        # Map the context vector with the input vector
        # (the vectors of caption) and then concatenate them
        x = tf.concat([tf.expand_dims(context_vector, 1),
                                    x], axis=-1)

        # Pass concatenated vector to the GRU
        output, state = self.gru(x)

        # shape == (batch_size, max_length, hidden_size)
        # Pass output of GRU layer through a Dense layer
        x = self.fc1(output)

        # x shape == (batch_size * max_length,
        #                       hidden_size)
        x = tf.reshape(x, (-1, x.shape[2]))

        # Add Dropout and BatchNorm Layers
        x = self.dropout(x)
        x = self.batchnormalization(x)
```

```python
        # output shape == (64 * 512)
        x = self.fc2(x)
        # shape : (64 * 8329(vocab))
        return x, state, attention_weights

    def reset_state(self, batch_size):
        return tf.zeros((batch_size, self.units))
        encoder = Inception_Encoder(embedding_dim)
        decoder = RNN_Decoder(embedding_dim,
                            units, vocab_size)

        tf.keras.utils.plot_model (encoder)

        tf.keras.utils.plot_model (decoder)

        optimizer = tf.keras.optimizers.Adam()
        loss_object = tf.keras.losses.
                        SparseCategoricalCrossentropy(
            from_logits=True, reduction='none')

    def loss_function(real, pred):
        mask = tf.math.logical_not(tf.math.equal(real, 0))
        loss_ = loss_object(real, pred)
        mask = tf.cast(mask, dtype=loss_.dtype)
        loss_ *= mask

        return tf.reduce_mean(loss_)

    checkpoint_path = "./checkpoints/train"
    ckpt = tf.train.Checkpoint(encoder=encoder,
                                decoder=decoder,
                                optimizer = optimizer)
    ckpt_manager = tf.train.CheckpointManager
                (ckpt, checkpoint_path, max_to_keep=5)

    start_epoch = 0

    if ckpt_manager.latest_checkpoint:
        start_epoch = int
            (ckpt_manager.latest_checkpoint.split('-')[-1])
    # restoring the latest checkpoint in checkpoint_path
    ckpt.restore(ckpt_manager.latest_checkpoint)

# if checkpoint is not restored run this code
ckpt.restore(tf.train.latest_checkpoint
                (checkpoint_path))

loss_plot = []

def train_step(img_tensor, target):
```

```python
    loss = 0

    # initialize the hidden state for each batch
    # because the captions are not related from image to image
    hidden = decoder.reset_state
                    (batch_size=target.shape[0])

    dec_input = tf.expand_dims([tokenizer.word_index
                                ['<start>']] *
                                BATCH_SIZE, 1)

    with tf.GradientTape() as tape:
        features = encoder(img_tensor)

        for i in range(1, target.shape[1]):
            # Pass the features through the decoder
            predictions, hidden, _ = decoder(dec_input,
                                    features, hidden)

            loss += loss_function(target[:, i],
                                  predictions)

            # Use teacher forcing
            dec_input = tf.expand_dims(target[:, i], 1)
    total_loss = (loss / int(target.shape[1]))
    trainable_variables = encoder.trainable_variables +
                          decoder.trainable_variables
    gradients = tape.gradient
                    (loss, trainable_variables)
    optimizer.apply_gradients(zip(gradients,
                                  trainable_variables))

    return loss, total_loss

for epoch in range(start_epoch, 20):
    start = time.time()

    total_loss_train = 0
    for (batch, (img_tensor, target))
                in enumerate(dataset):
        batch_loss, t_loss = train_step
                (img_tensor, target)
        total_loss_train += t_loss

    if epoch % 5 == 0:
    ckpt_manager.save()
    print ('Epoch {} Train-Loss {:.4f}'.format
                (epoch + 1,
                    (total_loss_train/num_steps)))
    print ('Time taken for this epoch {}
```

```python
                    sec\n'.format(time.time() - start))
def evaluate(image):
    hidden = decoder.reset_state(batch_size=1)
    temp_input = tf.expand_dims
                    (load_image(image)[0], 0)
    img_tensor_val = image_features_extract_model
                    (temp_input)
    img_tensor_val = tf.reshape(img_tensor_val,
                    (img_tensor_val.shape[0],
                    -1,
                    img_tensor_val.shape[3]))

    features = encoder(img_tensor_val)

    dec_input = tf.expand_dims([tokenizer.word_index
                                ['<start>']], 0)

    result = []

    for i in range(max_length):
        predictions, hidden, attention_weights =
                    decoder(dec_input,
                    features, hidden)

        predicted_id = tf.random.categorical
                    (predictions, 1)[0][0].numpy()
        result.append(tokenizer.index_word[predicted_id])
        if tokenizer.index_word[predicted_id] ==
                                '<end>':

            return result

        dec_input = tf.expand_dims([predicted_id], 0)

    return result

def predict(image_url , random_name):
    image_extension = image_url[-4:]
    image_path = tf.keras.utils.get_file
                    ('image'+ random_name +
                    image_extension,
                    origin=image_url)
    result = evaluate(image_path)
    print ('Prediction Caption:', ' '.join(result))
    Image.open(image_path)
    return image_path

image_url = 'https://tensorflow.org/images/surf.jpg'
path = predict(image_url , 'surfee')
Image.open(path)
```

```
image_url = 'https://farm4.staticflickr.com/3296/2765087292_535
6df67ce_z.jpg'
path = predict(image_url , 'baseball')
Image.open(path)

image_url = 'https://farm8.staticflickr.com/7139/8156048469_084
7c7ce15_z.jpg'
path = predict(image_url , 'dog')
Image.open(path)

image_url = 'https://farm5.staticflickr.com/4095/4910762818_
b1e9022005_z.jpg'
path = predict(image_url , 'tennis')
Image.open(path)

image_url = 'https://farm3.staticflickr.com/2690/4179330518_
b82897b153_z.jpg'
path = predict(image_url , 'competition')
Image.open(path)
```

总结

本章介绍了自然语言处理的另一个重要应用，即图像描述任务，用于为任何给定图像添加文字标注。图像描述包含两个重要部分，一个是提取图像特征，另一个是将图像特征映射为文本标注。提取图像特征是一项基本工作，有许多预训练网络可以使用，本章使用 InceptionV3 网络提取图像特征。为了将图像特征转换为文本标注，使用带有 Bahdanau 注意力的 循环神经网络。本章对 Bahdanau 注意力机制与实现过程进行了深入讲解。

在下一章中，将讲解另一项重要技术，即时间序列预测。

第 11 章
时间序列预测

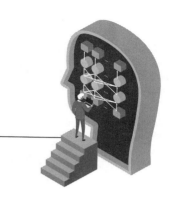

预测一直是每个人都感兴趣的话题。我的未来是什么样的？我会在未来 5 年内成为百万富翁吗？我什么时候结婚？诸如此类问题。这个世界上有人在做预测，至少在尝试为这些问题提供答案。到目前为止，神经网络在上述预测方面并不成功，但当过去数据中包含一些可发现未来的模式时，神经网络可以做出一些预测。本章的主题是学习如何训练神经网络来执行此类预测，这被称为时间序列预测。接下来，我们首先描述什么是时间序列，然后介绍如何对未来做出预测。

11.1 时间序列预测简介

11.1.1 什么是时间序列预测

对未来进行预测在经典统计学中称为外推法，我们现在使用的术语时间序列预测与外推法相同。预测包含构建适合历史数据的模型并使用该模型来预测未来——迄今为止尚未发生的事情，如"明天会很热吗？"。神经网络可以进行此类预测，并且比人类仅凭直觉做出的预测准确率更高。那么神经网络是如何做到这一点的呢？我们向神经网络提供过去几天、几个月甚至几年的温度数据，使神经网络挖掘这些数据中存在的时间序列模式，然后基于该模式进行预测。时间序列预测的另一个例子是预测股票的未来价格，如"一个月后 IBM 股票的价格是多少？"，可以通过一个训练好的神经网络来回答该问题。

以下是时间序列的更多示例。

① 根据脑电图（electroencephalogram, EEG）预测患者是否会癫痫发作。
② 预测 12 月份芭比娃娃产品的销量。
③ 预测服务器上每小时的带宽利用率需求。
④ 预测下一年的公司收益。
⑤ 预测某个城市的出生率。

我们可以发现很多和上述示例类似的令人感兴趣的问题，为了解决这些问题，人们已开发了许多基于统计的技术。如今，神经网络可以更好地解决这些问题。在讨论如何进行预测之前，我们首先描述预测中的一些重要问题。

11.1.2 预测中的问题

预测时的一些主要问题如下。
① 有多少过去的数据可以使用？
② 想要预测短期、中期还是长期结果？
③ 预测是否为静态？
④ 是否需要对给定问题进行定期预测？

时间序列预测的首个也是最重要的一点是我们有多少过去的数据。神经网络将分析过去的数据以找出其中的趋势，只有根据这些趋势，才能对未来做出预测。

短期预测通常较容易，但由于数据趋势可能会在一段时间段内发生变化，且存在很多不确定性，因此中期或长期预测较为困难。

当用户有预测需求时，他们希望以静态方式预测还是希望在条件发生变化时给出新的预测结果呢？动态预测的要求可能是模型设计的关注点之一。

有时，对同一问题的预测是以预先确定的频率进行的。预测频率的高低是影响数据采样的关键因素，进而影响着模型的构建。每天预测第二天的温度被认为是低频预测，而股票市场的高频交易被认为是高频预测。

接下来，将描述时间序列的各个组成部分。

11.1.3 时间序列组成

在分析用于建模的时间序列时，需要仔细观察以下几个时间序列的组成部分：整体或长期趋势、季节性变化、周期性变化、噪声（不规则扰动）。

整体或长期趋势是时间序列的主要组成部分。

遵循某种模式的时间序列可能会在一段时间内呈现上升或下降趋势，主要由社会经济和政治因素所影响。冰淇淋制造商在夏季的销售量增加而冬季销售量降低，这被称为销售趋势的季节性变动。假设我们拥有可用于分析的长期数据，如 5～12 年的数据，可能会观察到时间序列中的周期性波动，如在一个国家的经济数据中观察到的长周期。企业也会表现出长期波动，即众所周知的商业周期。最后，当出现 COVID-19（新型冠状病毒肺炎）大流行等不可预见的情况时，数据会出现噪声或不规则扰动。在开发用于时间序列分析的模型时，必须在分析中考虑到前面的各个部分。

接下来，我们介绍另一个重要问题，即时间序列的两个主要类别。

11.1.4 单变量与多变量

所有时间序列都可分为两大类：单变量和多变量。

具有随时间变化的单个变量的时间序列称为单变量时间序列。例如，在一年内测量一个家庭的电量消耗为单变量时间序列。此类时间序列最好使用统计技术建模，如自回归滑动平均模型 (autoregressive moving average, ARMA)。顾名思义，ARMA 模型将自回归技术与滑动平均技术相结合来预测单变量的未来值，该模型假设之前的观察值可为未来值提供良好的预测模型。

多变量时间序列具有多个随时间变化的变量，因此，预测结果不仅取决于多个变量的过去值，而且取决于变量之间的相互依赖性关系。向量自回归（vector autoregression, VAR）模型是一种用于对多变量时间序列建模的统计技术。该模型假设每个变量都是其自身过去值及所有相关变量过去值的函数。多变量时间序列的一个例子是空气污染，由多个相互依赖的因素决定。

接下来，将介绍用于预测单变量时间序列和多变量时间序列的神经网络模型。

11.2 单变量时间序列分析

本项目将构建一个预测能源消耗的模型。为此，我们使用 Kaggle 提供的数据集（www.kaggle.com/robikscube/hourly-energy-consumption）。本数据集由美国区域电网（PJM Interconnection LLC）提供，包含超过 10 年的每小时耗电量，以兆瓦表示。

11.2.1 创建项目

创建一个 Colab 项目并将其重命名为 Univariate – time series analysis，使用如下代码导入所需的库。

```
import tensorflow as tf
from tensorflow import keras
from tensorflow.keras import layers
import numpy as np
import pandas as pd
import matplotlib.pyplot as plt
import sklearn.preprocessing
from sklearn.metrics import r2_score
```

11.2.2 准备数据

从本书的下载页面将数据加载到项目中：

```
url = 'https://raw.githubusercontent.com/Apress/artificial-neural-networks-with-tensorflow-2/main/ch11/DOM_hourly.csv'
df = pd.read_csv(url)
```

通过将详细信息打印在屏幕上来查看数据，如图 11-1 所示。

可以看到，数据包含超过 100000 个电量读数，每隔一小时间隔一次。该数据从 2005 年开始记录，到 2018 年 1 月截止。使用以下语句从记录中生成每日数据。

```
# uncomment the following line for generating daily data
# df = df[df['Datetime'].str.contains("00:00:00")]
```

可以看到，在本书和可下载的代码中对上述代码进行了注释，因为我们首先将直接使用文件中提供的数据以每小时为单位进行实验。然后，我们将取消对上述代码的注释并重新运行该项目以实现以每周为单位的预测。我们将在本项目结束时解释这样做的目的。

在 datatime 字段上创建索引，使得 datatime 一列数据可以被视为日期而非对象，代码如下。

```
df['Datetime'] = pd.to_datetime(df.Datetime ,
format = '%Y-%m-%d %H:%M:%S')
df.index = df.Datetime
df.drop(['Datetime'], axis = 1,inplace = True)
```

使用 head() 方法查看创建索引后的数据：

```
df.head()
```

输出结果如图 11-2 所示。

图 11-1　每小时能耗量数据　　　　图 11-2　在 datatime 字段上创建索引后的数据

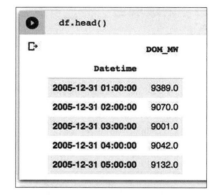

为了数据的完整性，使用以下代码检查数据中的空值。

```
#checking missing data
df.isna().sum()
```

输出结果显示数据中没有空值。通过绘制能耗与时间的关系图来查看能耗趋势，使用以下代码生成趋势图。

```
#@title Date Range
a = '2005-12-31'#@param {type:"date"}
b = '2018-01-31' #@param {type:"date"}

a = a+" 00:00:00"
b = b+" 00:00:00"
df.loc[a:b].plot(figsize = (16,4),legend = True)

plt.title('DOM hourly power consumption data')
plt.ylabel('Power consumption (MW)')

plt.show()
```

结果如图 11-3 所示。

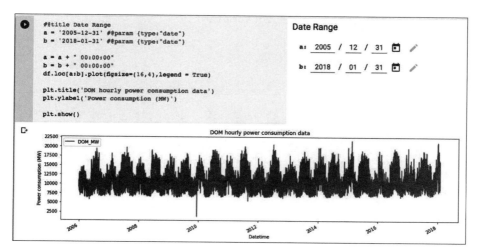

图 11-3　整个时期的能耗图

由图 11-3 可以看出,整个时期的能耗趋势几乎遵循一致的模式。在时间序列分析中,重要的是观察该模式的季节性变化。幸运的是,Colab 可以在运行时接收用户传递参数。本项目的绘图代码将开始日期和结束日期作为参数,因此可以尝试更改开始日期和结束日期以生成较小区域的图。图 11-4 显示了 2010 年 1 月的能耗图。

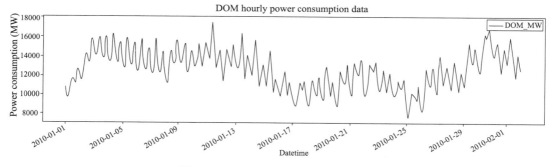

图 11-4　2010 年 1 月能耗趋势图

观察趋势后,假设我们的目标是预测 2018 年 2 月前两周的能耗。为此,我们需要创建一个模型,该模型将使用前 13 年的大量数据进行训练。

在将数据输入到模型进行训练之前,需要将数据缩放到 0 ～ 1 之间以进行归一化。使用 sklearn 的 MinMaxScaler 函数完成:

```
scaler = sklearn.preprocessing.MinMaxScaler()
df['DOM_MW'] = scaler.fit_transform
              (df['DOM_MW'].values.reshape(-1,1))
```

归一化之后,使用以下代码片段绘制归一化后的数据图。

```
df.plot(figsize = (16,4), legend = True)
```

```
plt.title('DOM hourly power consumption data -
                AFTER NORMALIZATION')
plt.ylabel('Normalized power consumption')
plt.show()
```

归一化后的数据图如图 11-5 所示。

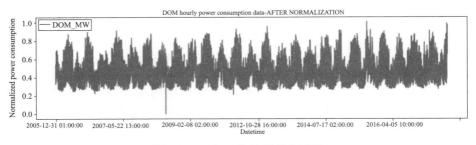

图 11-5　归一化的能耗曲线图

可以看到，现在所有 y 值的范围都在 0 ～ 1 之间。

至此，我们完成了数据的预处理。接下来，将创建训练集和测试集。

11.2.3　创建训练集和测试集

要创建训练集，需要在整个数据集中创建序列。假设我们定义序列长度为 20，那么前 20 个数据点将作为第一个序列，第 21 个点将作为预测的目标值。下一个序列从 2 开始到 21 结束，第 22 个数据点作为预测目标值，以此类推。图 11-6 中描述了创建序列的过程。

第一个图表显示了从 df[1] 到 df[20] 的 20 个数据点的序列，第 21 个数据点为预测目标。第二个序列包括从 df[2] 到 df[21] 的数据，第 22 个数据点为预测目标。第三个序列包括从 df[3] 到 df[22] 的数据，第 23 个数据点为预测目标。

定义一个名为 load_data 的函数来创建上述序列，如下所示。

```
def load_data(stock, seq_len):
```

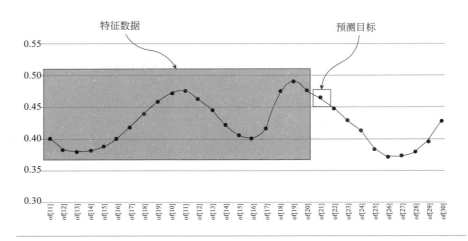

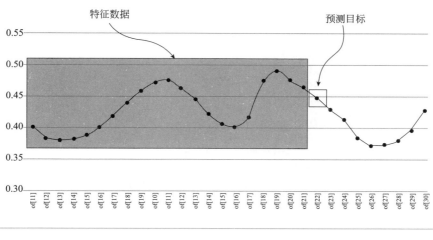

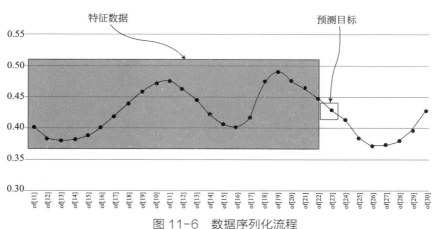

图 11-6　数据序列化流程

该函数的 stock 参数指定需要按序列拆分的数据集，seq_len 参数指定所需的序列长度。这里，我们将序列长度设置为一个参数，以便稍后可以尝试不同的序列范围来预测短期或长期趋势。使用以下 for 循环创建序列。

```
X_train = []
y_train = []
for i in range(seq_len, len(stock)):
    X_train.append(stock.iloc[i-seq_len : i, 0])
    y_train.append(stock.iloc[i, 0])
```

本数据集中有 116189 个数据点，我们使用前 90% 的数据进行训练，其余的数据用于测试：

```
X_test = X_train[int(0.9*(len(stock))):]
y_test = y_train[int(0.9*(len(stock))):]

X_train = X_train[:int(0.9*(len(stock)))]
y_train = y_train[:int(0.9*(len(stock)))]
```

接下来，将数据转换为 NumPy 数组：

```
# convert to numpy array
X_train = np.array(X_train)
y_train = np.array(y_train)
X_test = np.array(X_test)
y_test = np.array(y_test)
```

将 NumPy 数组转换为所需的数据维度：

```
# reshape data to input into RNN models
X_train = np.reshape(X_train,
           (int(0.9*(len(stock))), seq_len, 1))
X_test = np.reshape(X_test,
           (X_test.shape[0], seq_len, 1))
```

最后，将创建的数据集返回给函数调用者：

```
return [X_train, y_train, X_test, y_test]
```

完整的函数定义如清单 11-1 所示。

清单 11-1 load_data 函数完整代码

源码清单
链　接：https://pan.baidu.com/s/1NV0rimQ_8kRz22xfFHN-Cw
提取码：1218

```
def load_data(stock, seq_len):
    X_train = []
    y_train = []
    for i in range(seq_len, len(stock)):
        X_train.append(stock.iloc[i-seq_len : i, 0])
        y_train.append(stock.iloc[i, 0])

    X_test = X_train[int(0.9*(len(stock))):]
    y_test = y_train[int(0.9*(len(stock))):]

    X_train = X_train[:int(0.9*(len(stock)))]
    y_train = y_train[:int(0.9*(len(stock)))]

    # convert to numpy array
    X_train = np.array(X_train)
    y_train = np.array(y_train)

    X_test = np.array(X_test)
    y_test = np.array(y_test)

    # reshape data to input into RNN models
    X_train = np.reshape(X_train,
               (X_train.shape[0], seq_len, 1))
    X_test = np.reshape(X_test,
               (X_test.shape[0], seq_len, 1))

    return [X_train, y_train, X_test, y_test]
```

使用这个函数创建训练集和测试集，代码如下。

```
#create train, test data
seq_len = 20 #choose sequence length
X_train, y_train, X_test, y_test = load_data
                            (df, seq_len)
```

使用以下代码打印数据的 shape 属性，从而查看数据集的维度。

```
print('X_train.shape = ',X_train.shape)
print('y_train.shape = ', y_train.shape)
print('X_test.shape = ', X_test.shape)
print('y_test.shape = ',y_test.shape)
```

输出结果如下。

```
X_train.shape = (104570, 20, 1)
y_train.shape = (104570,)
X_test.shape = (11599, 20, 1)
y_test.shape = (11599,)
```

可以看到，X_train 和 X_test 都为包含 20 个点的序列数据。

11.2.4 创建输入张量

使用以下代码创建批量数据张量，用于输入预测模型。

```
batch_size = 256
buffer_size = 1000

train_data = tf.data.Dataset.from_tensor_slices
                    ((X_train , y_train))
train_data = train_data.cache().shuffle
            (buffer_size).batch(batch_size).repeat()

test_data = tf.data.Dataset.from_tensor_slices
                    ((X_test , y_test))
test_data = test_data.batch(batch_size).repeat()
```

如何在训练过程中打乱时间序列数据？实际上，我们打乱的是窗口的位置，而不是单个窗口内的数据。通常，当我们对时间序列这类训练数据的顺序进行打乱时，打乱的是这些序列输入模型的顺序，这样做将不会改变单个序列中数据的顺序。对于无状态网络，上述操作是可行的，因为网络的记忆不会跨序列持续存在。但对于一个有状态网络，即序列的评估需要记忆先前序列中的内容，这种操作是不可行的。就网络性质而言，所有 LSTM 在默认情况下都是有状态的。只有当我们决定是否让神经网络从一批数据到下一批数据保持状态时，才会出现在无状态和有状态之间做出选择的问题。因此，使用批量数据训练网络时，将给定的时间序列视为无状态还是有状态取决于我们自己的决定。

接下来，将构建神经网络模型。

11.2.5 构建模型

使用如下代码构建模型。

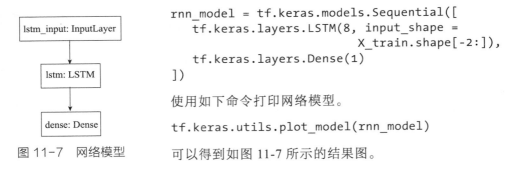

```
rnn_model = tf.keras.models.Sequential([
    tf.keras.layers.LSTM(8, input_shape =
                                X_train.shape[-2:]),
    tf.keras.layers.Dense(1)
])
```

使用如下命令打印网络模型。

```
tf.keras.utils.plot_model(rnn_model)
```

图 11-7 网络模型

可以得到如图 11-7 所示的结果图。

> **注意**
> 网络的第一层是一个具有 20 个节点的 LSTM 输入层,然后是具有 8 个节点的 LSTM 层,最后是一个输出层。

11.2.6 编译和训练

通过调用模型的 compile 方法编译模型:

```
rnn_model.compile(optimizer = 'adam', loss = 'mae')
```

通过调用模型的 fit 方法训练模型:

```
EVALUATION_INTERVAL = 200
EPOCHS = 10
rnn_model.fit(train_data, epochs = EPOCHS,
              steps_per_epoch = EVALUATION_INTERVAL,
                  validation_data = test_data,
                  validation_steps = 50)
```

模型训练完成后,即可对其性能进行评估。

11.2.7 评估

调用模型的 predcit 方法,并传入测试数据对模型的性能进行评估:

```
rnn_predictions = rnn_model.predict(X_test)
```

调用 sklearn_metrics 中的 r2_score 函数查看模型性能得分:

```
rnn_score = r2_score(y_test,rnn_predictions)
print("R2 Score of RNN model = 
            "+"{:.4f}".format(rnn_score));
```

得到评估结果如下。

```
R2 Score of RNN model = 0.9484
```

R2 被称为决定系数,是一种评价回归模型的分数,最好的分数为 1.0。当 R2 分数为 0.0 时表示模型恒定,它始终预测同一个 y 值,而忽略了输入特征。本项目中,R2 分数接近于 1.0,因此我们可以认为该预测模型训练得较好。

使用以下绘图代码绘制整个测试集中实际值与预测值的对比图。

```
#@title Data Range
a = 0 #@param {type:"slider", min:0,
            max:12000, step:1}
b = 12000 #@param {type:"slider", min:0,
            max:12000, step:1}
def plot_predictions(test, predicted, title):
  plt.figure(figsize = (16,4))
  plt.plot(test[a:b], color = 'blue',label =
                'Normalized power consumption')
  plt.plot(predicted[a:b], alpha = 0.7,
          color = 'orange',
          label = 'Predicted power consumption')
  plt.title(title)
  plt.xlabel('Time')
  plt.ylabel('Normalized power consumption')
  plt.legend()
  plt.show()
  plot_predictions(y_test, rnn_predictions,
"Predictions made by simple RNN model")
```

结果如图 11-8 所示。

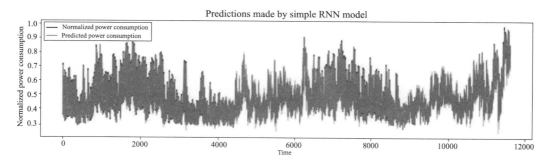

图 11-8　归一化尺度上的实际值与预测值对比图

可以看到,预测值接近实际值,这意味着 RNN 模型在预测能耗方面表现良好。

使用可视化平台提供的滑块可以选择数据的范围以查看放大的对比图。图 11-9 显示了一个缩小时间范围的对比图。

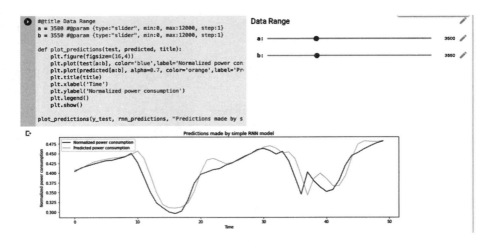

图 11-9　放大的实际值与预测值对比图

可以看到，预测曲线与预期目标值密切相关，因此，可以认为该模型即使在较小的数据范围内也具有较好的性能。使用以下代码对预测结果的尾部数据进行放大。

```
history_data = list(y_test[-40:])
plottingvalues = list(history_data)+list
                        (rnn_predictions[:50])
plt.figure(figsize = (16,4))
plt.plot(plottingvalues, color = 'orange',
         label = 'forecasted value',marker = 'o')
plt.plot(y_test[-40:], color = 'green',
         label = 'history',marker = 'x')
plt.xlabel('Time')
plt.ylabel('Normalized power consumption scale')
plt.legend()
plt.show()
```

输出结果如图 11-10 所示。

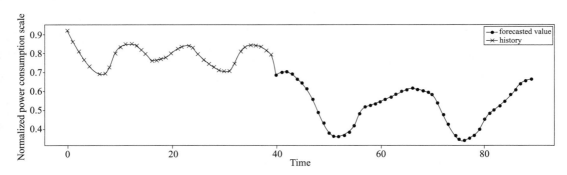

图 11-10　预测结果尾部数据的放大图

11.2.8　预测下一个数据点

我们将使用训练的模型预测下一个数据点，为此，从测试数据中提取最后一个数据点，并使用预测函数对其进行预测：

```
X = X_test[-1:]
rnn_predictions1 = rnn_model.predict(X)
```

可以通过在控制台上打印的方式查看预测值，命令及其输出结果如下。

```
rnn_predictions1
array([[0.798944]], dtype = float32)
```

这是我们的模型对时间戳 2018-01-02 00:00:00 所做的预测结果，因为我们数据集中的最后一个数据点为时间戳 2018-01-02 00:00:00。

为了以更好的方式可视化预测结果，我们使用如下代码片段将最后 40 个数据点及其预测值绘制为一个图。

```
history_data = list(y_test[-40:])
plottingvalues = list(history_data)+list
                    (rnn_predictions1)
plt.figure(figsize = (16,4))
plt.plot(plottingvalues, color = 'orange',
         label = 'forecasted value',marker = 'o')
plt.plot(y_test[-40:], color = 'green',
         label = 'history',marker = 'x')
plt.xlabel('Time')
plt.ylabel('Normalized power consumption scale')
plt.legend()
plt.show()
```

上述代码的执行结果如图 11-11 所示。

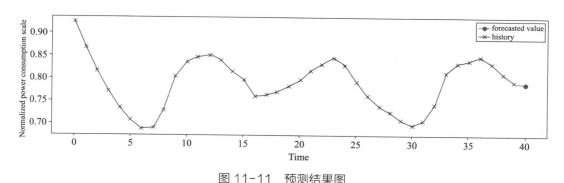

图 11-11　预测结果图

11.2.9　预测数据点区间

通常情况下，可能会对超出数据集中已有数据之后一段时间的功耗预测感兴趣。本数据集已有的最后一个数据点为 2018-01-02 00:00:00，假设我们想预测接下来的 25 个数据点。这就像一个多部分回归任务，生成这些预测数值的技巧是使用当前的预测结果作为下一个测试集的数据，并重复预测 25 次。接下来，我们将展示如何使用实际代码来执行此操作。首先，从测试集中提取最后 40 个数据点：

```
history_data = list(y_test[-40:])
```

然后，编写一个函数将最新的预测结果添加到数据集中并构建为一个新的数据集：

```
def make_data(X,rnn_predictions1):
    val = list(X[0][1:])+list
            (rnn_predictions1)
    X_new = []
    X_new.append(list(val))
    X_new = np.array(X_new)
    return X_new
```

创建一个列表变量来存储所有的预测结果：

```
forecast = list()
```

像以前一样提取最后一个测试数据点来进行下一个数据点的预测：

```
X = X_test[-1:]
```

接下来，定义一个循环来创建测试数据，对其进行预测，然后将其添加到预测列表中：

```
for i in range (25):
    X = make_data(X,rnn_predictions1)
    rnn_predictions1 = rnn_model.predict(X)
    forecast += list(rnn_predictions1)
```

最后，绘制由历史数据和接下来的 25 个预测数据组成的所有数据点：

```
plottingvalues = list(history_data)+list(forecast)
plt.figure(figsize=(16,4))
plt.plot(plottingvalues, color = 'orange',
         label = 'forecasted value',marker = 'o')
plt.plot(y_test[-40:], color = 'green',
         label = 'history',marker = 'x')
plt.xlabel('Time (ticks)')
plt.ylabel('Normalized power consumption scale')
plt.legend()
plt.show()
```

上述代码的运行结果如图 11-12 所示。

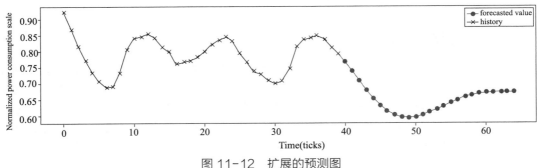

图 11-12 扩展的预测图

至此，我们已经可以预测接下来几个小时的能耗。但是，当我们拥有过去 13 年的数据时，可能会对预测下一周或接下来几周的能耗感兴趣。使用前面展示的外推法可能无法很好地完成这项工作。我们现在可以尝试在以天为单位的数据上训练模型，以便可以通过外推法对接下来 25 天的能耗进行预测。

回顾前文，我们已经编写了用于提取以天为单位的能耗数据的代码，只需对如下代码进行解注释即可。

```
# uncomment the following line for generating daily data
# df = df[df['Datetime'].str.contains("00:00:00")]
```

接下来，重置环境并运行整个项目。图 11-13 显示了 2018 年 1 月 1 日（数据集中的最后一个数据点）后一周的能耗预测结果。

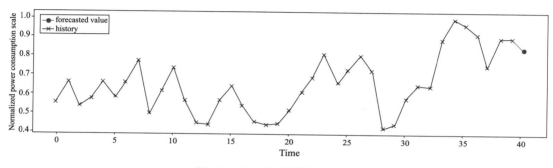

图 11-13 下一周的预测结果

图 11-14 显示了 2018 年 1 月 31 日之后 25 周的外推预测结果。

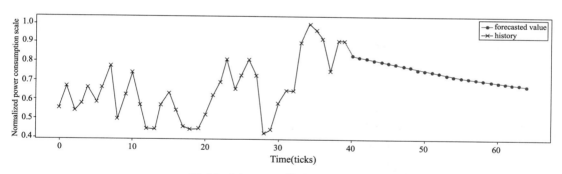

图 11-14 25 周的外推预测结果

要获得不同的预测结果，我们需要进行多个此类实验，包括更改 ANN 的配置（增加网络层数、使用 SimpleRNN、使用 dropouts 等）、训练更长的时间、尝试以周或者月为单位的数据点，等等。由于无法以一定的准确率真正预测未来，因此我们需要从上述实验中选择最合适的预测结果。

11.2.10 项目源码

清单 11-2 给出了本项目的完整代码。

源码清单
链　接：https://pan.baidu.com/s/1NV0rimQ_
8kRz22xfFHN-Cw
提取码：1218

清单 11-2 Univariate-time.ipynb

```
import tensorflow as tf
from tensorflow import keras
from tensorflow.keras import layers
import numpy as np
import pandas as pd
import matplotlib.pyplot as plt
import sklearn.preprocessing
from sklearn.metrics import r2_score

url = 'https://raw.githubusercontent.com/Apress/artificial
neural-networks-with-tensorflow-2/main/ch11/DOM_hourly.csv'
df = pd.read_csv(url)

df

# uncomment the following line for generating daily data
#df = df[df['Datetime'].str.contains("00:00:00")]

df['Datetime'] = pd.to_datetime(df.Datetime ,
                    format = '%Y-%m-%d %H:%M:%S')

df.index = df.Datetime
df.drop(['Datetime'], axis = 1,inplace = True)

df.head()

df

#checking missing data
df.isna().sum()

#@title Date Range
a = '2005-12-31' #@param {type:"date"}
b = '2018-01-31' #@param {type:"date"}

a = a + " 00:00:00"
b = b + " 00:00:00"
df.loc[a:b].plot(figsize = (16,4),legend = True)

plt.title('DOM hourly power consumption data')
plt.ylabel('Power consumption (MW)')

plt.show()

df.shape

scaler = sklearn.preprocessing.MinMaxScaler()
df['DOM_MW'] = scaler.fit_transform(df['DOM_MW'].
```

```python
                values.reshape(-1,1))

df.plot(figsize = (16,4), legend = True)
plt.title('DOM hourly power consumption data - 
                AFTER NORMALIZATION')
plt.ylabel('Normalized power consumption')
plt.show()

def load_data(stock, seq_len):
    X_train = []
    y_train = []
    for i in range(seq_len, len(stock)):
        X_train.append(stock.iloc[i-seq_len : i, 0])
        y_train.append(stock.iloc[i, 0])
    
    X_test = X_train[int(0.9*(len(stock))):]
    y_test = y_train[int(0.9*(len(stock))):]
    
    X_train = X_train[:int(0.9*(len(stock)))]
    y_train = y_train[:int(0.9*(len(stock)))]
    
    # convert to numpy array
    X_train = np.array(X_train)
    y_train = np.array(y_train)
    
    X_test = np.array(X_test)
    y_test = np.array(y_test)
    
    # reshape data to input into RNN models
    X_train = np.reshape(X_train, 
                (X_train.shape[0], seq_len, 1))
    X_test = np.reshape(X_test, 
                (X_test.shape[0], seq_len, 1))
    return [X_train, y_train, X_test, y_test]

#create train, test data
seq_len = 20 #choose sequence length

X_train, y_train, X_test, y_test = load_data
                                    (df, seq_len)

print('X_train.shape = ',X_train.shape)
print('y_train.shape = ', y_train.shape)
print('X_test.shape = ', X_test.shape)
print('y_test.shape = ',y_test.shape)

batch_size = 256
buffer_size = 1000

train_data = tf.data.Dataset.from_tensor_slices
```

```python
                    ((X_train , y_train))
train_data = train_data.cache().shuffle(buffer_size).
                    batch(batch_size).repeat()

test_data = tf.data.Dataset.from_tensor_slices
                    ((X_test , y_test))
test_data = test_data.batch(batch_size).repeat()

rnn_model = tf.keras.models.Sequential([
    tf.keras.layers.LSTM(8, input_shape =
                    X_train.shape[-2:]),
    tf.keras.layers.Dense(1)
])

tf.keras.utils.plot_model(rnn_model)

rnn_model.compile(optimizer = 'adam', loss = 'mae')

EVALUATION_INTERVAL = 200

EPOCHS = 10
rnn_model.fit(train_data, epochs=EPOCHS,
                    steps_per_epoch =
                    EVALUATION_INTERVAL,
                    validation_data = test_data,
                    validation_steps = 50)

rnn_predictions = rnn_model.predict(X_test)
rnn_score = r2_score(y_test,rnn_predictions)
print("R2 Score of RNN model =
                "+"{:.4f}".format(rnn_score));

#@title Data Range
a = 0 #@param {type:"slider", min:0,
                    max:12000, step:1}
b = 12000 #@param {type:"slider", min:0,
                    max:12000, step:1}

def plot_predictions(test, predicted, title):
    plt.figure(figsize = (16,4))
    plt.plot(test[a:b], color = 'blue',
            label = 'Normalized power consumption')
    plt.plot(predicted[a:b], alpha = 0.7,
            color = 'orange',
            label = 'Predicted power consumption')
    plt.title(title)
    plt.xlabel('Time')
    plt.ylabel('Normalized power consumption')
    plt.legend()
    plt.show()
```

```
plot_predictions(y_test, rnn_predictions,
            "Predictions made by simple RNN model")
history_data = list(y_test[-40:])
plottingvalues = list(history_data)+
                    list(rnn_predictions[:50])
plt.figure(figsize = (16,4))
plt.plot(plottingvalues, color = 'orange',
        label = 'forecasted value',marker = 'o')
plt.plot(y_test[-40:], color = 'green',
        label = 'history',marker = 'x')
plt.xlabel('Time')
plt.ylabel('Normalized power consumption scale')
plt.legend()
plt.show()

X = X_test[-1:]

rnn_predictions1 = rnn_model.predict(X)

rnn_predictions1

history_data = list(y_test[-40:])

plottingvalues = list(history_data)+
                    list(rnn_predictions1)
plt.figure(figsize = (16,4))
plt.plot(plottingvalues, color = 'orange',
        label = 'forecasted value',marker = 'o')
plt.plot(y_test[-40:], color = 'green',
        label = 'history',marker = 'x')
plt.xlabel('Time')
plt.ylabel('Normalized power consumption scale')
plt.legend()
plt.show()

history_data = list(y_test[-40:])

def make_data(X,rnn_predictions1):
    val = list(X[0][1:])+list(rnn_predictions1)
    X_new = []
    X_new.append(list(val))
    X_new = np.array(X_new)
    return X_new

forecast = list()
X = X_test[-1:]

for i in range (25):
    X = make_data(X,rnn_predictions1)
    rnn_predictions1 = rnn_model.predict(X)
```

```
        forecast += list(rnn_predictions1)

plottingvalues = list(history_data)+list(forecast)
plt.figure(figsize = (16,4))
plt.plot(plottingvalues, color = 'orange',
            label = 'forecasted value',marker = 'o')
plt.plot(y_test[-40:], color = 'green',
            label = 'history',marker = 'x')
plt.xlabel('Time (ticks)')
plt.ylabel('Normalized power consumption scale')
plt.legend()
plt.show()
```

接下来，将介绍如何开发用于多变量分析的模型。

11.3 多变量时间序列分析

本节将描述如何创建进行多变量时间序列分析的机器学习模型。为此，我们将使用 Kaggle 提供的伦敦共享单车数据集 (www.kaggle.com/hmavrodiev/london-bikesharing-dataset)。该数据集提供给定时间的共享单车数量及天气状况。据观察，对单车的需求也是季节性的。我们模型的目的是在预测单车的未来需求量时将数据集中的所有变量考虑在内。数据集中各个数据列的描述如下。

☑ timestamp：时间戳。

☑ cnt：共享单车数量。

☑ t1：实际摄氏温度。

☑ t2：体感摄氏温度。

☑ hum：湿度百分比。

☑ wind_speed：风速 (km/h)。

☑ weather_code：天气类别。

☑ is_holiday：布尔值，是否是假期，1 代表假期。

☑ is_weedend：布尔值，是否是周末，1 代表周末。

☑ season：季节类别：0 代表春季，1 代表夏季，2 代表秋季，3 代表冬季。

cnt 列将作为我们预测的目标值，其余所有数据列都可以用作特征。因此，这是一个多变量问题，预测目标结果取决于其他 9 个字段的值。

11.3.1 创建项目

创建一个 Colab 项目并将其重命名为 Multivariate time series analysis，并像往常一样导入所需的库：

```
import numpy as np
import pandas as pd
import tensorflow as tf
import matplotlib.pyplot as plt
import sklearn.preprocessing
import seaborn as sns
```

11.3.2　准备数据

使用如下代码将数据加载到项目中。

```
url = 'https://raw.githubusercontent.com/Apress/artificial-
neural-networks-with-tensorflow-2/main/ch11/london_merged.csv'
df = pd.read_csv(url,parse_dates=['timestamp'],
                                         index_col = "timestamp")
```

与前一个项目一样，我们将把时间戳一列处理为日期而不是对象。查看数据，如图 11-15 所示。

```
df
```

timestamp	cnt	t1	t2	hum	wind_speed	weather_code	is_holiday	is_weekend	season
2015-01-04 00:00:00	182	3.0	2.0	93.0	6.0	3.0	0.0	1.0	3.0
2015-01-04 01:00:00	138	3.0	2.5	93.0	5.0	1.0	0.0	1.0	3.0
2015-01-04 02:00:00	134	2.5	2.5	96.5	0.0	1.0	0.0	1.0	3.0
2015-01-04 03:00:00	72	2.0	2.0	100.0	0.0	1.0	0.0	1.0	3.0
2015-01-04 04:00:00	47	2.0	0.0	93.0	6.5	1.0	0.0	1.0	3.0
...	...	...	...	...	...	...	...	...	...
2017-01-03 19:00:00	1042	5.0	1.0	81.0	19.0	3.0	0.0	0.0	3.0
2017-01-03 20:00:00	541	5.0	1.0	81.0	21.0	4.0	0.0	0.0	3.0
2017-01-03 21:00:00	337	5.5	1.5	78.5	24.0	4.0	0.0	0.0	3.0
2017-01-03 22:00:00	224	5.5	1.5	76.0	23.0	4.0	0.0	0.0	3.0
2017-01-03 23:00:00	139	5.0	1.0	76.0	22.0	2.0	0.0	0.0	3.0

17414 rows × 9 columns

图 11-15　用于多变量时间序列分析的数据

可以看到，该数据集共有 17414 条记录（行）和 9 列。注意，我们已从数据集中提取出时间戳列并将其作为索引。通过调用 dftypes 方法查看数据类型，输出结果如图 11-16 所示。

由图 11-16 可知，该数据集中所有的特征列都为数字类型。但是需要注意，诸如 weather_code 和 season 之类的特征列使用了分类值，is_holiday 和 is_weekend 特征列包含了布尔值，当我们缩放特征数值以进行归一化时，不应该缩放这些特征列。

```
df.dtypes

cnt               int64
t1              float64
t2              float64
hum             float64
wind_speed      float64
weather_code    float64
is_holiday      float64
is_weekend      float64
season          float64
dtype: object
```

图 11-16　共享单车数据集中的数据类型

11.3.3　检查平稳性

我们将检查要分析的所有特征列是否具有平稳性。平稳序列是指属性（均值、方差和协方差）不随时间变化的序列。对于平稳的序列，其特征值的模应小于 1。Johansen 检验可用于最多 12 个时间序列之间的协整检验。在当前数据集中，我们有 9 个时间序列，因此可以应用 Johansen 检验，且无须对代码做任何调整。我们使用以下代码进行测试。

```
#checking stationarity
from statsmodels.tsa.vector_ar.vecm
                 import coint_johansen
johan_test_temp = df
coint_johansen(johan_test_temp,-1,1).eig
```

运行上述测试将得到所有 9 列特征的特征值，如下所示。

```
array([2.61219379e-01, 1.31970167e-01, 5.22046139e-02,
4.19830465e-02, 2.10126207e-02, 1.75450605e-02, 1.36518877e-02,
6.26085775e-04, 7.56291478e-05])
```

可以看到，所有特征值都小于 1，这意味着所有被测的时间序列都是平稳的。如果发现了一个非平稳序列，则需要检查列表以查看是哪个特征引起了问题。

11.3.4 探索数据

我们将探索一些特征列和对应的目标值以了解数据的分布。从代表共享单车数量的 cnt 列开始，它是我们即将预测的目标值。使用如下代码绘制 cnt 的分布。

```
plt.figure(figsize = (16,4))
plt.plot(df.index, df["cnt"]);
```

可以得到如图 11-17 所示的数据分布。

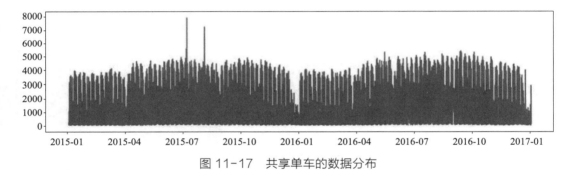

图 11-17　共享单车的数据分布

由图 11-17 可以看到，单车的需求在整个区间呈现均匀分布。接下来，我们检查单车需求分布是否有季节性变化，首先以每小时和每月为单位汇总数据并创建索引：

```
# create indexes
df['hour'] = df.index.hour
df['month'] = df.index.month
```

然后，使用前面的索引创建三个图表来观察周末、假期和季节性单车需求变化。使用以下代码生成图表。

```
fig,(ax1, ax2, ax3)= plt.subplots(nrows = 3)
fig.set_size_inches(16, 10)

sns.pointplot(data = df, x = 'month', y = 'cnt',
              hue = 'is_weekend', ax = ax1)
```

```
sns.pointplot(data = df, x = 'hour', y = 'cnt',
              hue = 'season', ax = ax2);
sns.pointplot(data = df, x = 'month', y = 'cnt',
              ax = ax3)
```

输出结果如图 11-18 所示。

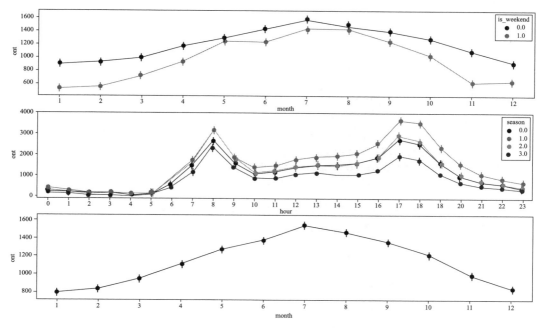

图 11-18　共享单车的需求变化

在图 11-18 中，第一个图表显示了一年中每个月周末和工作日的共享单车总量。可以清晰地看到，无论是周末还是工作日，每年 7 月份的单车需求量都最多。第二个图表显示了该数据集中四个季节（0- 春季、1- 夏季、2- 秋季、3- 冬季）的一天中每小时的单车需求量。可以观察到，每个季节上午 8 点和下午 5 点至下午 6 点的单车需求量更大。最后一个图表显示了不考虑周末、假期或季节的每月单车总需求量，可以看到，7 月份的单车需求量更多，而 1 月和 12 月期间需求量最低。

了解了单车需求模式后，接下来将继续进行数据清理和准备。

11.3.5　准备数据

使用 MinMaxScaler 函数缩放所有数字列，代码如下。

```
# scaling numeric columns
scaler = sklearn.preprocessing.MinMaxScaler()
df['t1'] = scaler.fit_transform(df['t1'].
                    values.reshape(-1,1))
df['t2'] = scaler.fit_transform(df['t2'].
                    values.reshape(-1,1))
df['hum'] = scaler.fit_transform(df['hum'].
                    values.reshape(-1,1))
```

```
df['wind_speed'] = scaler.fit_transform
                (df['wind_speed'].values.reshape(-1,1))
df['cnt'] = scaler.fit_transform
                (df['cnt'].values.reshape(-1,1))
```

注意

> 我们已经从预处理中排除了包含类别值和布尔值的数据列。

使用 90% 的数据进行训练，其余的数据用于测试：

```
# use 90% for training
train_size = int(len(df) * 0.9)
test_size = len(df) - train_size
train, test = df.iloc[0:train_size],
                    df.iloc[train_size:len(df)]
```

接下来，将创建张量以将数据输入到网络模型中。为此，定义一个名为 create_dataset 的函数，用于创建训练集和测试集，代码如下。

```
def create_dataset(X, y, time_steps = 1):
    Xs, ys = [], []
    for i in range(len(X) - time_steps):
        v = X.iloc[i:(i + time_steps)].values
        Xs.append(v)
        ys.append(y.iloc[i + time_steps])
    return np.array(Xs), np.array(ys)
```

使用如下代码创建训练集和测试集，并设置序列长度为 10。

```
time_steps = 10
X_train, y_train =
        create_dataset(train, train.cnt, time_steps)
X_test, y_test =
        create_dataset(test, test.cnt, time_steps)
```

对数据集进行拆分，并创建批量数据以进行训练，代码如下。

```
batch_size = 256
buffer_size = 1000

train_data = tf.data.Dataset.from_tensor_slices
                ((X_train , y_train))
train_data = train_data.cache().shuffle
            (buffer_size).batch(batch_size).repeat()

test_data = tf.data.Dataset.from_tensor_slices
                ((X_test , y_test))
test_data = test_data.batch(batch_size).repeat()
```

11.3.6 创建模型

使用如下代码定义神经网络模型。

```
simple_lstm_model = tf.keras.models.Sequential([
    tf.keras.layers.LSTM
            (8, input_shape = X_train.shape[-2:]),
    tf.keras.layers.Dense(1)
])
simple_lstm_model.compile
            (optimizer = 'adam', loss = 'mae')
```

该模型与前一个项目相似，因此这里不再解释。

11.3.7 训练

和往常一样，调用模型的 fit 方法来训练模型：

```
EVALUATION_INTERVAL = 200
EPOCHS = 10

history = simple_lstm_model.fit(
            train_data,
            epochs = EPOCHS,
            steps_per_epoch = EVALUATION_INTERVAL,
            validation_data = test_data,
            validation_steps = 50)
```

训练结束后，使用如下代码打印损失值。

```
# plot losses
plt.plot(history.history['loss'],
            label = 'train loss')
plt.plot(history.history['val_loss'],
            label = 'validation loss')
plt.xlabel("Epochs")
plt.ylabel("Loss")
plt.legend()
```

输出结果如图 11-19 所示。

可以看到，在第三个 epoch 之后，损失开始趋于平缓，且没有观察到过拟合。因此，模型的训练没有问题。接下来，我们将评估模型在测试数据上的性能。

11.3.8 评估

调用模型的 predict 方法，并输入测试数据评估模型的性能：

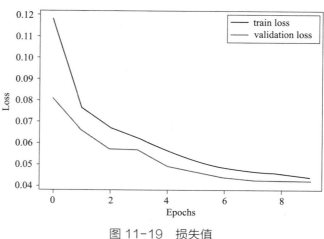

图 11-19　损失值

```
X_test,y_test = create_dataset(df,df.cnt,10)
y_pred = simple_lstm_model.predict(X_test)
```

接下来，绘制一些结果图以可视化预测结果。编写一个函数来创建一个指定长度的时间步长：

```
def create_time_steps(length):
    return list(range(-length, 0))
```

使用如下代码绘制结果图。

```
plt.figure(figsize = (16,4))
num_in = create_time_steps(91)
num_out = 28
plt.plot(num_in,y_train[15571:],label = 'history')
plt.plot(np.arange(num_out),
        y_test[15661:15689], 'b',label='Actual ')
plt.plot(np.arange(num_out),
        y_pred[15661:15689], 'r',label = 'Predicted')
plt.xlabel("Time")
plt.ylabel("bike shares (Normalized Value)")
plt.legend()
plt.show()
```

输出结果如图 11-20 所示。

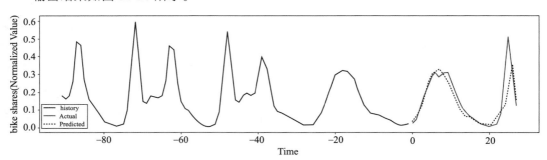

图 11-20　共享单车数量的实际值与预测值

11.3.9　预测未来点

模型的评估完成后，我们将预测未来的共享单车数量。用一个简单的语句来做预测：

```
y_pred = simple_lstm_model.predict(X_test[-1:])
```

打印预测结果值，代码如下。

```
# print value
y_pred
```

可以得到如下结果。

```
array([[0.03246454]], dtype = float32)
```

使用如下代码绘制预测结果图，以对数据点的拟合进行一些可视化。

```
# plot prediction

# plot prediction
plt.figure(figsize = (16,4))
num_in = create_time_steps(100)
num_out = 1
plt.plot(num_in,y_test[-100:])
plt.plot(np.arange(num_out),y_pred, 'ro',
                    label = 'Predicted')
plt.xlabel("Time")
plt.ylabel("bike shares (Normalized Value)")
plt.legend()
plt.show()
```

输出结果如图 11-21 所示。

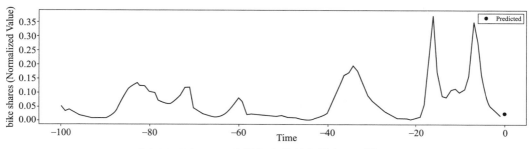

图 11-21　下一个数据点的共享单车预测结果

看起来模型对未来数据点的预测没有问题，那么可以对多个数据点进行预测吗？接下来让我们尝试一下。

11.3.10　预测数据点区间

测试集中的最后一个数据点为 2017-1-3 23:00:00，我们希望以每小时为单位预测 2017-1-4 00:00:00 以后的 100 个数据点。我们将需要这一时期的特征数据，因此，首先从测试集中挑选一些特征条目，代码如下。

```
df2 = df['2017-01-03 14:00:00':'2017-01-03 23:00:00']
```

在上述代码中，df2 将包含从 2017-1-3 14:00:00 到 2017-1-3 23:00:00 的测试数据。因此，我们选择了测试集的最后 10 个数据点。

然后，从原始数据集中提取 2016 年的特征数据：

```
df1 = df['2016-01-04 00:00:00':'2016-01-06 23:00:00']
```

　　上述代码从 2016-1-4 00:00:00 开始获取了接下来 3 天的数据。设置 df1 变量中的 cnt 字段为 0，代码如下。

```
df1['cnt'] = 0
```

通过将过去的数据附加到当前数据上来创建一个新的数据流：

```
df_future = df2.append(df1, sort = False)
```

通过打印 data_future 变量查看新建的数据：

```
df_future
```

输出结果如图 11-22 所示。

timestamp	cnt	t1	t2	hum	wind_speed	weather_code	is_holiday	is_weekend	season
2017-01-03 14:00:00	0.097328	0.211268	0.2000	0.666667	0.389381	3.0	0.0	0.0	3.0
2017-01-03 15:00:00	0.107506	0.211268	0.2000	0.635220	0.477876	4.0	0.0	0.0	3.0
2017-01-03 16:00:00	0.152799	0.211268	0.2000	0.635220	0.460177	4.0	0.0	0.0	3.0
2017-01-03 17:00:00	0.348855	0.211268	0.2000	0.666667	0.371681	3.0	0.0	0.0	3.0
2017-01-03 18:00:00	0.282443	0.183099	0.1750	0.761006	0.389381	2.0	0.0	0.0	3.0
...	...	...	...	...	...	...	...	...	...
2016-01-06 19:00:00	0.000000	0.239437	0.3000	0.874214	0.123894	2.0	0.0	0.0	3.0
2016-01-06 20:00:00	0.000000	0.239437	0.3000	0.836478	0.115044	2.0	0.0	0.0	3.0
2016-01-06 21:00:00	0.000000	0.211268	0.2625	0.911950	0.115044	2.0	0.0	0.0	3.0
2016-01-06 22:00:00	0.000000	0.211268	0.2500	0.911950	0.159292	2.0	0.0	0.0	3.0
2016-01-06 23:00:00	0.000000	0.225352	0.2500	0.874214	0.212389	2.0	0.0	0.0	3.0

82 rows × 9 columns

图 11-22　用于预测的新数据集

可以看到，新数据的时间戳存在不连续性，但这对于我们的分析来说无关紧要。可以简单地使用以下语句删除数据索引以使数据更加清洁。

```
#df_future = df_future.reset_index(drop = True)
```

接下来，我们在一个循环中完成预测，每次将新预测值添加到 df_future 数据流中。执行上述操作的 for 循环如下。

```
predictions = []

# make prediction in a loop every time adding the last
prediction
for i in range(50):
 X_f, y_f = create_dataset
           (df_future, df_future.cnt, time_steps)
 y_pred = simple_lstm_model.predict(X_f[i:i+1])
 df_future['cnt'][i+10] = y_pred
 predictions.append(float(y_pred[0][0]))
```

本书中我们只使用了 50 次迭代，读者可以遍历我们创建的整个数据集。

我们可以打印预测数组以检查所有预测值。更好的是使用以下代码绘制结果图可视化预测结果。

```
plt.figure(figsize = (16,4))
num_in = create_time_steps(100)
num_out = 50
plt.plot(num_in,y_test[-100:],label = 'History')
plt.plot(np.arange(num_out),predictions, 'r',
                    label = 'Predicted')
plt.xlabel("Time")
plt.ylabel("bike shares (Normalized Value)")
plt.legend()
plt.show()
```

上述代码生成的结果图如图 11-23 所示。

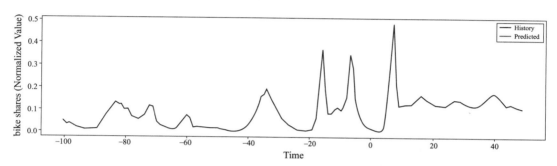

图 11-23 对共享单车数量未来几周的预测结果

至此，我们介绍完了如何创建用于多变量时间序列分析的深度神经网络模型。

> **注意**
>
> 该统计模型有可用的标准实现，如 statsmodels 库中有一个向量自回归模型（vector autoregression，VAR），代码如下。

```
from statsmodels.tsa.vector_ar.var_model import VAR
model = VAR(endog = train)
result = model.fit()
```

使用训练数据对模型进行拟合之后，就可以通过调用模型的 predict 方法来进行预测：

```
prediction = result.forecast(validation_data.y,
                    steps = len(valid))
```

随着深度学习技术的出现，在实际使用中了解复杂统计模型背后的数学原理不再是必须的，人工神经网络能够自动学习预设任务并做出预测。

11.3.11 项目源码

清单 11-3 给出了本项目的完整源码。

源码清单
链　接：https://pan.baidu.com/s/1NV0rimQ_8kRz22xfFHN-Cw
提取码：1218

清单 11-3 MultiVariate time.ipynb

```python
import numpy as np
import pandas as pd
import tensorflow as tf
import matplotlib.pyplot as plt
import sklearn.preprocessing
import seaborn as sns

url = 'https://raw.githubusercontent.com/Apress/artificialneural-networks-with-tensorflow-2/main/ch11/london_merged.csv'
df = pd.read_csv(url,parse_dates=['timestamp'],
                                  index_col="timestamp")

df

df.dtypes

#checking stationarity
from statsmodels.tsa.vector_ar.vecm
                 import coint_johansen
johan_test_temp = df
coint_johansen(johan_test_temp,-1,1).eig

plt.figure(figsize = (16,4))
plt.plot(df.index, df["cnt"]);

# create indexes
df['hour'] = df.index.hour
df['month'] = df.index.month

fig,(ax1, ax2, ax3) = plt.subplots(nrows = 3)
fig.set_size_inches(16, 10)

sns.pointplot(data = df, x = 'month',
              y = 'cnt', hue = 'is_weekend', ax = ax1)
sns.pointplot(data = df, x = 'hour',
              y = 'cnt', hue = 'season', ax = ax2);
sns.pointplot(data = df, x = 'month',
              y = 'cnt', ax = ax3)

# scaling numeric columns
scaler = sklearn.preprocessing.MinMaxScaler()
df['t1'] = scaler.fit_transform(df['t1'].values.reshape(-1,1))
df['t2'] = scaler.fit_transform
              (df['t2'].values.reshape(-1,1))
df['hum'] = scaler.fit_transform
              (df['hum'].values.reshape(-1,1))
df['wind_speed'] = scaler.fit_transform
              (df['wind_speed'].values.reshape(-1,1))
```

```python
df['cnt'] = scaler.fit_transform
            (df['cnt'].values.reshape(-1,1))

# use 90% for training
train_size = int(len(df) * 0.9)
test_size = len(df) - train_size
train, test = df.iloc[0:train_size],
              df.iloc[train_size:len(df)]

def create_dataset(X, y, time_steps = 1):
    Xs, ys = [], []
    for i in range(len(X) - time_steps):
        v = X.iloc[i:(i + time_steps)].values
        Xs.append(v)
        ys.append(y.iloc[i + time_steps])
    return np.array(Xs), np.array(ys)

# create input tensors
time_steps = 10
X_train, y_train = create_dataset
                   (train, train.cnt, time_steps)
X_test, y_test = create_dataset
                 (test, test.cnt, time_steps)

batch_size = 256
buffer_size = 1000

train_data = tf.data.Dataset.from_tensor_slices
                    ((X_train , y_train))
train_data = train_data.cache().shuffle
            (buffer_size).batch(batch_size).repeat()
test_data = tf.data.Dataset.from_tensor_slices
            ((X_test , y_test))
test_data = test_data.batch(batch_size).repeat()

simple_lstm_model = tf.keras.models.Sequential([
    tf.keras.layers.LSTM(8, input_shape =
                  X_train.shape[-2:]),
    tf.keras.layers.Dense(1)
])
simple_lstm_model.compile(optimizer = 'adam',
                          loss = 'mae')

EVALUATION_INTERVAL = 200
EPOCHS = 10

history = simple_lstm_model.fit(
                train_data,
                epochs = EPOCHS,
                steps_per_epoch = EVALUATION_INTERVAL,
```

```python
                    validation_data = test_data,
                    validation_steps = 50)
# plot losses
plt.plot(history.history['loss'],
                    label = 'train loss')
plt.plot(history.history['val_loss'],
                    label = 'validation loss')
plt.xlabel("Epochs")
plt.ylabel("Loss")
plt.legend()

X_test,y_test = create_dataset(df,df.cnt,10)
y_pred = simple_lstm_model.predict(X_test)

def create_time_steps(length):
  return list(range(-length, 0))

plt.figure(figsize = (16,4))
num_in = create_time_steps(91)
num_out = 28
plt.plot(num_in,y_train[15571:],label = 'history')
plt.plot(np.arange(num_out),
        y_test[15661:15689], 'b',label='Actual ')
plt.plot(np.arange(num_out),
       y_pred[15661:15689], 'r',label = 'Predicted')
plt.xlabel("Time")
plt.ylabel("bike shares (Normalized Value)")
plt.legend()
plt.show()

y_pred = simple_lstm_model.predict(X_test[-1:])

# print value
y_pred

# plot prediction
plt.figure(figsize = (16,4))
num_in = create_time_steps(100)
num_out = 1
plt.plot(num_in,y_test[-100:])
plt.plot(np.arange(num_out),y_pred, 'ro',
        label ='Predicted')
plt.xlabel("Time")
plt.ylabel("bike shares (Normalized Value)")
plt.legend()
plt.show()

df2 = df['2017-01-03 14:00:00':'2017-01-03 23:00:00']

df1 = df
```

```
        ['2016-01-04 00:00:00':'2016-01-06 23:00:00']

    df1['cnt'] = 0

    df_future = df2.append(df1, sort = False)

    # dropping index is not truly required.
    #df_future = df_future.reset_index(drop = True)

    df_future

    predictions = []

    # make prediction in a loop every time adding the last
    prediction
    for i in range(50):
      X_f, y_f = create_dataset
              (df_future, df_future.cnt, time_steps)
      y_pred = simple_lstm_model.predict(X_f[i:i+1])
      df_future['cnt'][i+10] = y_pred
      predictions.append(float(y_pred[0][0]))

    predictions

    plt.figure(figsize = (16,4))
    num_in = create_time_steps(100)
    num_out = 50
    plt.plot(num_in,y_test[-100:],label = 'History')
    plt.plot(np.arange(num_out),predictions, 'r',
            label = 'Predicted')
    plt.xlabel("Time")
    plt.ylabel("bike shares (Normalized Value)")
    plt.legend()
    plt.show()
```

总结

本章，我们学习了如何使用神经网络对单变量时间序列和多变量时间序列进行建模。在单变量时间序列中，目标值仅依赖于一个自变量；而在多变量时间序列中，目标值依赖于多个自变量。对于这两种类型的时间序列分析，都有广泛的基于统计的技术可供使用。然而，随着深度神经网络的出现，神经网络可以为我们节省大量时间和精力来学习这些统计模型的理论及代码实现。当然，有一些统计模型的标准库可用于时间序列预测，但要实现自定义统计模型，需要了解其背后的全部数学原理。神经网络使得我们能够轻松实现时间序列预测模型，还允许我们使用不同的模型配置来试验数据，以达到可接受的预测性能。

下一章，我们将学习风格迁移。

第 12 章
风格迁移

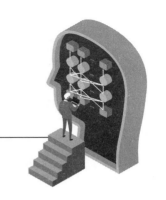

你是否曾希望能像毕加索或印度著名画家 M.F. 侯赛因一样绘画？现在，神经网络已经让这个愿望成为现实。本章，我们将学习一种新技术，该技术使用神经网络将选择的图像转换为著名艺术家的绘画风格，或者更确切地说是转换为我们自己选择的风格，这种技术称为风格迁移（style transfer）。Leon A. Gatys 在其论文 *A Neural Algorithm of Artistic Style* 中概述了风格迁移。

风格迁移是一种优化技术，可将一幅图像的内容与另一幅图像的风格相融合。我们在之前章节使用的 TensorFlow Hub 中包含一个用于风格迁移的预训练模型。本章将首先展示如何使用此预训练模型进行快速风格迁移学习，接下来介绍一个自定义风格迁移示例，该示例介绍如何从两幅不同的图像中提取内容和风格，然后对提取的内容图像执行风格迁移，以创建另一个风格的图像。

图 12-1 展示了我们将要实现的目标。

图 12-1 内容、风格和风格化图像

可以看到，左侧第一幅图为内容图像，中间为风格图像，右侧为风格迁移后的图像，即中间图像的风格被应用于左侧图像的内容以生成新的风格化图像。

风格迁移背后的理论较为简单,我们将在本章的自定义风格迁移示例中介绍。首先,让我们开始介绍一个快速风格迁移的应用。

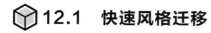 12.1 快速风格迁移

TensorFlow Hub 提供了一个可以进行快速风格迁移的预训练模型,该预训练模型的名称为 "arbitrary-image-stylization-v1-256/2"。与从头开始搭建神经网络进行风格迁移的工作相比,该模型执行了快速风格迁移操作,且可以适用于任意的绘画风格。该模型基于 Golnaz 等人在论文 *Exploring the structure of a realtime, arbitrary neural artistic stylization network*(https://arxiv.org/abs/1705.06830)中提出的风格迁移技术,所提出的模型结合了艺术风格化神经网络算法的灵活性和快速风格迁移算法的速度,促进了使用任意内容图像与风格图像对进行实时风格迁移的发展。

接下来,我们将使用 TensorFlow Hub 上提供的这个预训练模型对风格迁移进行快速了解。

12.1.1 创建项目

创建一个新的 Colab 项目并将其重命名为 TFHubStyleTransfer。导入所需的库,代码如下。

```
import tensorflow as tf
import re
import urllib
import numpy as np
import matplotlib.pyplot as plt
import PIL.Image
import tensorflow_hub as hub
from tensorflow.keras.preprocessing.image
import load_img, img_to_array
from matplotlib import gridspec
from IPython import display
from PIL import Image
```

12.1.2 下载图像

该项目在任何给定实例中都需要两幅图像——内容图像和风格图像。内容图像将被改变以适应风格图像中提供的风格。为了测试需要,本书提供的存储库中已上传了一些图像,每幅图像的 URL 采用以下形式 "https://raw.githubusercontent.com/Apress/artificial-neuralnetworks-with-tensorflow-2/main/ch12/ferns.jpg"。

编写函数提取图像文件名,并下载文件及设置图像的新路径,函数定义如清单 12-1 所示。

清单 12-1 用于创建图像 URL 的函数

源码清单
链 接:https://pan.baidu.com/s/1NV0rimQ_8kRz22xfFHN-Cw
提取码:1218

```
def download_image_from_URL(imageURL):
    imageName = re.search('[a-z0-9\-]+\.
                         (jpe?g|png|gif|bmp|JPG)',
```

```
                        imageURL, re.IGNORECASE)
    imageName = imageName.group(0)
    urllib.request.urlretrieve(imageURL, imageName)
    imagePath = "./" + imageName
    return imagePath
```

调用此函数以创建内容图像的路径：

```
# This is the path to the image you want to transform.
target_url = "https://raw.githubusercontent.com/Apress/
artificial-neural-networks-with-tensorflow-2/main/ch12/ferns.jpg"
target_path = download_image_from_URL(target_url)
```

同样地，下载风格图像并设置其路径：

```
# This is the path to the style image.
style_url = "https://raw.githubusercontent.com/Apress/artificial-
neural-networks-with-tensorflow-2/main/ch12/on-the-road.jpg"
style_path = download_image_from_URL(style_url)
```

在 Colab notebook 中看到的用户界面如图 12-2 所示。

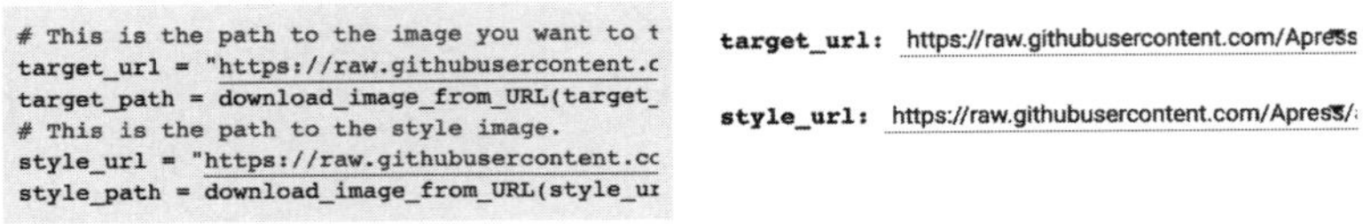

图 12-2　用于选择图像文件的 Colab 界面

可以从单元格右侧的下拉列表中选择所需的内容图像和风格图像。

使用 matplotlib imshow 函数和以下代码片段显示两幅选定的图像。

```
content = Image.open(target_path)
style = Image.open(style_path)

plt.figure(figsize=(10, 10))

plt.subplot(1, 2, 1)
plt.imshow(content)
plt.title('Content Image')

plt.subplot(1, 2, 2)
plt.imshow(style)
plt.title('Style Image')

plt.tight_layout()
plt.show()
```

输出结果如图 12-3 所示。

图 12-3　内容图像和风格图像

可以看到，我们获取了两幅具有不同尺寸的图像，分别为内容图像和风格图像。

12.1.3　准备模型输入图像

TensorFlow Hub 中执行图像转换的模型需要采用特定格式的图像以获得良好效果。首先，使用以下函数将风格图像转换为张量。

```
def image_to_tensor_style(path_to_img):
    img = tf.io.read_file(path_to_img)
    img = tf.image.decode_image
            (img, channels=3, dtype=tf.float32)
    img = tf.image.resize(img, [256,256])
    img = img[tf.newaxis, :]
    return img
```

该函数通过调用 read_file 方法读取图像数据，通过调用 decode_image 将图像解码为 RGB 三个通道，通过调用 resize 方法将图像尺寸调整为 256×256，因为预训练模型使用此特定尺寸进行风格转换。最后，将图像数据维度转换为 (1, 256, 256, 3)，并返回给函数调用者。可以看到，这里为图像添加了一个新维度，稍后在图像批处理时将使用该维度。

同样，编写一个函数将内容图像转换为张量，代码如下所示。

```
def image_to_tensor_target(path_to_img, image_size):
    img = tf.io.read_file(path_to_img)
    img = tf.image.decode_image
            (img, channels=3, dtype=tf.float32)
    img = tf.image.resize(img, [image_size,image_size],
                    preserve_aspect_ratio=True)
    img = img[tf.newaxis, :]
    return img
```

该函数除了图像路径之外还需要一个额外的参数：用户定义的图像尺寸。

大尺寸图像在处理过程中会占用大量内存，因此该函数添加了一个参数，以便可以在保持图像纵横比的情况下缩小尺寸。图像纵横比如 resize 方法中的 preserve_aspect_ratio 参数所示，为图像的宽度与高度的比率。假设需要将维度为 (1, 1200, 1600, 3) 的图像尺寸调整为 400，由于需要保持图像纵横比，如果将图像高度设置为 400，则宽度将以相应的图像纵横比成比例地减少。

调用上述定义的两个函数,将内容图像和风格图像转换为张量:

```
output_image_size = 400
target_image = image_to_tensor_target
                (target_path,output_image_size)
style_image = image_to_tensor_style(style_path)
```

12.1.4 执行风格迁移

为了将新风格应用于内容图像,需要从 Tensor Flow Hub 加载预训练模型,使用以下语句。

```
hub_module = hub.load('https://tfhub.dev/google/magenta/
arbitrary-image-stylization-v1-256/2')
```

模型名称为"arbitrary-image-stylization-v1-256/2"。加载模型后,只需将两个张量输入到模型即可执行风格迁移,代码如下。

```
outputs = hub_module(tf.constant(target_image),
                    tf.constant(style_image))
stylized_image = outputs[0]
```

输出是一个维度为 (1, 300, 400, 3) 的张量,这是我们进行风格迁移后的图像数据。

> **注意** 该图像的尺寸为 300×400,这是原始图像 1600×1200 的缩小版。

12.1.5 显示输出

为了显示图像,使用以下代码将张量转换为图像。

```
tensor = stylized_image*256
tensor = np.array(tensor, dtype=np.uint8)
tensor = tensor[0]
PIL.Image.fromarray(tensor)
```

由于该图像的数值范围为 0 ~ 1,因此我们需要将图像乘以 256 以方便显示。输出图像如图 12-4 所示。

如果要显示如图 12-5 所示的缩小图像,只需调用 imshow 方法:

```
plt.imshow(tensor)
```

12.1.6 更多结果

对项目中加载的图像进行更多的风格迁移,结果如图 12-6 所示。

图 12-4 风格化图像

图 12-5 缩小尺度的风格化图像

图 12-6 更多的风格迁移

12.1.7 项目源码

TFHubStyleTransfer 项目的完整源码如清单 12-2 所示。

清单 12-2 TFHubStyleTransfer 的完整源码

源码清单
链 接：https://pan.baidu.com/s/1NV0rimQ_
8kRz22xfFHN-Cw
提取码：1218

```python
import tensorflow as tf
import re
import urllib
import numpy as np
import matplotlib.pyplot as plt
import PIL.Image
import tensorflow_hub as hub
from tensorflow.keras.preprocessing.image
import load_img, img_to_array
from matplotlib import gridspec
from IPython import display
from PIL import Image

def download_image_from_URL(imageURL):
    imageName = re.search('[a-z0-9\-]+\.'
                          '(jpe?g|png|gif|bmp|JPG)',
                          imageURL, re.IGNORECASE)
    imageName = imageName.group(0)
    urllib.request.urlretrieve(imageURL, imageName)
    imagePath = "./" + imageName
    return imagePath

# This is the path to the image you want to transform.
target_url = "https://raw.githubusercontent.com/Apress/
artificial-neural-networks-with-tensorflow-2/main/ch12/ferns.jpg"
target_path = download_image_from_URL(target_url)
# This is the path to the style image.
style_url = "https://raw.githubusercontent.com/Apress/
artificial-neural-networks-with-tensorflow-2/main/ch12/on-the-
road.jpg"
style_path = download_image_from_URL(style_url)

content = Image.open(target_path)
style = Image.open(style_path)

plt.figure(figsize=(10, 10))

plt.subplot(1, 2, 1)
plt.imshow(content)
plt.title('Content Image')

plt.subplot(1, 2, 2)
plt.imshow(style)
plt.title('Style Image')
```

```python
plt.tight_layout()
plt.show()

def image_to_tensor_style(path_to_img):
    img = tf.io.read_file(path_to_img)
    img = tf.image.decode_image
            (img, channels=3, dtype=tf.float32)
    img = tf.image.resize(img, [256,256])
    img = img[tf.newaxis, :]
    return img

def image_to_tensor_target(path_to_img, image_size):
    img = tf.io.read_file(path_to_img)
    img = tf.image.decode_image
            (img, channels=3, dtype=tf.float32)
    img = tf.image.resize(img,
                    [image_size,image_size],
                    preserve_aspect_ratio=True)
    img = img[tf.newaxis, :]
    return img

output_image_size = 400

target_image = image_to_tensor_target
                    (target_path,output_image_size)
style_image = image_to_tensor_style(style_path)

hub_module = hub.load('https://tfhub.dev/google/magenta/arbitrary-image-stylization-v1-256/2')

outputs = hub_module(tf.constant(target_image),
                    tf.constant(style_image))
stylized_image = outputs[0]

tensor = stylized_image*256
tensor = np.array(tensor, dtype=np.uint8)
tensor = tensor[0]
PIL.Image.fromarray(tensor)

plt.imshow(tensor)
```

12.2 自定义风格迁移

学习了如何进行快速风格迁移后，我们将介绍风格迁移背后的技术原理。风格迁移是指提取一幅图像的风格（如一幅名画），并将其应用于另一幅图像的内容中。因此，风格迁移需要两幅输入图像，即内容图像和风格图像，新生成的图像一般称为风格化图像。生成的图像包含与内容图像相同的内容，但获得了与风格图像相似的风格。显然，这不是通过简单地叠加图像来完成的。因此，我们的程序必须能够区分给定图像的内容和风格。我们将使用 VGG16 预训练模型来提取上述信息并构建自定义的神经网络，以根据输入图像创建风格化图像。一些安卓应用程序如 Prisma 和 Lucid 会进行这种风格迁移，虽然本项目不会介绍开发类似的安卓应用程序，但将介绍此类应用程序的内部结构。

首先，我们介绍 VGG16 结构，以理解如何提取一幅图像的内容和风格。

12.2.1　VGG16 结构

Gatys 等人 (2015) 提出了风格迁移的核心思想，该核心思想认为用于图像分类的预训练卷积神经网络结构（convolutional neural network, CNN）可以对图像的感知和语义信息进行编码。目前有很多这样的预训练 CNN 结构，我们将使用 VGG16 模型来提取图像特征，然后分别处理其内容和风格。Simonyan 和 Zisserman(2015) 的原始论文使用了 19 层 VGG 网络模型。具有 16 层的 VGG16 模型架构如图 12-7 所示。

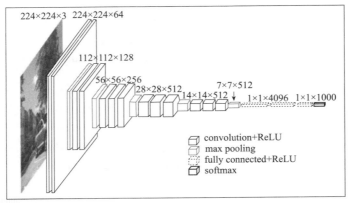

图 12-7　VGG16 架构

由于我们不需要进行图像分类，并且只对特征提取感兴趣，因此不需要 VGG 模型的全连接层及最后的 softmax 分类器，而只需要模型的一部分。那么，如何只提取模型的特定部分呢？幸运的是，这是一项非常简单的任务，因为 Keras 提供了一个预训练的 VGG16 模型，我们可以直接分离其中的某些层。Keras 还提供了包括 VGG19 在内的许多其他模型。为了移除最顶层的全连接层，我们需要在提取模型层时将 include_top 变量的值设置为 False。

12.2.2　创建项目

创建一个新的 Colab 项目并将其重命名为 CustomStyleTransfer。安装以下两个包：

```
!pip install keras==2.3.1
!pip install tensorflow==2.1.0
```

在撰写本书时，预训练的 VGG16 模型需要使用 Keras 和 tensorflow 版本一起运行，尚不支持较新的版本。

导入所需的库，代码如下。

```
import tensorflow as tf
import re
import urllib
from tensorflow.keras.preprocessing.image
import load_img, img_to_array
from matplotlib import pyplot as plt
from IPython import display
from PIL import Image
import numpy as np
from tensorflow.keras.applications import vgg16
from tensorflow.keras import backend as K
from keras import backend as K
from scipy.optimize import fmin_l_bfgs_b
```

12.2.3 下载图像

与上一个项目相同，我们编写一个下载函数用来下载项目所需的图像，代码如清单12-3所示。

清单 12-3 下载函数

源码清单
链　接：https://pan.baidu.com/s/1NV0rimQ_8kRz22xfFHN-Cw
提取码：1218

```
def download_image_from_URL(imageURL):
    imageName = re.search
            ('[a-z0-9\-]+\.(jpe?g|png|gif|bmp|JPG)',
            ImageURL, re.IGNORECASE)
    imageName = imageName.group(0)
    urllib.request.urlretrieve(imageURL, imageName)
    imagePath = "./" + imageName
    return imagePath

# This is the path to the image you want to transform.
target_url = "https://raw.githubusercontent.com/Apress/
artificial-neural-networks-with-tensorflow-2/main/ch12/blank-
sign.jpg"
target_path = download_image_from_URL(target_url)
# This is the path to the style image.
style_url = "https://raw.githubusercontent.com/Apress/
artificial-neural-networks-with-tensorflow-2/main/ch12/road.jpg"
```

将目标图像的高度缩放为400像素，为了保持图像纵横比，重新计算图像宽度，代码如下。

```
width, height = load_img(target_path).size
img_height = 400
img_width = int(width * img_height / height)
```

12.2.4 显示图像

为了显示内容图像和风格图像，我们使用和上一个项目类似的代码，如清单 12-4 所示。

清单 12-4 显示内容图像和风格图像

```
content = Image.open(target_path)
style = Image.open(style_path)

plt.figure(figsize=(10, 10))

plt.subplot(1, 2, 1)
plt.imshow(content)
plt.title('Content Image')

plt.subplot(1, 2, 2)
plt.imshow(style)
plt.title('Style Image')

plt.tight_layout()
plt.show()
```

源码清单
链 接：https://pan.baidu.com/s/1NV0rimQ_8kRz22xfFHN-Cw
提取码：1218

输出结果如图 12-8 所示。

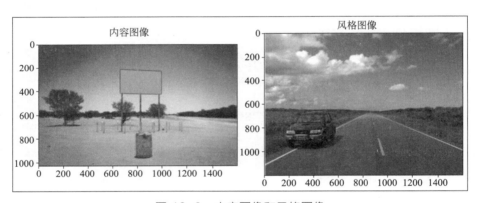

图 12-8　内容图像和风格图像

12.2.5 图像预处理

如前所述，我们将使用 VGG16 模型提取图像特征，因此需要按照 VGG 的训练过程处理图像数据。幸运的是，Keras 不仅为 VGG16 模型提供了图像预处理程序，还为许多其他流行模型提供了数据预处理程序，如 ResNet、Inception、DenseNet 等。Keras 库提供了一个名为 preprocess_input 的函数，该函数将批量图像的张量或 NumPy 数组作为输入，并返回一个预处理过的 numpy 数组或张量，输出数据类型为 float 32。该函数还将图像从 RGB（red、green、blue）通道转换为 BGR (blue、green、red) 通道，并将每个通道置为零均值。

> **注意**
>
> VGG 模型需要使用特定格式的图像进行训练，这些图像的色彩通道顺序需要为 BGR，并且每个通道均按均值 [103.939, 116.779, 123.68] 进行归一化。上述对图像进行预处理的过程如清单 12-5 所示。

清单 12-5 为 VGG16 模型准备预处理图像

源码清单
链 接：https://pan.baidu.com/s/1NV0rimQ_8kRz22xfFHN-Cw
提取码：1218

```
def preprocess_image(image_path):
    img = load_img(image_path,
        target_size=(img_height, img_width))
    img = img_to_array(img)
    img = np.expand_dims(img, axis=0)
    img =
    tf.keras.applications.vgg16.preprocess_input(img)
    return img
```

查看输出数据时，我们需要进行反向预处理。此外，还需要将 0 ～ 255 范围内的所有数值裁剪出来。使用以下函数执行上述操作。

```
def deprocess_image(x):
    # Remove zero-center by mean pixel
    x[:, :, 0] += 103.939
    x[:, :, 1] += 116.779
    x[:, :, 2] += 123.68
    # 'BGR'->'RGB'
    x = x[:, :, ::-1]
    x = np.clip(x, 0, 255).astype('uint8')
    return x
```

接下来，将基于 VGG16 预训练模型构建风格迁移模型。

12.2.6 构建模型

为了构建风格迁移模型，将图像张量数据输入 VGG16 模型，并提取特征图、内容和风格等特征表达。该模型将加载在 ImageNet 数据集上预训练的权重。模型构建代码如下。

```
target = K.constant(preprocess_image(target_path))
style = K.constant(preprocess_image(style_path))

# This placeholder will contain our generated image
combination_image = K.placeholder
                    ((1, img_height, img_width, 3))

# We combine the 3 images into a single batch
```

```
input_tensor = K.concatenate([target,
                              style,
                              combination_image],
                             axis=0)

# Build the VGG16 network with our batch of 3 images as input.
model = vgg16.VGG16(input_tensor=input_tensor,
                    weights='imagenet',
                    include_top=False)
```

在这段代码中，我们首先调用之前定义的预处理方法为内容图像和风格图像构建输入张量；然后为风格化图像创建一个占位符（placeholder），调用 concatenate 方法为三幅图像合并一个张量；最后将合并后的图像输入 VGG16 模型中获取想要的风格迁移模型。

通过如下命令查看模型的摘要信息。

```
model.summary()
```

输出结果如图 12-9 所示。

```
Model: "vgg16"
Layer (type)                 Output Shape              Param #
=================================================================
input_4 (InputLayer)         [(3, 400, 533, 3)]        0
block1_conv1 (Conv2D)        (3, 400, 533, 64)         1792
block1_conv2 (Conv2D)        (3, 400, 533, 64)         36928
block1_pool (MaxPooling2D)   (3, 200, 266, 64)         0
block2_conv1 (Conv2D)        (3, 200, 266, 128)        73856
block2_conv2 (Conv2D)        (3, 200, 266, 128)        147584
block2_pool (MaxPooling2D)   (3, 100, 133, 128)        0
block3_conv1 (Conv2D)        (3, 100, 133, 256)        295168
block3_conv2 (Conv2D)        (3, 100, 133, 256)        590080
block3_conv3 (Conv2D)        (3, 100, 133, 256)        590080
block3_pool (MaxPooling2D)   (3, 50, 66, 256)          0
block4_conv1 (Conv2D)        (3, 50, 66, 512)          1180160
block4_conv2 (Conv2D)        (3, 50, 66, 512)          2359808
block4_conv3 (Conv2D)        (3, 50, 66, 512)          2359808
block4_pool (MaxPooling2D)   (3, 25, 33, 512)          0
block5_conv1 (Conv2D)        (3, 25, 33, 512)          2359808
block5_conv2 (Conv2D)        (3, 25, 33, 512)          2359808
block5_conv3 (Conv2D)        (3, 25, 33, 512)          2359808
block5_pool (MaxPooling2D)   (3, 12, 16, 512)          0
=================================================================
Total params: 14,714,688
Trainable params: 14,714,688
Non-trainable params: 0
```

图 12-9　模型摘要信息

12.2.7 内容损失

我们将在一些网络层中计算内容损失,并将所有损失值相加。在每次迭代,我们将图像输入模型,该模型计算出所有内容损失,且由于 Eager 执行,所有梯度也会被计算出来。内容损失表示随机生成的噪声图像与内容图像之间的相似程度,其计算过程如下。

假设我们在预训练的 VGG 模型中选择一个隐藏层(L)来计算损失。令 P 和 F 代表原始内容图像和生成的风格化图像,F[l] 和 P[l] 分别为第 L 层各图像的特征表达,则内容损失的定义如下。

$$L_{\text{content}}\left(\vec{P},\vec{X},l\right) = \frac{1}{2}\sum_{ij}\left(F_{ij}^{l} - P_{ij}^{l}\right)^{2}$$

使用如下代码计算内容损失。

```
def content_loss(base, combination):
    return K.sum(K.square(combination - base))
```

12.2.8 风格损失

为了计算风格损失,首先需要计算格拉姆矩阵(Gram matrix)。计算格拉姆矩阵是一个额外的预处理步骤,为了找到不同通道之间的相关性,用于衡量学习到的风格信息。

使用如下代码计算格拉姆矩阵。

```
def gram_matrix(x):
    features = K.batch_flatten
                (K.permute_dimensions(x, (2, 0, 1)))
    gram = K.dot(features, K.transpose(features))
    return gram
```

风格损失计算风格图像和生成的风格化图像的格拉姆矩阵,然后将损失返回给调用者。损失为风格图像的格拉姆矩阵与风格化图像的格拉姆矩阵差的平方。在数学上,可表示为:

$$L_{\text{GM}}(S,G,l) = \frac{1}{4N_l^2 M_l^2}\sum_{ij}\left(\text{GM}[l](S)_{ij} - \text{GM}[l](G)_{ij}\right)^{2}$$

风格损失函数的定义如下。

```
def style_loss(style, combination):
    S = gram_matrix(style)
    C = gram_matrix(combination)
    channels = 3
    size = img_height * img_width
    return K.sum(K.square(S - C)) /
            (4. * (channels ** 2) * (size ** 2))
```

12.2.9 全变分损失

为了对生成的风格化图像的平滑度进行约束,我们定义了全变分(total variation)损失,代码如下。

```
def total_variation_loss(x):
    a = K.square(
        x[:, :img_height - 1, :img_width - 1,
            :] - x[:, 1:, :img_width - 1, :])
    b = K.square(
        x[:, :img_height - 1, :img_width - 1,
            :] - x[:, :img_height - 1, 1:, :])
    return K.sum(K.pow(a + b, 1.25))
```

12.2.10 计算内容和风格损失

我们首先选择用于计算损失的 VGG16 内容和风格层。这里使用了 Johnson 等 (2016) 定义的网络层，而不是 Gatys 等人定义的网络层，因为前者可以产生更好的结果。

将所有层映射到字典中：

```
# Dict mapping layer names to activation tensors
outputs_dict = dict([(layer.name, layer.output)
                    for layer in model.layers])
```

使用如下代码提取出内容层。

```
# Name of layer used for content loss
content_layer = 'block5_conv2'
```

使用如下代码提取出风格层。

```
# Name of layers used for style loss;
style_layers = ['block1_conv1',
                'block2_conv1',
                'block3_conv1',
                'block4_conv1',
                'block5_conv1']
```

定义一些权重变量，用于计算不同损失分量的加权平均值。这些权重视为风格层和内容层的超参数，它们决定了在最终模型中赋予每一层的权值代码如下。

```
total_variation_weight = 1e-4
style_weight = 10.
content_weight = 0.025
```

将所有损失加在一起以计算整体损失，代码如下。

```
# Define the loss by adding all components to a `loss` variable
loss = K.variable(0.)
layer_features = outputs_dict[content_layer]
target_features = layer_features[0, :, :, :]
combination_features = layer_features[2, :, :, :]
loss = loss + content_weight *
            content_loss(target_features,
                        combination_features)
for layer_name in style_layers:
    layer_features = outputs_dict[layer_name]
```

```
            style_reference_features = layer_features
                                       [1, :, :, :]
            combination_features = layer_features
                                   [2, :, :, :]
            sl = style_loss(style_reference_features,
                            combination_features)
            loss += (style_weight / len(style_layers)) * sl
        loss += total_variation_weight *
                total_variation_loss(combination_image)
```

12.2.11 Evaluator 类

最后，定义一个名为 Evaluator 的类来计算一次传播的损失和梯度：

```
grads = K.gradients(loss, combination_image)[0]
# Function to fetch the values of the current loss and the
current gradients
fetch_loss_and_grads =
                K.function([combination_image],
                           [loss, grads])
class Evaluator(object):

    def __init__(self):
        self.loss_value = None
        self.grads_values = None

    def loss(self, x):
        assert self.loss_value is None
        x = x.reshape((1, img_height, img_width, 3))
        outs = fetch_loss_and_grads([x])
        loss_value = outs[0]
        grad_values =
                outs[1].flatten().astype('float64')
        self.loss_value = loss_value
        self.grad_values = grad_values
        return self.loss_value

    def grads(self, x):
        assert self.loss_value is not None
        grad_values = np.copy(self.grad_values)
        self.loss_value = None
        self.grad_values = None
        return grad_values

evaluator = Evaluator()
```

12.2.12 生成输出图像

我们已经准备好了所有的相关函数，接下来将生成一幅风格化图像。从像素的随机组合（一幅随机图像）开始，并使用 L-BFGS（limited memory broyden- fletchergoldfarbshanno)算法进行优化。该算法使用二阶导数来最小化或最大化函数,并且明显快于标准梯度下降法。训练循环如下。

```
iterations = 50

x = preprocess_image(target_path)
x = x.flatten()
for i in range(1, iterations):
    x, min_val, info = fmin_l_bfgs_b
                       (evaluator.loss,
                        x,
                        fprime=evaluator.grads,
                                    maxfun=10)
    print('Iteration %0d, loss: %0.02f' %
                        (i, min_val))
img = x.copy().reshape((img_height, img_width, 3))
img = deprocess_image(img)
```

训练结束后,将最终输出图像复制到一个变量中并对其进行重新处理以显示图像。

12.2.13 显示图像

使用如下代码显示三幅图像。

```
plt.figure(figsize=(50, 50))

plt.subplot(3,3,1)
plt.imshow(load_img(target_path, target_size=(img_height,
                                              img_width)))

plt.subplot(3,3,2)
plt.imshow(load_img(style_path, target_size=(img_height,
                                             img_width)))

plt.subplot(3,3,3)
plt.imshow(img)

plt.show()
```

输出结果如图 12-10 所示。

图 12-10　内容图像、风格图像及风格化图像

12.2.14 项目源码

CustomStyleTransfer 项目的完整源码如清单 12-6 所示。

清单 12-6 CustomStyleTransfer 完整源码

源码清单
链　接：https://pan.baidu.com/s/1NV0rimQ_8kRz22xfFHN-Cw
提取码：1218

```python
!pip install keras==2.3.1
!pip install tensorflow==2.1.0

import tensorflow as tf
import re
import urllib
from tensorflow.keras.preprocessing.image
import load_img, img_to_array
from matplotlib import pyplot as plt
from IPython import display
from PIL import Image
import numpy as np
from tensorflow.keras.applications import vgg16
from tensorflow.keras import backend as K
from keras import backend as K
from scipy.optimize import fmin_l_bfgs_b

def download_image_from_URL(imageURL):
    imageName = re.search(
            ('[a-z0-9\-]+\.(jpe?g|png|gif|bmp|JPG)',
            imageURL, re.IGNORECASE)
    imageName = imageName.group(0)
    urllib.request.urlretrieve(imageURL, imageName)
    imagePath = "./" + imageName
    return imagePath

# This is the path to the image you want to transform.
target_url = "https://raw.githubusercontent.com/Apress/artificial-neural-networks-with-tensorflow-2/main/ch12/blank-sign.jpg"
target_path = download_image_from_URL(target_url)
# This is the path to the style image.
style_url = "https://raw.githubusercontent.com/Apress/artificial-neural-networks-with-tensorflow-2/main/ch12/road.jpg"
style_path = download_image_from_URL(style_url)

# Dimensions for the generated picture.
width, height = load_img(target_path).size
img_height = 400
img_width = int(width * img_height / height)

content = Image.open(target_path)
style = Image.open(style_path)

plt.figure(figsize=(10, 10))
```

```python
plt.subplot(1, 2, 1)
plt.imshow(content)
plt.title('Content Image')

plt.subplot(1, 2, 2)
plt.imshow(style)
plt.title('Style Image')

plt.tight_layout()
plt.show()

# Preprocess the data as per VGG16 requirements
def preprocess_image(image_path):
    img = load_img(image_path,
                target_size=(img_height, img_width))
    img = img_to_array(img)
    img = np.expand_dims(img, axis=0)
    img = 
  tf.keras.applications.vgg16.preprocess_input(img)
    return img

def deprocess_image(x):
    # Remove zero-center by mean pixel
    x[:, :, 0] += 103.939
    x[:, :, 1] += 116.779
    x[:, :, 2] += 123.68
    # 'BGR'->'RGB'
    x = x[:, :, ::-1]
    x = np.clip(x, 0, 255).astype('uint8')
    return x

target = K.constant(preprocess_image(target_path))
style = K.constant(preprocess_image(style_path))

# This placeholder will contain our generated image
combination_image = K.placeholder
                    ((1, img_height, img_width, 3))
# We combine the 3 images into a single batch
input_tensor = K.concatenate([target, style, combination_
                            image], axis=0)

# Build the VGG16 network with our batch of 3 images as input.
model = vgg16.VGG16(input_tensor=input_tensor,
                weights='imagenet',
                include_top=False)

model.summary()

# compute content loss for the generated image
def content_loss(base, combination):
```

```python
        return K.sum(K.square(combination - base))

def gram_matrix(x):
    features = K.batch_flatten
              (K.permute_dimensions(x, (2, 0, 1)))
    gram = K.dot(features, K.transpose(features))
    return gram

def style_loss(style, combination):
    S = gram_matrix(style)
    C = gram_matrix(combination)
    channels = 3
    size = img_height * img_width
    return K.sum(K.square(S - C)) /
           (4. * (channels ** 2) * (size ** 2))

def total_variation_loss(x):
    a = K.square(
        x[:, :img_height - 1, :img_width - 1,
            :] - x[:, 1:, :img_width - 1, :])
    b = K.square(
        x[:, :img_height - 1, :img_width - 1,
            :] - x[:, :img_height - 1, 1:, :])
    return K.sum(K.pow(a + b, 1.25))

# Dict mapping layer names to activation tensors
outputs_dict = dict([(layer.name, layer.output)
                    for layer in model.layers])

# Name of layer used for content loss
content_layer = 'block5_conv2'

# Name of layers used for style loss;
style_layers = ['block1_conv1',
                'block2_conv1',
                'block3_conv1',
                'block4_conv1',
                'block5_conv1']

# Weights in the weighted average of the loss components
total_variation_weight = 1e-4
style_weight = 10.
content_weight = 0.025

# Define the loss by adding all components to a `loss` variable
loss = K.variable(0.)
layer_features = outputs_dict[content_layer]
target_features = layer_features[0, :, :, :]
combination_features = layer_features[2, :, :, :]
loss = loss + content_weight *
```

```python
                content_loss(target_features,
                        combination_features)
for layer_name in style_layers:
    layer_features = outputs_dict[layer_name]
    style_reference_features = layer_features
                                [1, :, :, :]
    combination_features = layer_features
                                [2, :, :, :]
    sl = style_loss(style_reference_features,
                    combination_features)
    loss += (style_weight / len(style_layers)) * sl
    loss += total_variation_weight *
        total_variation_loss(combination_image)

grads = K.gradients(loss, combination_image)[0]

# Function to fetch the values of the current loss and the
current gradients
fetch_loss_and_grads =
                K.function([combination_image],
                            [loss, grads])

class Evaluator(object):

    def __init__(self):
        self.loss_value = None
        self.grads_values = None

    def loss(self, x):
        assert self.loss_value is None
        x = x.reshape((1, img_height, img_width, 3))
        outs = fetch_loss_and_grads([x])
        loss_value = outs[0]
        grad_values =
            outs[1].flatten().astype('float64')
        self.loss_value = loss_value
        self.grad_values = grad_values

        return self.loss_value

    def grads(self, x):
        assert self.loss_value is not None
        grad_values = np.copy(self.grad_values)
        self.loss_value = None
        self.grad_values = None
        return grad_values

evaluator = Evaluator()

iterations = 50
```

```
x = preprocess_image(target_path)
x = x.flatten()
for i in range(1, iterations):
    x, min_val, info = fmin_l_bfgs_b
                        (evaluator.loss,
                         x,
                         fprime=evaluator.grads,
                                      maxfun=10)
    print('Iteration %0d, loss: %0.02f' %
                        (i, min_val))
img = x.copy().reshape((img_height, img_width, 3))
img = deprocess_image(img)

plt.figure(figsize=(50, 50))

plt.subplot(3,3,1)
plt.imshow(load_img(target_path, target_size=(img_height,
                                              img_width)))

plt.subplot(3,3,2)
plt.imshow(load_img(style_path, target_size=(img_height,
                                             img_width)))

plt.subplot(3,3,3)
plt.imshow(img)

plt.show()
```

总结

本章介绍了神经网络中的另一项重要技术：风格迁移。该技术允许我们将所选图像的内容的风格转换为另一幅图像的风格。我们学习了以两种不同的方式进行风格迁移。第一种方法使用 TensorFlow Hub 中提供的预训练模型来执行快速风格迁移。使用这种方法进行风格迁移很快，而且做得很好。第二种方法是构建自定义的神经网络模型来进行风格迁移。我们使用了预训练的 VGG16 图像分类模型来提取图像的内容和风格。然后创建了一个模型，该模型在多次迭代中学会了将特性风格应用于给定的内容图像。这种方法允许我们在自己的实验上进行风格迁移。

下一章，将介绍如何使用生成对抗网络（generative adversarial networks, GAN）生成一幅图像。

第 13 章
图像生成

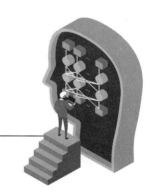

您有没有想过神经网络可以用来生成复杂的彩色图像？如动漫海报、明星宣传画、生活照等。这是不是听起来很有趣？所有这些都可以通过神经网络中最有趣的设计实现，即生成对抗网络（generative adversarial networks，GAN）。这个想法是由 Ian J. Goodfellow 在 2014 年提出的。通过 GAN 创建的图像看起来非常真实，以至于几乎无法区分假图像和真实图像。本章将讲解 GAN 的工作原理。

13.1 GAN（生成对抗网络）

在 GAN 中，有两个神经网络模型通过对抗过程同时训练。一个模型称为生成器，另一个称为判别器。生成器可以理解为艺术家，创建看起来真实的图像。判别器可以理解为评论家，学会区分真实图像和假图像。因此，上述两个模型相互竞争，试图相互击败。通过训练，将实现生成器超越判别器作为最终目标。

13.2 GAN 如何工作

如上所说，GAN 由两个模型组成，训练 GAN 需要两步。

① 保持生成器不变，训练判别器。在真实图像上训练多轮判别器，判断它是否可以正确地将真实图像预测为真实的。在同一训练阶段，使用由生成器生成的假图像训练判别器，判断它是否可以将生成图像预测为假图像。

② 保持判别器不变，训练生成器。使用判别器对假图像的预测结果来改进生成器的输出图像。

将上述步骤重复多次，并手动检查结果以查看它们是否接近真实图像。如果人工已经无法分别图像真假，即可停止训练。如果不是，继续前两个步骤，直到假图像看起来接近真实图像。整个过程如图 13-1 所示。

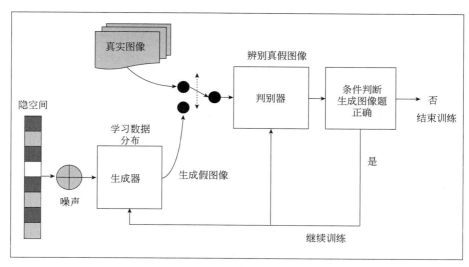

图 13-1 GAN 的工作流程

13.3 生成器

生成器的工作原理如图 13-2 所示。

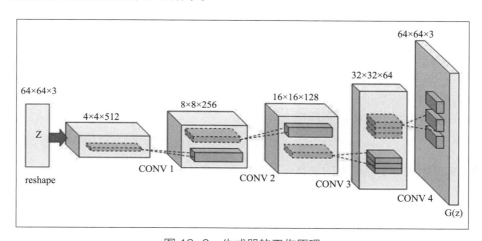

图 13-2 生成器的工作原理

生成器首先生成某个随机噪声向量，在示例中使用维度为 100 的噪声向量。从这个随机向量中，生成一个 64×64×3 的图像。该图像通过卷积层的一系列计算被放大。每个卷积层之后是批量归一化和 Leaky ReLU 激活层，并在每个卷积层中使用 strides 避免训练不稳定。最终通过若干个卷积层后，图像被放大到 64×64×3。

13.4 判别器

判别器的工作原理如图 13-3 所示。

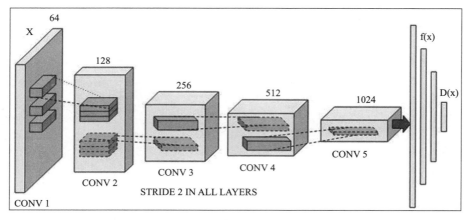

图 13-3 判别器的工作原理

判别器同意使用卷积层缩小给定图像的大小，并进行评估。

13.5 数学公式

GAN 的工作可以用如下简单的数学方程表示。

$$\min_{G} \max_{D} V(D,G) = \min_{G} \max_{D} E_{z \sim P_{\text{data}}(x)} \log D(x) + E_{z \sim P_{z}(z)} \left[\log \left(1 - D(G(z))\right) \right]$$

其中，G 代表生成器，D 代表判别器。data(x) 表示真实数据的分布，$p_z(z)$ 表示生成或假数据的分布。x 代表来自真实数据的样本，z 代表来自生成数据的样本。$D(x)$ 代表判别器模型，$G(z)$ 代表生成器模型。

在真实数据上训练时的判别器损失表示为：

$$L(D(x),1) = \log(D(x)) \tag{13.1}$$

训练来自生成器的假数据时的判别器损失表示为：

$$L(D(G(z)),0) = \log(1 - D(G(z))) \tag{13.2}$$

对于真实数据，判别器的预测结果应该接近 1。因为 log 函数是一个递增函数，因此，应最大化方程 1 以使 $D(x)$ 接近 1。式（13.1）是真实数据上的判别器损失，应该最大化以使 $D(G(z))$ 接近 1。因为式（13.2）是假数据上的判别器损失，所以也应该最大化。

式（13.2）中，对生成的假数据的判别结果应该接近于 0。为了最大化式（13.2），必须将 $D(G(z))$ 的值最小化为 0。因此，需要最大化判别器的两个损失，判别器的总损失是式（13.1）和式（13.2）给出的两个损失的和。最终组合总损失也将最大化。

生成器的损失表示为：

$$L^{(G)} = \min \left[\log(D(x)) + \log(1 - D(G(z))) \right] \tag{13.3}$$

训练生成器时，需要将上述损失最小化。

13.6 数字生成

在本节项目中，将使用 Kaggle 提供的 MNIST 数据集。如之前介绍，该数据集由手写数字图像组成。本节将创建一个 GAN 模型用于生成与真实数字图像看起来相同的假图像。此过程有助于在未来项目中进行训练数据集扩增。

13.6.1 创建项目

创建一个 Colab 项目并将其重命名为 DigitGen-GAN。导入所需的库，代码如下。

```
import matplotlib.pyplot as plt
import tensorflow as tf
import numpy as np
import time
from tensorflow import keras
from tensorflow.keras import layers
import os
```

13.6.2 加载数据集

可以使用以下语句将数据集加载到代码中。

```
(train_images,train_labels),
    (test_images,test_labels) =
        tf.keras.datasets.mnist.load_data()
```

训练和测试数据集被加载到单独的 numpy 数组中。由于本节项目生成单个数字，如 9，因此，从训练数据集中提取包含 9 的所有图像。

```
digit9_images = []
for i in range(len(train_images)):
    if train_labels[i] == 9:
        digit9_images.append(train_images[i])
train_images = np.array(digit9_images)

train_images.shape
```

train_image 变量形状是 (5949, 28, 28)，表示有 5949 张大小为 28×28 的图像。这是一个巨大的数据库。本节模型将尝试生成这种尺寸的图像，并与真实图像的外观相匹配。

使用以下代码在终端上打印一些图像来验证变量值是否只有数字 9 的图像。

```
n = 10
f = plt.figure()
for i in range(n):
    f.add_subplot(1, n, i + 1)
    plt.subplot(1, n, i+1 ).axis("off")
    plt.imshow(train_images[i])
plt.show()
```

上述代码输出结果如图 13-4 所示。

图 13-4　图像示例

接下来，准备用于训练的数据集。

13.6.3　准备数据集

首先，使用以下语句进行数据重采样。

> 本节项目数据集中的图像大小为 28×28 像素。

```
train_images = train_images.reshape (
    train_images.shape[0], 28, 28, 1).astype
                    ('float32')
```

由于图像中的每个颜色范围为 0～256，将颜色范围标准化为 –1～1 之间，以便模型更好地学习样本分布。原始图像像素均值为 127.5，使用以下等式将对 –1～1 范围内的像素值进行归一化。

```
train_images = (train_images - 127.5) / 127.5
```

通过调用 from_tensor_slices 方法创建用于训练的批处理数据集：

```
train_dataset = tf.data.Dataset.from_tensor_slices(
    train_images).shuffle
        (train_images.shape[0]).batch(32)
```

接下来是本项目的重要部分，即定义生成器模型。

13.6.4　定义生成器模型

生成器的目的是创建包含数字 9 的图像，并使生成图像看起来与训练数据集中的真实图像相似。使用 Keras 序列模型来创建生成器：

```
gen_model = tf.keras.Sequential()
```

添加 Keras Dense 层作为第一层，或者可以将 Conv2D 层作为第一层：

```
gen_model.add(tf.keras.layers.Dense(7*7*256, use_bias=False,

input_shape=(100,)))
```

因为稍后将使用维度为 100 的噪声向量作为此 GAN 模型的输入，因此将此层的输入指定为 100。从 7×7 的图像尺寸开始，将其放大到最终目标尺寸 28×28。z 维度指定用于生成图像的滤波器，初始为 256 通道，最终将图像转换为 3 通道。

接下来，为模型添加一个批量归一化层以维持摩擦的稳定性。

```
gen_model.add(tf.keras.layers.BatchNormalization())
gen_model.add(tf.keras.layers.LeakyReLU())
```

使用 Leaky ReLu 为激活函数，并将输出结果调整为 7×7×256：

```
gen_model.add(tf.keras.layers.Reshape((7, 7, 256)))
```

现在使用 Conv2D 层对生成图像进行放大：

```
gen_model.add(tf.keras.layers.Conv2DTranspose
                (128, (5, 5),
                 Strides=(1, 1),
                 padding='same',
                 use_bias=False))
```

第一个参数是输出维度，即卷积中输出滤波器的数量。第二个参数 kernel_size 指定了卷积滤波器的高度和宽度。第三个参数指定卷积计算沿高度和宽度的步长。步长可理解为在图像中从左到右，从上到下移动一个过滤器，一次移动 1 个像素。上述过程步长为 (1, 1)。步长为 (2, 2) 时，过滤器在每侧移动 2 个像素，将图像放大 2×2。由于步长指定为 (1, 1)，因此该层的输出尺寸将与其输入相同，即大小为 7×7 的图像。Padding 参数用于确保输出尺寸不变。use_bias 参数设置为 False 值表示该层不使用偏置向量。该层之后是批量归一化和激活层，代码如下。

```
gen_model.add(tf.keras.layers.BatchNormalization())
gen_model.add(tf.keras.layers.LeakyReLU())
```

接下来，添加步长设置为 (2, 2) 的 Conv2DTranspose 层及批量归一化和激活层。代码如下。

```
gen_model.add(tf.keras.layers.Conv2DTranspose
                (64, (5, 5),
                 strides=(2, 2),
                 padding='same',
                 use_bias=False))
gen_model.add(tf.keras.layers.BatchNormalization())
gen_model.add(tf.keras.layers.LeakyReLU())
```

此时，输出图像大小为 14×14。

接下来，添加步长为 (2, 2) 的最后一个 Conv2D 层，从而进一步将图像放大到 28×28，即最终想要的图像尺寸。代码如下。

```
gen_model.add(tf.keras.layers.Conv2DTranspose
                (1, (5, 5),
                 strides=(2, 2),
                 padding='same',
                 use_bias=False,
                 activation='tanh'))
```

最后一层使用 tanh 激活函数，map 参数值为 1，为模型提供单个输出图像。
使用 plot 函数生成的模型图如图 13-5 所示。
模型概要如图 13-6 所示。

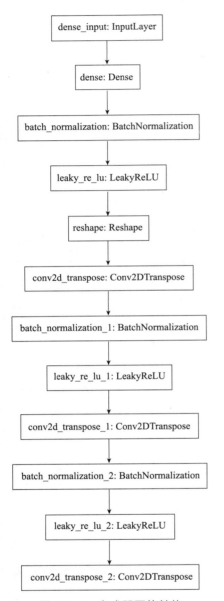

图 13-5　生成器网络结构　　　　图 13-6　生成器模型概要

13.6.5　测试生成器

使用随机向量测试生成器，并使用以下代码显示生成器输出结果。

```
noise = tf.random.normal([1, 100])
#giving random input vector
```

```
generated_image = gen_model(noise, training=False)
plt.imshow(generated_image[0, :, :, 0], cmap='gray')
```

生成图像如图 13-7 所示。

使用如下代码检查图像尺寸。

```
generated_image.shape
```

图像尺寸为：

```
TensorShape([1, 28, 28, 1])
```

输出图像的尺寸为 28×28，符合本项目模型的要求。

接下来，将定义鉴别器。

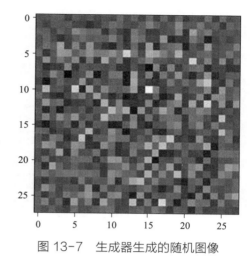

图 13-7　生成器生成的随机图像

13.6.6　定义判别器模型

使用如下语句定义判别器模型。

```
discri_model = tf.keras.Sequential()

discri_model.add(tf.keras.layers.Conv2D
                    (64, (5, 5),
                     strides=(2, 2),
                     padding='same',
                     input_shape=[28, 28, 1]))
discri_model.add(tf.keras.layers.LeakyReLU())
discri_model.add(tf.keras.layers.Dropout(0.3))

discri_model.add(tf.keras.layers.Conv2D
                    (128, (5, 5),
                     strides=(2, 2),
                     padding='same'))
discri_model.add(tf.keras.layers.LeakyReLU())
discri_model.add(tf.keras.layers.Dropout(0.3))

discri_model.add(tf.keras.layers.Flatten())
discri_model.add(tf.keras.layers.Dense(1))
```

判别器仅使用两个卷积层。最后一个卷积层的输出为"（批量大小、高度、宽度、过滤器）"。网络中的 Flatten 层将输出展平，并将其输入到网络的最后一个 Dense 层。

判别器模型图如图 13-8 所示。

判别器模型概要如图 13-9 所示。

> **注意**
>
> 判别器只有约 200000 个可训练参数。

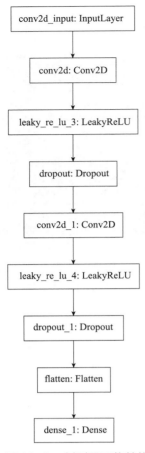

```
Model: "sequential_1"
_____
Layer (type)                 Output Shape              Param #
=================================================================
conv2d (Conv2D)              (None, 14, 14, 64)        1664
_____
leaky_re_lu_3 (LeakyReLU)    (None, 14, 14, 64)        0
_____
dropout (Dropout)            (None, 14, 14, 64)        0
_____
conv2d_1 (Conv2D)            (None, 7, 7, 128)         204928
_____
leaky_re_lu_4 (LeakyReLU)    (None, 7, 7, 128)         0
_____
dropout_1 (Dropout)          (None, 7, 7, 128)         0
_____
flatten (Flatten)            (None, 6272)              0
_____
dense_1 (Dense)              (None, 1)                 6273
=================================================================
Total params: 212,865
Trainable params: 212,865
Non-trainable params: 0
```

图 13-8　判别器网络结构　　　　图 13-9　判别器模型概要

13.6.7　测试判别器

使用之前生成的图像测试判别器：

`decision = discri_model(generated_image)`

如果图像是假的，判别器将给出一个负值；如果它是真实的，则给出正值。使用如下语句打印判别器对该图像的测试结果。

`print (decision)`

测试结果为：

`tf.Tensor([[0.0033829]], shape=(1, 1), dtype=float32)`

测试结果为正数 0.0033829，表示图像是真实的。如果测试结果为最大值 1，表示模型确定此图像是真实的。如果使用生成器生成的另一张图像测试判别器，可能会输出负数，表示判别为假。产生上述结果的原因是还没有使用真实数据集训练生成器和判别器。

13.6.8 定义损失函数

接下来，为生成器和判别器定义损失函数。代码如下。

```
cross_entropy = tf.keras.losses.BinaryCrossentropy
                            (from_logits=True)
```

使用 Keras 中的二元交叉熵函数作为损失函数。

> 本项目模型预测两个类别，真实图像表示为 1，假图像表示为 0。由此可见本项目研究的是二分类问题，因此使用二元交叉熵函数。

定义生成器损失函数，代码如下。

```
def generator_loss(generated_output):
    return cross_entropy(tf.ones_like
              (generated_output),generated_output)
```

该函数的返回值是对生成器欺骗判别器能力的量化。如果生成器性能很好，判别器将假图像分类为真实图像，并返回函数结果 1。

定义判别器损失函数，代码如下。

```
def discriminator_loss(real_output,
                        generated_output):
# compute loss considering the image is real [1,1,...,1]
    real_loss = cross_entropy(tf.ones_like
              (real_output),real_output)

# compute loss considering the image is fake[0,0,...,0]
    generated_loss = cross_entropy(tf.zeros_like
              (generated_output),
               generated_output)

# compute total loss
total_loss = real_loss + generated_loss

return total_loss
```

首先让判别器考虑给定的图像是真实的，计算一组损失函数；然后让判别器认为图像是假的，计算关于零数组的损失函数。判别器的总损失是这两个损失之和。

对生成器和判别器均使用 Adam 优化器进行优化：

```
gen_optimizer = tf.optimizers.Adam(1e-4)
discri_optimizer = tf.optimizers.Adam(1e-4)
```

接下来，编写一些在训练期间使用的函数。

13.6.9 定义新训练函数

首先，声明变量：

```
epoch_number = 0
EPOCHS = 100
noise_dim = 100
seed = tf.random.normal([1, noise_dim])
```

本项目将对模型进行 100 轮训练，可以根据时间需求更改此变量。更高数量的训练次数将生成更加逼真的图像。噪声维度设置为 100，用于为生成器的第一个输入创建随机图像。种子设置为一张图像的随机数据。

1. 设置 Checkpoint

由于训练可能需要很长时间，在训练中设置了 Checkpoing 功能，以便将生成器和判别器的中间状态保存到本地文件中。代码如下

```
checkpoint_dir ='/content/drive/My Drive/GAN1/Checkpoint'
checkpoint_prefix = os.path.join(checkpoint_dir, "ckpt")
checkpoint = tf.train.Checkpoint
            (generator_optimizer = gen_optimizer,
discriminator_optimizer=discri_optimizer,
             generator= gen_model,
             discriminator = discri_model)
```

如果服务器断开连接，可以从最后一个 checkpoint 处继续训练。

2. 设置驱动器

Checkpoint 的数据保存在 Google Drive 中名为"GAN1/Checkpoint"的文件夹中，因此，在运行代码之前，需确保已在 Google 云端硬盘中创建了此文件夹结构。

使用以下代码在本项目中安装驱动器。

```
from google.colab import drive
drive.mount('/content/drive')
```

将当前文件夹更改为新位置，以便 checkpoint 文件存储在该位置：

```
cd '/content/drive/My Drive/GAN1'
```

接下来，编写一个 gradient_tuning 函数进行模型训练。

3. 模型训练步骤

GAN 中的生成器模型和判别器模型都将分几个步骤进行训练，下面为训练步骤编写函数。使用梯度磁带 (tf.GradientTape) 对生成器和判别器进行自动微分，然后使用反向模式微分来计算新的梯度。

在每一步中，给函数提供一批图像作为输入。使用判别器为训练生成的图像预测输出。本项目将训练输出称为真实结果，将生成图像输出称为生成结果，分别计算生成器损失和判别器损失。完整的函数定义及每行的注释如清单 13-1 所示。

清单 13-1 梯度调整函数

源码清单

链　接：https://pan.baidu.com/s/1NV0rimQ_8kRz22xfFHN-Cw

提取码：1218

```python
def gradient_tuning(images):
    # create a noise vector.
    noise = tf.random.normal([16, noise_dim])

    # Use gradient tapes for automatic
    # differentiation
    with tf.GradientTape() 
            as generator_tape, tf.GradientTape() 
            as discriminator_tape:

        # ask genertor to generate random images
        generated_images = gen_model(noise, training=True)

        # ask discriminator to evalute the real images and
        generate its output
        real_output = discri_model(images,
                        training = True)
        # ask discriminator to do the evlaution on generated
        (fake) images
        fake_output = discri_model(generated_images,
                        training = True)
        # calculate generator loss on fake data
        gen_loss = generator_loss(fake_output)

        # calculate discriminator loss as defined earlier
        disc_loss = discriminator_loss(real_output,
                        fake_output)

    # calculate gradients for generator
    gen_gradients = generator_tape.gradient
            (gen_loss, gen_model.trainable_variables)

    # calculate gradients for discriminator
    discri_gradients = 
            discriminator_tape.gradient(disc_loss,
                discri_model.trainable_variables)

    # use optimizer to process and apply gradients to variables
    gen_optimizer.apply_gradients(zip(gen_gradients,
                    gen_model.trainable_variables))

    # same as above to discriminator
    discri_optimizer.apply_gradients(
        zip(discri_gradients,
            discri_model.trainable_variables))
```

接下来，编写一个函数来生成数字 9 的图像，并计算输出，将其保存到 Google Drive 中。

```python
def generate_and_save_images(model, epoch,
                             test_input):
    global epoch_number
    epoch_number = epoch_number + 1

    # set training to false to ensure inference mode
    predictions = model(test_input,
                        training = False)
    # display and save image
    fig = plt.figure(figsize=(4,4))
    for i in range(predictions.shape[0]):
        plt.imshow(predictions[i, :, :, 0] *
                   127.5 + 127.5, cmap='gray')
        plt.axis('off')
    plt.savefig('image_at_epoch_{:01d}.png'.format
                (epoch_number))
    plt.show()
```

该函数使用全局变量 epoch_number 跟踪训练过程，以防断开连接，并保留断点续传功能。模型中 test_input 为初始化随机种子，将训练模式关闭，即将 training 设置为 False，使用批量归一化对该种子进行预测。然后，在用户的控制台上显示图像，将其保存到驱动器中，并以 epoch_number 命名文件。编写上述函数后，编写代码进行模型训练并生成预测结果。

4. 模型训练

设置一个简单的 for 循环训练生成器模型和判别器模型，如清单 13-2 所示。train 方法接收真实图像的数据集作为其第一个参数。使用参数化的图像传递方式能够测试不同的图像数据集或尺寸不同的数据集。第二个参数代表模型训练次数。为数据集中的批量数据调用 gradient_tuning。在每轮训练结束时，生成预测结果并将其保存到用户的驱动器中。此外，保存网络状态作为 checkpoint 节点，以便在与最后一个 checkpoint 断开连接的情况下继续训练。每轮训练所花费的时间被记录并打印在用户控制台上。

清单 13-2 模型训练函数

源码清单
链　接：https://pan.baidu.com/s/1NV0rimQ_8kRz22xfFHN-Cw
提取码：1218

```python
def train(dataset, epochs):
  for epoch in range(epochs):
    start = time.time()

    for image_batch in dataset:
      gradient_tuning(image_batch)

    # Produce images as we go
    generate_and_save_images(gen_model,
                             epoch + 1,
                             seed)

    # save checkpoint data
    checkpoint.save(file_prefix = checkpoint_prefix)
    print ('Time for epoch {} is {} sec'.format
```

```
                           (epoch + 1,
                           time.time()-start))
```

通过调用此训练方法开始模型训练:

```
train(train_dataset, EPOCHS)
```

随着训练的进行，保存 checkpoint 文件与每轮训练的生成图像，将图像显示在控制台上并保存到 Google Drive 中。代码运行 100 次训练后的输出如图 13-10 所示。

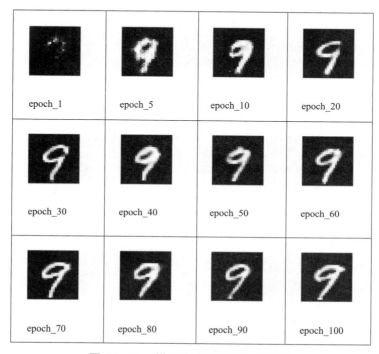

图 13-10　模型生成的数字 9 的图像

从输出结果中可以看出，模型在第 20～30 轮次之后得到可接受的输出结果，并且在第 70 轮次后，生成图像的质量最好。

在训练期间如果出现断开连接的情况，可以从先前保存的 checkpoint 文件中恢复网络状态，并继续训练，使用如下语句实现。

```
#run this code only if there is a runtime disconnection
try:
    checkpoint.restore(tf.train.latest_checkpoint
                        (checkpoint_dir))
except Exception as error:
    print("Error loading in model :
            {}".format(error))
train(train_dataset, EPOCHS)
```

使用 GPU 运行本项目应用程序时，完成每轮训练大约需要 10s。很多时候，生成更复杂的图像可能需要几个小时才能得到理想的输出结果。在这种情况下，使用 checkpoint

功能有助于重新开始训练。

13.6.10 项目源码

生成手写数字图像的完整源码如清单 13-3 所示。

清单 13-3 DigitGen-GAN.ipynb

源码清单
链　接：https://pan.baidu.com/s/1NV0rimQ_8kRz22xfFHN-Cw
提取码：1218

```python
import matplotlib.pyplot as plt
import tensorflow as tf
import numpy as np
import time
from tensorflow import keras
from tensorflow.keras import layers
import os

(train_images,train_labels), 
    (test_images,test_labels) = 
            tf.keras.datasets.mnist.load_data()
digit9_images = []
for i in range(len(train_images)):
    if train_labels[i] == 9:
        digit9_images.append(train_images[i])
train_images = np.array(digit9_images)

train_images.shape

n = 10
f = plt.figure()
for i in range(n):
    f.add_subplot(1, n, i + 1)
    plt.subplot(1, n, i+1 ).axis("off")
    plt.imshow(train_images[i])
plt.show()

train_images = train_images.reshape (
    train_images.shape[0], 28, 28, 1).astype
                        ('float32')
train_images = (train_images - 127.5) / 127.5
train_dataset = tf.data.Dataset.from_tensor_slices(
    train_images).shuffle
            (train_images.shape[0]).batch(32)

gen_model = tf.keras.Sequential()

# Feed network with a 7x7 random image
gen_model.add(tf.keras.layers.Dense(7*7*256, 
                                    use_bias=False, 
input_shape=(100,)))
# Add batch normalization for stability
gen_model.add(tf.keras.layers.BatchNormalization())
```

```python
gen_model.add(tf.keras.layers.LeakyReLU())
# reshape the output
gen_model.add(tf.keras.layers.Reshape((7, 7, 256)))

# Apply (5x5) filter and shift of (1,1).
# The image output is still 7x7.
gen_model.add(tf.keras.layers.Conv2DTranspose
                 (128, (5, 5),
                    strides=(1, 1),
                    padding='same',
                    use_bias=False))
gen_model.add(tf.keras.layers.BatchNormalization())
gen_model.add(tf.keras.layers.LeakyReLU())

# apply stride of (2,2). The output image is now 14x14.
gen_model.add(tf.keras.layers.Conv2DTranspose
                 (64, (5, 5),
                    strides=(2, 2),
                    padding='same',
                    use_bias=False))
gen_model.add(tf.keras.layers.BatchNormalization())
gen_model.add(tf.keras.layers.LeakyReLU())

# another shift upscales the image to 28x28, whihch is our final size.
gen_model.add(tf.keras.layers.Conv2DTranspose
                 (1, (5, 5),
                    strides=(2, 2),
                    padding='same',
                    use_bias=False,
                    activation='tanh'))
gen_model.summary()
tf.keras.utils.plot_model(gen_model)

noise = tf.random.normal([1, 100])
#giving random input vector
generated_image = gen_model(noise, training=False)
plt.imshow(generated_image[0, :, :, 0], cmap='gray')

generated_image.shape

discri_model = tf.keras.Sequential()

discri_model.add(tf.keras.layers.Conv2D
                    (64, (5, 5),
                       strides=(2, 2),
                       padding='same',
                       input_shape=[28, 28, 1]))
discri_model.add(tf.keras.layers.LeakyReLU())
discri_model.add(tf.keras.layers.Dropout(0.3))
```

```python
    discri_model.add(tf.keras.layers.Conv2D
                     (128, (5, 5),
                      strides=(2, 2),
                      padding='same'))
    discri_model.add(tf.keras.layers.LeakyReLU())
    discri_model.add(tf.keras.layers.Dropout(0.3))

    discri_model.add(tf.keras.layers.Flatten())
    discri_model.add(tf.keras.layers.Dense(1))
    discri_model.summary()

tf.keras.utils.plot_model(discri_model)

decision = discri_model(generated_image)
print (decision)
cross_entropy = tf.keras.losses.BinaryCrossentropy
                                (from_logits=True)
 #creating loss function

def generator_loss(generated_output):
    return cross_entropy(tf.ones_like
              (generated_output),generated_output)

def discriminator_loss(real_output,
                        generated_output):
# compute loss considering the image is real [1,1,...,1]
    real_loss = cross_entropy(tf.ones_like
              (real_output),real_output)

# compute loss considering the image is fake[0,0,...,0]
    generated_loss = cross_entropy(tf.zeros_like
              (generated_output),
                generated_output)

    # compute total loss
    total_loss = real_loss + generated_loss

    return total_loss

gen_optimizer = tf.optimizers.Adam(1e-4)
discri_optimizer = tf.optimizers.Adam(1e-4)

epoch_number = 0
EPOCHS = 100
noise_dim = 100
seed = tf.random.normal([1, noise_dim])

checkpoint_dir =
        '/content/drive/My Drive/GAN1/Checkpoint'
checkpoint_prefix =
            os.path.join(checkpoint_dir, "ckpt")
```

```python
checkpoint = tf.train.Checkpoint
            (generator_optimizer = gen_optimizer,
                            discriminator_
optimizer=discri_optimizer,
            generator= gen_model,
            discriminator = discri_model)

from google.colab import drive
drive.mount('/content/drive')

cd '/content/drive/My Drive/GAN1'

def gradient_tuning(images):
    # create a noise vector.
    noise = tf.random.normal([16, noise_dim])

    # Use gradient tapes for automatic differentiation
    with tf.GradientTape()
        as generator_tape, tf.GradientTape()
        as discriminator_tape:

      # ask genertor to generate random images
      generated_images = gen_model(noise,
                        training=True)

      # ask discriminator to evalute the real images and
generate its output
      real_output = discri_model(images,
                        training = True)

      # ask discriminator to do the evlaution on generated
(fake) images
      fake_output = discri_model(generated_images,
                        training = True)
    # calculate generator loss on fake data
    gen_loss = generator_loss(fake_output)

    # calculate discriminator loss as defined earlier
    disc_loss = discriminator_loss(real_output,
                    fake_output)
# calculate gradients for generator
gen_gradients = generator_tape.gradient
        (gen_loss, gen_model.trainable_variables)

# calculate gradients for discriminator
discri_gradients =
      discriminator_tape.gradient(disc_loss,
          discri_model.trainable_variables)

# use optimizer to process and apply gradients to variables
```

```python
        gen_optimizer.apply_gradients(zip(gen_gradients,
                    gen_model.trainable_variables))

        # same as above to discriminator
        discri_optimizer.apply_gradients(
            zip(discri_gradients,
                discri_model.trainable_variables))

    def generate_and_save_images(model, epoch,
                                  test_input):

        global epoch_number
        epoch_number = epoch_number + 1

        # set training to false to ensure inference mode
        predictions = model(test_input,
                            training = False)
        # display and save image
        fig = plt.figure(figsize=(4,4))
        for i in range(predictions.shape[0]):
            plt.imshow(predictions[i, :, :, 0] *
                        127.5 + 127.5, cmap='gray')
            plt.axis('off')
        plt.savefig('image_at_epoch_{:01d}.png'.format
                        (epoch_number))
        plt.show()

    def train(dataset, epochs):
      for epoch in range(epochs):
        start = time.time()

        for image_batch in dataset:
          gradient_tuning(image_batch)

        # Produce images as we go
        generate_and_save_images(gen_model,
                                  epoch + 1,
                                  seed)

        # save checkpoint data
        checkpoint.save(file_prefix = checkpoint_prefix)
        print ('Time for epoch {} is {} sec'.format
                                  (epoch + 1,
                                   time.time()-start))

    train(train_dataset, EPOCHS)

    #run this code only if there is a runtime disconnection
    try:
        checkpoint.restore(tf.train.latest_checkpoint
```

```
                    (checkpoint_dir))
except Exception as error:
    print("Error loading in model :
                    {}".format(error))
train(train_dataset, EPOCHS)
```

本项目训练了一个用于生成手写数字图像的 GAN 模型。下一节将展示如何创建手写字符。

13.7 字母生成

与提供数字数据集的 Kaggle 文件类似，手写字母数据集在另一个名为 extrakeras-datasets 的文件中。本节使用此数据集生成手写字母。生成器模型、判别器模型及训练和测试过程都与上一节相同。因此，本节仅提供如何从 Kaggle 站点加载字母数据集的代码及生成输出图像的代码。完整的项目源码命名为 emnist-GAN，可在本书的代码库中找到。

13.7.1 下载数据

本项目的数据集在一个单独的包中，可以通过运行 pip 语句安装该数据包。代码如下。

```
pip install extra-keras-datasets
```

将此包导入到项目中。代码如下。

```
from extra_keras_datasets import emnist
```

将数据加载到项目中并使用如下代码显示一张图像和相应的标签。

```
(train_images,train_labels),
    (test_images,test_labels) =
        emnist.load_data(type='letters')
plt.imshow(train_images[1])
print ("label: ", train_labels[1])
```

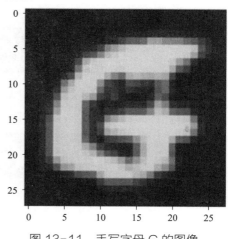

图 13-11 手写字母 G 的图像

上述代码的输出结果如图 13-11 所示。

与上一节数据库一样，每张图像的尺寸都是 28×28，因此，能够使用之前的生成器模型来生成该维度的图像。本项目每个字母的标签都取其在字母集中的位置值。例如，字母 a 的标签值为 1，字母 b 的标签值为 2，以此类推。

13.7.2 创建单字母数据集

就像上一节示例中一样，本项目将训练 GAN 模型用于生成单个字母图像。因此，需要创建一个仅包含所需字母表的图像数据集，使用如下代码实现。

```
letter_G_images = []
for i in range(len(train_images)):
    if train_labels[i] == 7:
        letter_G_images.append(train_images[i])
train_images = np.array(letter_G_images)
```

运行一个 for 循环验证是否只提取了 G 字母的图像，代码如下。

```
n = 10
f = plt.figure()
for i in range(n):
    f.add_subplot(1, n, i + 1)
    plt.subplot(1, n, i+1 ).axis("off")
    plt.imshow(train_images[i])
plt.show()
```

上述代码的输出结果如图 13-12 所示。

图 13-12　手写字母 G 的图像示例

目前为止，数据集已准备就绪。用于预处理数据、定义模型、损失函数、优化器、训练等的其余代码与上一节相同，因此，本项目只显示代码的最终输出。

13.7.3　输出结果

本项目各训练轮次的输出结果如图 13-13 所示。

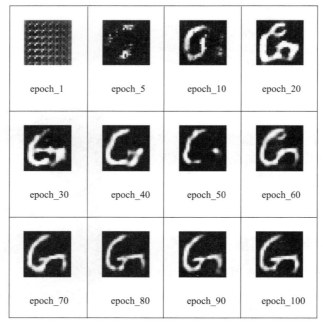

图 13-13　不同训练轮次生成的图像

与上一节的情况一样，可以注意到模型在前几轮中快速收敛，100 轮训练结束后，获得高质量的输出结果。

13.7.4 项目源码

本项目完整源码如清单 13-4 所示。

清单 13-4 emnist-GAN.ipynb

源码清单

链　接：https://pan.baidu.com/s/1NV0rimQ_8kRz22xfFHN-Cw

提取码：1218

```python
import matplotlib.pyplot as plt
import tensorflow as tf
import numpy as np
import time
from tensorflow import keras
from tensorflow.keras import layers
import os

pip install extra-keras-datasets
from extra_keras_datasets import emnist

(train_images,train_labels),
        (test_images,test_labels) = 
         emnist.load_data(type='letters')
plt.imshow(train_images[1])
print ("label: ", train_labels[1])

letter_G_images = []
for i in range(len(train_images)):
    if train_labels[i] == 7:
        letter_G_images.append(train_images[i])
train_images = np.array(letter_G_images)
n = 10
f = plt.figure()
for i in range(n):
    f.add_subplot(1, n, i + 1)
    plt.subplot(1, n, i+1 ).axis("off")
    plt.imshow(train_images[i])
plt.show()

train_images = train_images.reshape (
    train_images.shape[0], 28, 28, 1).astype('float32')
train_images = (train_images - 127.5) / 127.5
train_dataset = tf.data.Dataset.from_tensor_slices(
    train_images).shuffle(
        (train_images.shape[0]).batch(32)

gen_model = tf.keras.Sequential()

# Feed network with a 7x7 random image
gen_model.add(tf.keras.layers.Dense
                      (7*7*256,
```

```python
                                    use_bias=False,
                                    input_shape=(100,)))
# Add batch normalization for stability
gen_model.add(tf.keras.layers.BatchNormalization())
gen_model.add(tf.keras.layers.LeakyReLU())

# reshape the output
gen_model.add(tf.keras.layers.Reshape((7, 7, 256)))

# Apply (5x5) filter and shift of (1,1).
# The image output is still 7x7.
gen_model.add(tf.keras.layers.Conv2DTranspose
                                (128, (5, 5),
                                    strides=(1, 1),
                                    padding='same',
                                    use_bias=False))
gen_model.add(tf.keras.layers.BatchNormalization())
gen_model.add(tf.keras.layers.LeakyReLU())
# apply stride of (2,2). The output image is now 14x14.
gen_model.add(tf.keras.layers.Conv2DTranspose
                                (64, (5, 5),
                                    strides=(2, 2),
                                    padding='same',
                                    use_bias=False))
gen_model.add(tf.keras.layers.BatchNormalization())
gen_model.add(tf.keras.layers.LeakyReLU())

# another shift upscales the image to 28x28, whihch is our
final size.
gen_model.add(tf.keras.layers.Conv2DTranspose
                                (1, (5, 5),
                                    strides=(2, 2),
                                    padding='same',
                                    use_bias=False,
                                    activation='tanh'))

gen_model.summary()

tf.keras.utils.plot_model(gen_model)

noise = tf.random.normal([1, 100])#giving random input vector
generated_image = gen_model(noise, training=False)
plt.imshow(generated_image[0, :, :, 0], cmap='gray')

generated_image.shape

discri_model = tf.keras.Sequential()

discri_model.add(tf.keras.layers.Conv2D
                                (64, (5, 5),
```

```python
                            strides=(2, 2),
                            padding='same',
                            input_shape=[28, 28, 1]))
discri_model.add(tf.keras.layers.LeakyReLU())
discri_model.add(tf.keras.layers.Dropout(0.3))

discri_model.add(tf.keras.layers.Conv2D
                            (128, (5, 5),
                             strides=(2, 2),
                             padding='same'))
discri_model.add(tf.keras.layers.LeakyReLU())
discri_model.add(tf.keras.layers.Dropout(0.3))

discri_model.add(tf.keras.layers.Flatten())
discri_model.add(tf.keras.layers.Dense(1))

discri_model.summary()

tf.keras.utils.plot_model(discri_model)

decision = discri_model(generated_image)
print (decision)

cross_entropy = tf.keras.losses.BinaryCrossentropy(from_logits=True)
#creating loss function

def generator_loss(generated_output):
    return cross_entropy(tf.ones_like(generated_output),
                         generated_output)

def discriminator_loss(real_output,
                       generated_output):
    # compute loss considering the image is real [1,1,...,1]
    real_loss = cross_entropy(tf.ones_like
                    (real_output),real_output)

    # compute loss considering the image is fake[0,0,...,0]
    generated_loss = cross_entropy
                (tf.zeros_like(generated_output),
                             generated_output)
    # compute total loss
    total_loss = real_loss + generated_loss

    return total_loss

gen_optimizer = tf.optimizers.Adam(1e-4)
discri_optimizer = tf.optimizers.Adam(1e-4)

epoch_number = 0
```

```python
EPOCHS = 100
noise_dim = 100
seed = tf.random.normal([1, noise_dim])

checkpoint_dir = 
    '/content/drive/My Drive/GAN2/Checkpoint'
checkpoint_prefix = os.path.join
                 (checkpoint_dir, "ckpt")
checkpoint = tf.train.Checkpoint
            (generator_optimizer = gen_optimizer,
             discriminator_optimizer = 
                       discri_optimizer,
             generator= gen_model,
             discriminator = discri_model)
from google.colab import drive
drive.mount('/content/drive')

cd '/content/drive/My Drive/GAN2'

def gradient_tuning(images):
    # create a noise vector.
    noise = tf.random.normal([16, noise_dim])
    # Use gradient tapes for automatic differentiation
    with tf.GradientTape() 
        as generator_tape, tf.GradientTape() 
        as discriminator_tape:

      # ask genertor to generate random images
      generated_images = gen_model
                   (noise, training = True)

      # ask discriminator to evalute the real images and
generate its output
      real_output = discri_model
                    (images, training = True)

      # ask discriminator to do the evlaution on generated
(fake) images
      fake_output = discri_model
              (generated_images, training = True)

      # calculate generator loss on fake data
      gen_loss = generator_loss(fake_output)

      # calculate discriminator loss as defined earlier
      disc_loss = discriminator_loss
                  (real_output, fake_output)
    # calculate gradients for generator
    gen_gradients = generator_tape.gradient
                 (gen_loss,
```

```python
                    gen_model.trainable_variables)
    
    # calculate gradients for discriminator
    discri_gradients = discriminator_tape.gradient
                (disc_loss,
                    discri_model.trainable_variables)
    # use optimizer to process and apply gradients to variables
    gen_optimizer.apply_gradients(zip(gen_gradients,
                gen_model.trainable_variables))
    
    # same as above to discriminator
    discri_optimizer.apply_gradients(
        zip(discri_gradients,
            discri_model.trainable_variables))
    def generate_and_save_images
            (model, epoch, test_input):
        global epoch_number
        epoch_number = epoch_number + 1
        
        # set training to false to ensure inference mode
        predictions = model(test_input,
                        training = False)
        # display and save image
        fig = plt.figure(figsize=(4,4))
        for i in range(predictions.shape[0]):
            plt.imshow(predictions
                [i, :, :, 0] * 127.5 + 127.5,
                        cmap='gray')
            plt.axis('off')
        plt.savefig('image_at_epoch_
                {:01d}.png'.format(epoch_number))
        plt.show()

def train(dataset, epochs):
    for epoch in range(epochs):
        start = time.time()
        for image_batch in dataset:
            gradient_tuning(image_batch)
        
        # Produce images as we go
        generate_and_save_images(gen_model,
                            epoch + 1,
                            seed)
        
        # save checkpoint data
        checkpoint.save(file_prefix =
                        checkpoint_prefix)
        print ('Time for epoch {} is {} sec'.format
                        (epoch + 1,
                        time.time()-start))
```

```
train(train_dataset, EPOCHS)

#run this code only if there is a runtime disconnection
try:
    checkpoint.restore(tf.train.latest_checkpoint
                        (checkpoint_dir))
except Exception as error:
    print("Error loading in model :
                        {}".format(error))
train(train_dataset, 100)
```

13.8 印刷体到手写体

通过前面两节的学习，现在可以通过输入自定义的 a～z、A～Z 和数字 0～9 的手写图像来训练 GAN 模型。完成模型训练后，可以使用该模型将任何印刷文本转换为用户个性化手写文本。本节使用上述模型将单词 "tensor" 转换为手写体写，如图 13-14 所示。

图 13-14　通过组合图像创建的示例文本

接下来创建更复杂的图像。

13.9 生成彩色卡通图像

到目前为止，本章已经创建了手写数字图像和字母图像，如何创建像卡通这样的复杂彩色图像？目前学到的技术已经可用于创建复杂的彩色图像。本项目将展示相关内容。

13.9.1 下载数据集

Kaggle 网站上有大量的动漫角色数据集，本书的下载站点上保留了该项目的数据集供读者使用。使用 wget 语句将数据下载到本项目中：

```
! wget --no-check-certificate -r 'https://drive.google.com/uc?exp
ort=download&id=1z7rXRIFtRBFZHt-Mmti4HxrxHqUfG3Y8' -O tf-book.zip
```

使用如下语句解压缩下载文件。

```
!unzip tf-book.zip
```

13.9.2 创建数据集

编写创建数据集函数，代码如下。

```
def load_dataset(batch_size, img_shape,
                    data_dir = None):
    # Create a tuple of size(30000,64,64,3)
    sample_dim = (batch_size,) + img_shape
    # Create an uninitialized array of shape (30000,64,64,3)
    sample = np.empty(sample_dim, dtype=np.float32)
```

```
    # Extract all images from our file
    all_data_dirlist = list(glob.glob(data_dir))

    # Randomly select an image file from our data list
    sample_imgs_paths = np.random.choice
                    (all_data_dirlist,batch_size)

    for index,img_filename in enumerate
                        (sample_imgs_paths):
        # Open the image
        image = Image.open(img_filename)
        # Resize the image
        image = image.resize(img_shape[:-1])
        # Convert the input into an array
        image = np.asarray(image)
        # Normalize data
        image = (image/127.5) -1
        # Assign the preprocessed image to our sample
        sample[index,...] = image
    print("Data loaded")
    return sample
```

以上代码附有完整注释，现在调用此函数来创建数据集：

```
x_train=load_dataset(30000,(64,64,3),
    "/content/tf-book/chapter13/anime/data/*.png")
BUFFER_SIZE = 30000
BATCH_SIZE = 256
train_dataset = tf.data.Dataset.from_tensor_slices
        (x_train).shuffle(BUFFER_SIZE).batch
                (BATCH_SIZE)
```

13.9.3 显示图像

从集合中打印部分图像检查数据集是否正确加载：

```
n = 10
f = plt.figure(figsize=(15,15))
for i in range(n):
    f.add_subplot(1, n, i + 1)
    plt.subplot(1, n, i+1 ).axis("off")
    plt.imshow(x_train[i])
plt.show()
```

图像示例如图 13-15 所示。

图 13-15　动漫图像示例

使用如下语句检查训练数据形状。

x_train.shape

训练数据形状如下。

(30000, 64, 64, 3)

输出结果表明有 30000 个 RGB 图像，每个图像的大小为 64×64。接下来使用此数据集进行模型训练与测试。本项目其余代码与前两个项目完全相同，因此不在这里重复介绍，仅展示不同训练轮次的输出结果。

13.9.4　输出结果

模型不同时期的生成图像如图 13-16 所示。

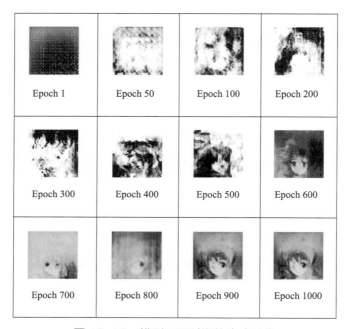

图 13-16　模型不同时期的生成图像

可以看到，大约 1000 轮训练后，模型重现了很多原始卡通图像内容。想要模型生成更加逼真的图像，需要将代码运行 10000 次或更多。每轮训练在 GPU 上大约需要运行 16s。本项目说明 GAN 模型能够生成更加复杂的图像。

13.9.5　项目源码

用于生成动漫图像的完整源码如清单 13-5 所示。

清单 13-5 CS-Anime.ipynb

```
import tensorflow as tf
import numpy as np
import sys
```

源码清单
链　接：https://pan.baidu.com/s/1NV0rimQ_8kRz22xfFHN-Cw
提取码：1218

```python
import os
import cv2
import glob
from PIL import Image
import matplotlib.pyplot as plt
import time
from tensorflow import keras
from tensorflow.keras import layers
from keras.layers import UpSampling2D, Conv2D

! wget --no-check-certificate -r 'https://drive.google.com/uc?exp
ort=download&id=1z7rXRIFtRBFZHt-Mmti4HxrxHqUfG3Y8' -O tf-book.zip

!unzip tf-book.zip

def load_dataset(batch_size, img_shape,
                 data_dir = None):
    # Create a tuple of size(30000,64,64,3)
    sample_dim = (batch_size,) + img_shape
    # Create an uninitialized array of shape (30000,64,64,3)
    sample = np.empty(sample_dim, dtype=np.float32)
    # Extract all images from our file
    all_data_dirlist = list(glob.glob(data_dir))

    # Randomly select an image file from our data list
    sample_imgs_paths = np.random.choice
                    (all_data_dirlist,batch_size)

    for index,img_filename in enumerate
                            (sample_imgs_paths):
        # Open the image
        image = Image.open(img_filename)
        # Resize the image
        image = image.resize(img_shape[:-1])
        # Convert the input into an array
        image = np.asarray(image)
        # Normalize data
        image = (image/127.5) -1
        # Assign the preprocessed image to our sample
        sample[index,...] = image
    print("Data loaded")
    return sample

x_train=load_dataset(30000,(64,64,3),
        "/content/tf-book/chapter13/anime/data/*.png")
BUFFER_SIZE = 30000
BATCH_SIZE = 256
train_dataset = tf.data.Dataset.from_tensor_slices
            (x_train).shuffle(BUFFER_SIZE).batch
                    (BATCH_SIZE)
```

```python
n = 10
f = plt.figure(figsize=(15,15))
for i in range(n):
    f.add_subplot(1, n, i + 1)
    plt.subplot(1, n, i+1 ).axis("off")
    plt.imshow(x_train[i])
plt.show()

x_train.shape

gen_model = tf.keras.Sequential()

# seed image of size 4x4
gen_model.add(tf.keras.layers.Dense
                        (64*4*4,
                         use_bias=False,
                         input_shape=(100,)))
gen_model.add(tf.keras.layers.BatchNormalization())
gen_model.add(tf.keras.layers.LeakyReLU())

gen_model.add(tf.keras.layers.Reshape((4,4,64)))

# size of output image is still 4x4
gen_model.add(tf.keras.layers.Conv2DTranspose
                        (256, (5, 5),
                          strides=(1, 1),
                          padding='same',
                          use_bias=False))
gen_model.add(tf.keras.layers.BatchNormalization())
gen_model.add(tf.keras.layers.LeakyReLU())
# size of output image is 8x8
gen_model.add(tf.keras.layers.Conv2DTranspose
                        (128, (5, 5),
                          strides=(2, 2),
                          padding='same',
                          use_bias=False))
gen_model.add(tf.keras.layers.BatchNormalization())
gen_model.add(tf.keras.layers.LeakyReLU())

# size of output image is 16x16
gen_model.add(tf.keras.layers.Conv2DTranspose
                        (64, (5, 5),
                          strides=(2, 2),
                          padding='same',
                          use_bias=False))
gen_model.add(tf.keras.layers.BatchNormalization())
gen_model.add(tf.keras.layers.LeakyReLU())

# size of output image is 32x32
gen_model.add(tf.keras.layers.Conv2DTranspose
```

```python
                    (32, (5, 5),
                    strides=(2, 2),
                    padding='same',
                    use_bias=False))
gen_model.add(tf.keras.layers.BatchNormalization())
gen_model.add(tf.keras.layers.LeakyReLU())

# size of output image is 64x64
gen_model.add(tf.keras.layers.Conv2DTranspose
                    (3, (5, 5),
                    strides=(2, 2),
                    padding='same',
                    use_bias=False,
                    activation='tanh'))

gen_model.summary()
noise = tf.random.normal([1, 100])
generated_image = gen_model(noise, training=False)
plt.imshow(generated_image[0, :, :, 0] )
discri_model = tf.keras.Sequential()
discri_model.add(tf.keras.layers.Conv2D
                    (128, (5, 5), strides=(2, 2),
                    padding='same',
                    input_shape=[64,64,3]))
discri_model.add(tf.keras.layers.LeakyReLU())
discri_model.add(tf.keras.layers.Dropout(0.3))

discri_model.add(tf.keras.layers.Conv2D(
                    256, (5, 5), strides=(2, 2),
                    padding='same'))
discri_model.add(tf.keras.layers.LeakyReLU())
discri_model.add(tf.keras.layers.Dropout(0.3))

discri_model.add(tf.keras.layers.Flatten())
discri_model.add(tf.keras.layers.Dense(1))
discri_model.summary()

tf.keras.utils.plot_model(discri_model)

decision = discri_model(generated_image)
#giving the generated image to discriminator,the discriminator
will give negative value if it is fake,while if it is real then
it will give positive value.
print (decision)

cross_entropy = tf.keras.losses.BinaryCrossentropy
                    (from_logits=True)

def generator_loss(generated_output):
    return cross_entropy(tf.ones_like(generated_output),
```

```python
                            generated_output)

    def discriminator_loss(real_output,
                           generated_output):
        # compute loss considering the image is real [1,1,...,1]
        real_loss = cross_entropy
                        (tf.ones_like(real_output),
                         real_output)
        # compute loss considering the image is fake[0,0,...,0]
        generated_loss = cross_entropy
                        (tf.zeros_like
                         (generated_output),
                         generated_output)

        # compute total loss
        total_loss = real_loss + generated_loss

        return total_loss

gen_optimizer = tf.optimizers.Adam(1e-4)
discri_optimizer = tf.optimizers.Adam(1e-4)

epoch_number = 0
EPOCHS = 10000
noise_dim = 100
seed = tf.random.normal([1, noise_dim])

checkpoint_dir =
        '/content/drive/My Drive/GAN3/Checkpoint'
checkpoint_prefix = os.path.join
                        (checkpoint_dir, "ckpt")
checkpoint = tf.train.Checkpoint
            (generator_optimizer=gen_optimizer,
             discriminator_optimizer=discri_optimizer,
             generator= gen_model,
             discriminator = discri_model)

from google.colab import drive
drive.mount('/content/drive')

cd '/content/drive/My Drive/GAN3'

def gradient_tuning(images):
    # create a noise vector.
    noise = tf.random.normal([16, noise_dim])

    # Use gradient tapes for automatic differentiation
    with tf.GradientTape()
            as generator_tape, tf.GradientTape()
            as discriminator_tape:
```

```python
# ask genertor to generate random images
generated_images = gen_model
                (noise, training=True)
# ask discriminator to evalute the real images and
generate its output
real_output = discri_model(images,
                    training=True)

# ask discriminator to do the evlaution on generated
(fake) images
fake_output = discri_model(generated_images,
                    training=True)

# calculate generator loss on fake data
gen_loss = generator_loss(fake_output)

# calculate discriminator loss as defined earlier
disc_loss = discriminator_loss(real_output,
                    fake_output)
# calculate gradients for generator
gen_gradients = generator_tape.gradient
            (gen_loss,
             gen_model.trainable_variables)

# calculate gradients for discriminator
discri_gradients = discriminator_tape.gradient
            (disc_loss,
             discri_model.trainable_variables)
# use optimizer to process and apply gradients to variables
gen_optimizer.apply_gradients(zip(gen_gradients,
            gen_model.trainable_variables))

# same as above to discriminator
discri_optimizer.apply_gradients(
    zip(discri_gradients,
        discri_model.trainable_variables))

def generate_and_save_images(model, epoch,
                        test_input):
    global epoch_number
    epoch_number = epoch_number + 1

    # set training to false to ensure inference mode
    predictions = model(test_input,
                        training=False)

    # display and save image
    fig = plt.figure(figsize=(4,4))
    for i in range(predictions.shape[0]):
        plt.imshow(predictions[i, :, :, 0]
```

```
                                    * 127.5 + 127.5, cmap='gray')
            plt.axis('off')

        plt.savefig('image_at_epoch_
                {:01d}.png'.format(epoch_number))
        plt.show()
def train(dataset, epochs):
    for epoch in range(epochs):
        start = time.time()

        for image_batch in dataset:
            gradient_tuning(image_batch)

        # Produce images as we go
        generate_and_save_images(gen_model,
                                 epoch + 1,
                                 seed)

        # save checkpoint data
        checkpoint.save(file_prefix = checkpoint_prefix)
        print ('Time for epoch {} is {} sec'.format
                                (epoch + 1,
                                 time.time()-start))

train(train_dataset, EPOCHS)

#run this code only if there is a runtime disconnection
try:
      checkpoint.restore(tf.train.latest_checkpoint
                                (checkpoint_dir))
except Exception as error:
     print("Error loading in model :
                                  {}".format(error))
train(train_dataset, EPOCHS)
```

总结

　　GAN 提供了一种模仿已知图像的新想法。GAN 由两个模型组成——生成器和判别器，两个模型都通过对抗过程同时训练。本章讲解了构建 GAN，并用于创建手写数字图像、字母图像甚至动漫图像。训练 GAN 需要大量训练图像，才能达到满意的结果。如今，GAN 已成功应用于许多应用中。除了本章示例外，GAN 还可用于从照片创建表情符号、图像到图像和文本到图像的翻译等应用。通过本章学习，读者可使用 GAN 完成实际应用的具体任务。

第 14 章

图像转换

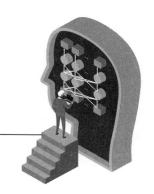

当我们需要给旧黑白照片涂上彩色时，可能会联系精通 Photoshop 的设计师完成这项工作，但我们需要向他们支付高额费用并等待几天或几周的时间。现在这项工作可以使用深度神经网络完成。本章将介绍将黑白图像转换为彩色图像的深度学习技术，该技术很简单，使用 AutoEncoders 网络架构实现。首先，让我们先学习 AutoEncoders。

14.1 自动编码器

AutoEncoders 由两部分组成：编码器和解码器，如图 14-1 所示。

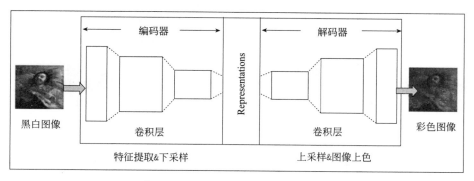

图 14-1　AutoEncoders 结构图

图 14-1 左侧，有一张黑白图像输入到神经网络中。图 14-1 右侧，网络输出与输入图像内容相同的彩色图像。计算过程如下：首先编码器通过一系列卷积层提取图像特征并对特征图下采样以学习输入图像的降维表示，然后解码器将图像传递给另一系列卷积层重新生成彩色图像。

要了解如何为图像着色，需要了解颜色空间。

14.2 色彩空间

彩色图像由给定颜色空间中的颜色和光照强度定义。不同颜色使用原色（如红色、绿色、蓝色）创建，整个颜色范围称为颜色空间，如 RGB。在数学术语中，颜色空间是一种抽象的数学模型，将颜色范围使用数字元组描述。每种颜色可以由一个点表示。

本章将描述三种最流行的颜色空间：RGB、YCbCr、Lab。

RGB 是最常用的颜色空间，包含三个通道，即红色 (R)、绿色 (G) 和蓝色 (B)。每个通道由 8 位表示，最大值为 256。整个 RGB 图像空间可以代表超过 1600 万种颜色。

YCbCr 颜色空间的图像通常使用 JPEG 和 MPEG 格式存储。与 RGB 相比，YCbCr 在数字传输和存储方面更高效。Y 通道表示灰度图像的亮度，数值范围为 16～235。Cb 和 Cr 分别代表蓝色和红色差异色度分量，数值范围为 16～240。

> **注意**
>
> 所有这些通道的组合值可能无法代表有效颜色，因此在本章的图像着色应用中，不使用 YCbCr 颜色空间。

Lab 颜色空间由国际照明委员会 (CIE) 设计。此颜色空间的视觉表示如图 14-2 所示。

Lab 颜色空间大于计算机显示器和打印机的色域。Lab 颜色空间的位图中每个颜色需要更多的数据才能获得与 RGB 或 CMYK 相同的精度。因此，Lab 颜色空间通常用作中介而不是最终颜色空间。

L 通道表示亮度，取值范围为 0～100。a 通道代码从绿色 (-) 到红色 (+)，b 通道代码从蓝色 (-) 到黄色 (+)。使用 8 位数据描述 a 通道和 b 通道，需要使用 -127～127 范围内的数值。Lab 颜色空间近似于人类视觉，空间内不同数值组合与人类视觉感知范围大致相同。

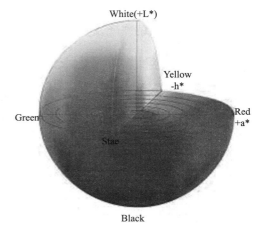

图 14-2 Lab 颜色空间

本项目使用 Lab 颜色空间。通过分离代表亮度的灰度分量，模型只需要学习余下两个通道的数值进行着色，这有助于减小模型规模并加快收敛速度。

接下来，讨论本项目中 AutoEncoders 的不同网络拓扑结构。

14.3 网络配置

AutoEncoder 可以通过三种不同的方式进行配置：Vanilla、Merged、使用预训练的 Merged 模型。

接下来介绍上述三种模型。

14.3.1　Vanilla 模型

vanilla 模型结构与如图 14-1 所示的结构相同，其中编码器包含一系列卷积层，用于提取特征及特征图下采样，解码器中的卷积层用于还原特征图尺寸与图像着色。在这种自动编码器中，编码器的深度不足以提取图像的全局特征。全局特征能够帮助我们确定如何对图像的某些区域进行着色。如果使编码器网络加深，特征图尺寸会继续缩小，以至于解码器无法还原原始图像。因此，本项目编码器需要两条路径，一条获取全局特征，另一条获取图像的细节特征，使用接下来介绍的两个模型实现。

14.3.2　Merged 模型

此模型由 Lizuka 等人提出，发表于论文 *Let there be Color!* 中（http://iizuka.cs.tsukuba.ac.jp/projects/colorization/data/colorization_sig2016.pdf）。Merged 模型结构如图 14-3 所示。

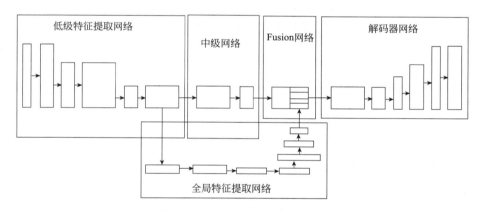

图 14-3　Merged 模型结构图

Merged 模型使用含有 8 个卷积层的编码器来提取中级特征，第 6 层的输出连接另一个 7 层网络以提取全局特征，另一个 Fusion 网络连接前两个输出并将它们输入到解码器。

14.3.3　使用预训练的 Merged 模型

此模型由 Baldassarre 等人提出，发表在论文 *Deep Koalarization: Image Colorization using CNNs and Inception-Resnet-v2* 中（https://arxiv.org/pdf/1712.03400.pdf）。使用预训练的 Merged 模型的结构如图 14-4 所示。

特征提取由预训练的 ResNet 完成。

在本章的第二个项目中，展示了如何使用预训练模型进行特征提取。

以上完成了对 AutoEncoders 及其组件的介绍，接下来介绍具体实现细节。

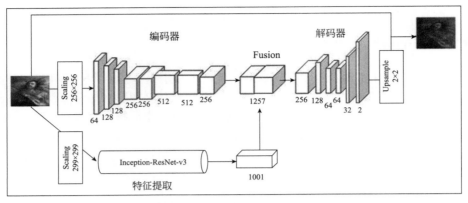

图 14-4　使用预训练的 Merged 模型

14.4　自动编码器

在本项目中，使用 vanilla 自动编码器。

打开一个新的 Colab 项目并将其重命名为 AutoEncoder–Custom。通过如下代码添加附加依赖库。

```
import numpy as np
import pandas as pd
import os

import matplotlib.pyplot as plt

from tqdm import tqdm
from itertools import chain
import skimage
from skimage.io import imread, imshow
from skimage.transform import resize
from skimage.util import crop, pad
from skimage.morphology import label
from skimage.color import import rgb2gray, gray2rgb,
                          rgb2lab, lab2rgb
from sklearn.model_selection import train_test_split

import tensorflow as tf
from tensorflow.keras.models
          import Model, load_model,Sequential
from tensorflow.keras.preprocessing.image
          import ImageDataGenerator
from tensorflow.keras.layers import Input, Dense,
          UpSampling2D, RepeatVector, Reshape
from tensorflow.keras.layers import Dropout, Lambda
from tensorflow.keras.layers import Conv2D,
                                    Conv2DTranspose
from tensorflow.keras.layers import MaxPooling2D
```

```
from tensorflow.keras import backend as K
```

14.4.1 加载数据

此项目使用 Kaggle 站点上提供的数据集，网址为 www.kaggle.com/thedownhill/art-images-drawingspainting-sculpture-engraving，提供约 9000 张图像数据集，包含五种类型的艺术图像。如果拥有 Kaggle 账户，可以使用以下代码下载数据集。

```
#!pip install -q kaggle
#!mkdir ~/.kaggle
#!touch ~/.kaggle/kaggle.json
#api_token = {"username":"Your UserName",
                          "key":"Your key"}

#import json

#with open('/root/.kaggle/kaggle.json', 'w') as file:
#    json.dump(api_token, file)

#!chmod 600 ~/.kaggle/kaggle.json
#!kaggle datasets download -d thedownhill/art-images-drawings-painting-
sculpture-engraving
```

或者，可以从本书的下载站点获得数据集，使用 wget 下载数据集到本项目中。代码如下。

```
!wget --no-check-certificate -r 'https://drive.google.com/
uc?export=download&id=1CKs7s_MZMuZFBXDchcL_AgmCxgPBTJXK' -O
art-images-drawings-painting-sculpture-engraving.zip
```

下载数据文件后，将其解压缩：

```
!unzip art-images-drawings-painting-sculpture-engraving.zip
```

解压缩文件后，计算机上将存储大量图像，并按特定文件夹结构排列。图像有不同的大小，将所有训练图像转换为 256×256 的固定大小。

定义了一些变量用于创建训练数据集，代码如下。

```
IMG_WIDTH = 256
IMG_HEIGHT = 256
TRAIN_PATH =
'/content/dataset/dataset_updated/training_set/painting/'
train_ids = next(os.walk(TRAIN_PATH))[2]
```

os.walk 语句获取文件夹中存在的所有文件名。

首先，检查代码中是否存在任何不可读图像，并将其从数据集中删除。使用如下语句实现。

```
missing_count = 0
for n, id_ in tqdm(enumerate(train_ids),
                              total=len(train_ids)):
```

```
        path = TRAIN_PATH + id_+''
        try:
            img = imread(path)
        except:
            missing_count += 1

print("\n\nTotal missing: "+ str(missing_count))
```

运行上述代码,将会发现图像集合中有 86 个不可读图像。接下来,将创建训练集,使用如下语句实现。

```
X_train = np.zeros((len(train_ids)-missing_count,
        IMG_HEIGHT, IMG_WIDTH, 3), dtype=np.uint8)
missing_images = 0
for n, id_ in tqdm(enumerate(train_ids),
                    total=len(train_ids)):
    path = TRAIN_PATH + id_+''
    try:
        img = imread(path)
        img = resize(img, (IMG_HEIGHT, IMG_WIDTH),
            mode='constant', preserve_range=True)
        X_train[n-missing_images] = img
    except:
        missing_images += 1

X_train = X_train.astype('float32') / 255.
```

可以使用以下语句检查图像。

```
plt.imshow(X_train[5])
```

输出图像如图 14-5 所示。

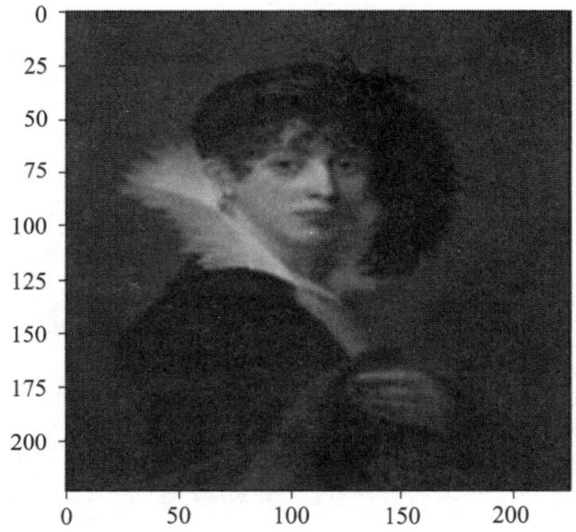

14.4.2　创建训练、测试数据集

接下来,从之前创建的数据集中保留一些图像用于测试:

图 14-5　图像示例

```
x_train, x_test = train_test_split(X_train,
                    test_size=20)
```

train_test_split 函数中的 test_size 参数指定预留 20 张图片进行测试。

14.4.3　准备训练数据

为了训练模型,将图像从 RGB 转换为 Lab 格式。如前所述,L 通道的灰度值代表图像亮度,a 通道是绿色和红色之间的颜色平衡,b 通道是蓝色和黄色之间的颜色平衡。

首先,从 Keras 库中创建一个 ImageDataGenerator 实例,将图像转换为像素数组,最后将它们组合成一个巨大的向量。代码如下。

```
datagen = ImageDataGenerator(
```

```
            shear_range=0.2,
            zoom_range=0.2,
            rotation_range=20,
            horizontal_flip=True)
```

shear_range 将图像向左或向右倾斜,其他参数分别代表缩放、旋转、水平翻转。

现在,将编写一个函数来创建用于训练的批量数据,使用如下代码实现。

```
def create_training_batches(dataset=X_train,
                            batch_size = 20):
    # iteration for every image
    for batch in datagen.flow(dataset, batch_size=batch_size):
        # convert from rgb to grayscale
        X_batch = rgb2gray(batch)
        # convert rgb to Lab format
        lab_batch = rgb2lab(batch)
        # extract L component
        X_batch = lab_batch[:,:,:,0]
        # reshape

        X_batch = X_batch.reshape(X_batch.shape+(1,))
        # extract a and b features of the image
        Y_batch = lab_batch[:,:,:,1:] / 128
        yield X_batch, Y_batch
```

该函数首先通过调用 rgb2gray 方法将给定的图像从 RGB 转换为灰度图像,然后通过调用 rgb2lab 方法将图像转换为 Lab 格式。采用 Lab 颜色空间后,只需要预测两个分量,而其他颜色空间则需要预测三个或四个分量。这有助于减小模型规模并加快收敛速度。最后,从图像中提取 L、a 和 b 分量。

14.4.4 定义模型

接下来,将定义本项目的自动编码器模型。模型配置基于论文 *Let there be Color!* 中的建议 (http://iizuka.cs.tsukuba.ac.jp/projects/colorization/data/colorization_sig2016.pdf)。

```
# the input for the encoder layer
inputs1 = Input(shape=(IMG_WIDTH, IMG_HEIGHT, 1,))

# encoder

# Using Conv2d to reduce the size of feature maps and image
size
# convert image to 128x128
encoder_output = Conv2D(64, (3,3), activation='relu',
                    padding='same', strides=2)(inputs1)
encoder_output = Conv2D(128, (3,3),
                    activation='relu',
                    padding='same')(encoder_output)
# convert image to 64x64
encoder_output = Conv2D(128, (3,3),
                    activation='relu', padding='same',
```

```
                          strides=2)(encoder_output)
encoder_output = Conv2D(256, (3,3),
                        activation='relu',
                        padding='same')(encoder_output)
# convert image to 32x32
encoder_output = Conv2D(256, (3,3),
                        activation='relu', padding='same',
                        strides=2)(encoder_output)
encoder_output = Conv2D(512, (3,3),
                        activation='relu', padding='same')
                        (encoder_output)

# mid-level feature extractions
encoder_output = Conv2D(512, (3,3),
                        activation='relu',
                        padding='same')(encoder_output)
encoder_output = Conv2D(256, (3,3),
                        activation='relu',
                        padding='same')(encoder_output)

# decoder

# Adding colors to the grayscale image and upsizing it
decoder_output = Conv2D(128, (3,3),
                        activation='relu',
                        padding='same')(encoder_output)
decoder_output = UpSampling2D((2, 2))(decoder_output)
# image size 64x64
decoder_output = Conv2D(64, (3,3), activation='relu',
                        padding='same')(decoder_output)
decoder_output = Conv2D(64, (3,3), activation='relu',
                        padding='same')(decoder_output)
decoder_output = UpSampling2D((2, 2))(decoder_output)
# image size 128x128
decoder_output = Conv2D(32, (3,3), activation='relu',
                        padding='same')(decoder_output)
decoder_output = Conv2D(2, (3, 3), activation='tanh',
                        padding='same')(decoder_output)
decoder_output = UpSampling2D((2, 2))(decoder_output)
# image size 256x256
```

编码器和解码器都包含少量的 Conv2D 卷积层。编码器通过一系列层运算对图像进行下采样完成图像特征提取，解码器使用一组层对特征图的各个点上采样重新生成原始图像并向灰度图像中添加颜色，创建大小为 256×256 的最终图像。最后一个解码器层使用 tanh 激活函数将网络输出映射到 $-1 \sim 1$ 之间。

在定义了编码器和解码器后，构建模型并使用其 compile 方法进行编译，结合损失函数和 Adam 优化器训练网络。代码如下。

```
model = Model(inputs=inputs1, outputs=decoder_output)
model.compile(loss='mse', optimizer='adam',
```

```
                    metrics=['accuracy'])
print(model.summary())
```

模型概要如图 14-6 所示。

可以通过 plot_model 函数对模型进行可视化:

```
tf.keras.utils.plot_model(model)
```

模型可视化结果如图 14-7 所示。

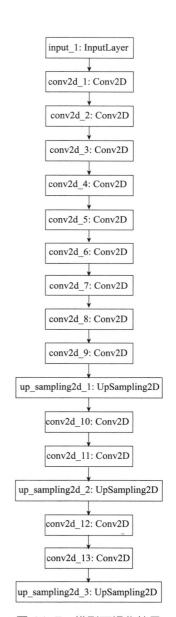

图 14-6　编码器模型概要

图 14-7　模型可视化结果

14.4.5 模型训练

通过调用 fit 方法训练模型：

```
BATCH_SIZE = 20
model.fit_generator(create_training_batches
                    (X_train,BATCH_SIZE),
            epochs= 100,
            verbose=1,
            steps_per_epoch=X_train.shape[0]/BATCH_SIZE)
```

在 GPU 上训练模型时，每个 epoch 需要约 1mim 的时间。通过使用预训练模型，训练时间缩短到每个 epoch 约 1s。本章中下一项目将使用此训练方法。

14.4.6 测试

现在，可以在之前创建的测试数据集上检查模型性能。

> **注意**
> 对于测试图像，与测试数据不同，不会对它们进行预处理，只需将测试图像转换为 Lab 格式并进行预测。

对测试图像进行模型预测的代码如下。

```
test_image = rgb2lab(x_test)[:,:,:,0]
test_image = test_image.reshape
                (test_image.shape+(1,))
output = model.predict(test_image)
output = output * 128

# making the output image array
generated_images = np.zeros
                    ((len(output),256, 256, 3))

for i in range(len(output)):
#iterating for the output
    cur = np.zeros((256, 256, 3))
# dummy array
    cur[:,:,0] = test_image[i][:,:,0]
#assigning the gray scale component
    cur[:,:,1:] = output[i]
#assigning the a and b component
#converting from lab to rgb format as plt only work for rgb mode
    generated_images[i] = lab2rgb(cur)
```

使用以下代码片段显示生成的图像和原始图像。

```
plt.figure(figsize=(20, 6))
for i in range(10):
    # grayscale
    plt.subplot(3, 10, i + 1)
    plt.imshow(rgb2gray(x_test)[i].reshape(256, 256))
    plt.gray()
    plt.axis('off')
    # recolorization
    plt.subplot(3, 10, i + 1 +10)
    plt.imshow(generated_images[i].reshape
                                    (256, 256,3))
    plt.axis('off')

    # original
    plt.subplot(3, 10, i + 1 + 20)
    plt.imshow(x_test[i].reshape(256, 256,3))
    plt.axis('off')

plt.tight_layout()
plt.show()
```

输出如图 14-8 所示。

图 14-8　模型测试结果

第一行是从第三行给出的原始彩色图像创建的一组灰度图像，中间一行显示了模型生成的图像。如图 14-8 所示，该模型能够生成与原始图像足够接近的图像。

411

接下来，将展示如何使用此模型预测不同尺寸的新图像。

14.4.7 未知图像预测

选择使用未知的新图像测试上述模型的性能需要在本书的网站上下载一个示例图像，可以使用 wget 下载，如下所示。

```
!wget https://raw.githubusercontent.com/Apress/artificialneural-
networks-with-tensorflow-2/main/ch14/mountain.jpg
```

示例图像如图 14-9 所示。
使用以下代码运行模型测试。

> **注意** 需要在将图像输入网络之前更改图像大小。

```
img = resize(img, (IMG_HEIGHT, IMG_WIDTH),
             mode='constant', preserve_range=True)
img = img.astype('float32') / 255.

test_image = rgb2lab(img)[:,:,0]
test_image = test_image.reshape
                    ((1,)+test_image.shape+(1,))
output = model.predict(test_image)
output = output * 128

plt.imshow(img)
plt.axis('off')
```

生成图像如图 14-10 所示。

图 14-9　示例图像

图 14-10　通过自定义自动编码器模型生成的彩色图像

14.4.8 项目源码

本项目的完整源码如清单 14-1 所示。

清单 14-1 AutoEncoder 模型源码

源码清单
链　接：https://pan.baidu.com/s/1NV0rimQ_8kRz22xfFHN-Cw
提取码：1218

```python
import numpy as np
import pandas as pd
import os

import matplotlib.pyplot as plt

from tqdm import tqdm
from itertools import chain
import skimage
from skimage.io import imread, imshow
from skimage.transform import resize
from skimage.util import crop, pad
from skimage.morphology import label
from skimage.color import rgb2gray, gray2rgb,
                          rgb2lab, lab2rgb
from sklearn.model_selection import train_test_split

import tensorflow as tf
from tensorflow.keras.models
          import Model, load_model,Sequential
from tensorflow.keras.preprocessing.image
          import ImageDataGenerator
from tensorflow.keras.layers import Input, Dense,
          UpSampling2D, RepeatVector, Reshape
from tensorflow.keras.layers import Dropout, Lambda
from tensorflow.keras.layers import Conv2D,
                                    Conv2DTranspose
from tensorflow.keras.layers import MaxPooling2D
from tensorflow.keras import backend as K

#!pip install -q kaggle
#!mkdir ~/.kaggle
#!touch ~/.kaggle/kaggle.json

#api_token = {"username":"Your UserName",
                          "key":"Your key"}
#import json

#with open('/root/.kaggle/kaggle.json', 'w') as file:
#    json.dump(api_token, file)

#!chmod 600 ~/.kaggle/kaggle.json
#!kaggle datasets download -d thedownhill/art-images-drawings-painting-sculpture-engraving
```

```python
!wget --no-check-certificate -r 'https://drive.google.com/
uc?export=download&id=1CKs7s_MZMuZFBXDchcL_AgmCxgPBTJXK' -O
art-images-drawings-painting-sculpture-engraving.zip

!unzip art-images-drawings-painting-sculpture-engraving.zip

IMG_WIDTH = 256
IMG_HEIGHT = 256
TRAIN_PATH =
'/content/dataset/dataset_updated/training_set/painting/'
train_ids = next(os.walk(TRAIN_PATH))[2]

missing_count = 0
for n, id_ in tqdm(enumerate(train_ids),
                    total=len(train_ids)):
    path = TRAIN_PATH + id_+''
    try:
        img = imread(path)
    except:
        missing_count += 1
print("\n\nTotal missing: "+ str(missing_count))

X_train = np.zeros((len(train_ids)-missing_count,
        IMG_HEIGHT, IMG_WIDTH, 3), dtype=np.uint8)
missing_images = 0
for n, id_ in tqdm(enumerate(train_ids),
                    total=len(train_ids)):
    path = TRAIN_PATH + id_+''
    try:
        img = imread(path)
        img = resize(img, (IMG_HEIGHT, IMG_WIDTH),
            mode='constant', preserve_range=True)
        X_train[n-missing_images] = img
    except:
        missing_images += 1

X_train = X_train.astype('float32') / 255.

plt.imshow(X_train[5])

x_train, x_test = train_test_split(X_train,
                                    test_size=20)

datagen = ImageDataGenerator(
        shear_range=0.2,
        zoom_range=0.2,
        rotation_range=20,
        horizontal_flip=True)

def create_training_batches(dataset=X_train,
                                batch_size = 20):
```

```python
        # iteration for every image
        for batch in datagen.flow(dataset, batch_size=batch_size):
            # convert from rgb to grayscale
            X_batch = rgb2gray(batch)
            # convert rgb to Lab format
            lab_batch = rgb2lab(batch)
            # extract L component
            X_batch = lab_batch[:,:,:,0]
            # reshape
            X_batch = X_batch.reshape(X_batch.shape+(1,))
            # extract a and b features of the image
            Y_batch = lab_batch[:,:,:,1:] / 128
            yield X_batch, Y_batch

# the input for the encoder layer
inputs1 = Input(shape=(IMG_WIDTH, IMG_HEIGHT, 1,))

# encoder

# Using Conv2d to reduce the size of feature maps and image
size
# convert image to 128x128
encoder_output = Conv2D(64, (3,3), activation='relu',
                 padding='same', strides=2)(inputs1)
encoder_output = Conv2D(128, (3,3),
                 activation='relu',
                 padding='same')(encoder_output)
# convert image to 64x64
encoder_output = Conv2D(128, (3,3),
                 activation='relu', padding='same',
                 strides=2)(encoder_output)
encoder_output = Conv2D(256, (3,3),
                 activation='relu',
                 padding='same')(encoder_output)
# convert image to 32x32
encoder_output = Conv2D(256, (3,3),
                 activation='relu', padding='same',
                 strides=2)(encoder_output)
encoder_output = Conv2D(512, (3,3),
                 activation='relu', padding='same')
                 (encoder_output)

# mid-level feature extractions
encoder_output = Conv2D(512, (3,3),
                 activation='relu',
                 padding='same')(encoder_output)
encoder_output = Conv2D(256, (3,3),
                 activation='relu',
                 padding='same')(encoder_output)
```

```python
# decoder

# Adding colors to the grayscale image and upsizing it
decoder_output = Conv2D(128, (3,3),
                   activation='relu',
                   padding='same')(encoder_output)
decoder_output = UpSampling2D((2, 2))(decoder_output)
# image size 64x64
decoder_output = Conv2D(64, (3,3), activation='relu',
                   padding='same')(decoder_output)
decoder_output = Conv2D(64, (3,3), activation='relu',
                   padding='same')(decoder_output)
decoder_output = UpSampling2D((2, 2))(decoder_output)
# image size 128x128
decoder_output = Conv2D(32, (3,3), activation='relu',
                   padding='same')(decoder_output)
decoder_output = Conv2D(2, (3, 3), activation='tanh',
                   padding='same')(decoder_output)
decoder_output = UpSampling2D((2, 2))(decoder_output)
# image size 256x256

# compiling model
model = Model(inputs=inputs1, outputs=decoder_output)
model.compile(loss='mse', optimizer='adam',
                        metrics=['accuracy'])
print(model.summary())

tf.keras.utils.plot_model(model)

BATCH_SIZE = 20
model.fit_generator(create_training_batches
                   (X_train,BATCH_SIZE),
          epochs= 100,
          verbose=1,
          steps_per_epoch=X_train.shape[0]/BATCH_SIZE)

test_image = rgb2lab(x_test)[:,:,:,0]
test_image = test_image.reshape
              (test_image.shape+(1,))
output = model.predict(test_image)
output = output * 128

# making the output image array
generated_images = np.zeros
                    ((len(output),256, 256, 3))
for i in range(len(output)):
#iterating for the output
    cur = np.zeros((256, 256, 3))
# dummy array
    cur[:,:,0] = test_image[i][:,:,0]
```

```
    #assigning the gray scale component
        cur[:,:,1:] = output[i]
    #assigning the a and b component
    #converting from lab to rgb format as plt only work for rgb
    mode
        generated_images[i] = lab2rgb(cur)

plt.figure(figsize=(20, 6))
for i in range(10):
    # grayscale
    plt.subplot(3, 10, i + 1)
    plt.imshow(rgb2gray(x_test)[i].reshape(256, 256))
    plt.gray()
    plt.axis('off')

    # recolorization
    plt.subplot(3, 10, i + 1 +10)
    plt.imshow(generated_images[i].reshape
                            (256, 256,3))
    plt.axis('off')

    # original
    plt.subplot(3, 10, i + 1 + 20)
    plt.imshow(x_test[i].reshape(256, 256,3))
    plt.axis('off')

plt.tight_layout()
plt.show()
!wget https://raw.githubusercontent.com/Apress/artificialneural-
networks-with-tensorflow-2/main/ch14/mountain.jpg

img = imread("mountain.jpg")
plt.imshow(img)

img = resize(img, (IMG_HEIGHT, IMG_WIDTH),
            mode='constant', preserve_range=True)
img = img.astype('float32') / 255.

test_image = rgb2lab(img)[:,:,0]
test_image = test_image.reshape
                    ((1,)+test_image.shape+(1,))
output = model.predict(test_image)
output = output * 128
plt.imshow(img)
plt.axis('off')
```

接下来，将展示如何使用预先训练好的模型进行特征提取，从而节省大量训练时间并提供更好的特征提取。

14.5 编码器的预训练模型

有多种预训练模型可用于图像处理,如 VGG16 模型。使用预训练模型可以提取图像特征,与第 13 章程序中自定义的编码器效果相同。那么为什么不使用 VGG16 预训练模型代替编码器来使用迁移学习呢?答案将在此节中揭晓。与自定义的编码器相比,使用预训练模型肯定会提供更好的结果,训练速度也有所提升。

14.5.1 项目简介

本项目将使用与上一项目相同的图像数据集,因此,数据加载和预处理代码将保持不变,不同的是模型定义和测试部分。因此,此处将只讲解变化的部分。完整的项目源代码可在本书的下载网站中找到,也在本节末尾提供。本项目名称为"AutoEncoder-TransferLearning"。

由于 VGG16 是在大小为 224×224 的图像上训练的,因此需要将如下两个常量值进行更改。

```
IMG_WIDTH = 224
IMG_HEIGHT = 224
```

14.5.2 定义模型

VGG16 的模型架构已经在第 12 章(图 12-6)中进行了介绍。VGG 模型的前 18 层用于提取图像特征,因此,本项目将使用这些层并丢弃后续的全连接层。使用如下代码片段创建一个新的模型。

```
vggmodel = tf.keras.applications.vgg16.VGG16()
newmodel = Sequential()
num = 0
for i, layer in enumerate(vggmodel.layers):
    if i<19:
        newmodel.add(layer)
newmodel.summary()
for layer in newmodel.layers:
  layer.trainable=False
```

将所有这些层的可训练参数设置为 False,以便使用预训练模型进行特征提取。模型概要如图 14-11 所示。

14.5.3 提取特征

下面,将使用创建的新模型提取训练数据集图像中的特征,通过网络传递每个训练图像并记录第 19 层的预测特征。

```
vggfeatures = []
for sample in x_train:
  sample = gray2rgb(sample)
  sample = sample.reshape((1,224,224,3))
  prediction = newmodel.predict(sample)
```

```
            prediction = prediction.reshape((7,7,512))
            vggfeatures.append(prediction)
    vggfeatures = np.array(vggfeatures)
```

```
Model: "sequential_2"
_____
Layer (type)                 Output Shape              Param #
=================================================================
block1_conv1 (Conv2D)        (None, 224, 224, 64)      1792
block1_conv2 (Conv2D)        (None, 224, 224, 64)      36928
block1_pool (MaxPooling2D)   (None, 112, 112, 64)      0
block2_conv1 (Conv2D)        (None, 112, 112, 128)     73856
block2_conv2 (Conv2D)        (None, 112, 112, 128)     147584
block2_pool (MaxPooling2D)   (None, 56, 56, 128)       0
block3_conv1 (Conv2D)        (None, 56, 56, 256)       295168
block3_conv2 (Conv2D)        (None, 56, 56, 256)       590080
block3_conv3 (Conv2D)        (None, 56, 56, 256)       590080
block3_pool (MaxPooling2D)   (None, 28, 28, 256)       0
block4_conv1 (Conv2D)        (None, 28, 28, 512)       1180160
block4_conv2 (Conv2D)        (None, 28, 28, 512)       2359808
block4_conv3 (Conv2D)        (None, 28, 28, 512)       2359808
block4_pool (MaxPooling2D)   (None, 14, 14, 512)       0
block5_conv1 (Conv2D)        (None, 14, 14, 512)       2359808
block5_conv2 (Conv2D)        (None, 14, 14, 512)       2359808
block5_conv3 (Conv2D)        (None, 14, 14, 512)       2359808
block5_pool (MaxPooling2D)   (None, 7, 7, 512)         0
=================================================================
Total params: 14,714,688
Trainable params: 14,714,688
Non-trainable params: 0
_____
```

图 14-11　预训练的编码器模型概要

14.5.4　定义网络

现在，定义本项目的编码器和解码器架构代码如下。

```
#Encoder
encoder_input = Input(shape=(7, 7, 512,))
#Decoder
decoder_output = Conv2D(256, (3,3),
                 activation='relu', padding='same')
                       (encoder_input)
decoder_output = Conv2D(128, (3,3),
                 activation='relu', padding='same')
                       (decoder_output)
```

```
decoder_output = UpSampling2D((2, 2))(decoder_output)
decoder_output = Conv2D(64, (3,3), activation='relu',
                padding='same')(decoder_output)
decoder_output = UpSampling2D((2, 2))(decoder_output)
decoder_output = Conv2D(32, (3,3), activation='relu',
                padding='same')(decoder_output)
decoder_output = UpSampling2D((2, 2))(decoder_output)
decoder_output = Conv2D(16, (3,3), activation='relu',
                padding='same')(decoder_output)
decoder_output = UpSampling2D((2, 2))(decoder_output)
decoder_output = Conv2D(2, (3, 3), activation='tanh',
                padding='same')(decoder_output)
decoder_output = UpSampling2D((2, 2))(decoder_output)
model = Model(inputs=encoder_input,
            outputs=decoder_output)
model.summary()
```

对编码器使用指定的输入。解码器架构与之前的示例相同，完成放大特征图并为图像添加颜色的功能。

模型概要如图 14-12 所示。

```
Model: "model_2"
_____
Layer (type)                 Output Shape              Param #
=================================================================
input_4 (InputLayer)         (None, 7, 7, 512)         0
_____
conv2d_7 (Conv2D)            (None, 7, 7, 256)         1179904
_____
conv2d_8 (Conv2D)            (None, 7, 7, 128)         295040
_____
up_sampling2d_6 (UpSampling2 (None, 14, 14, 128)       0
_____
conv2d_9 (Conv2D)            (None, 14, 14, 64)        73792
_____
up_sampling2d_7 (UpSampling2 (None, 28, 28, 64)        0
_____
conv2d_10 (Conv2D)           (None, 28, 28, 32)        18464
_____
up_sampling2d_8 (UpSampling2 (None, 56, 56, 32)        0
_____
conv2d_11 (Conv2D)           (None, 56, 56, 16)        4624
_____
up_sampling2d_9 (UpSampling2 (None, 112, 112, 16)      0
_____
conv2d_12 (Conv2D)           (None, 112, 112, 2)       290
_____
up_sampling2d_10 (UpSampling (None, 224, 224, 2)       0
=================================================================
Total params: 1,572,114
Trainable params: 1,572,114
Non-trainable params: 0
```

图 14-12　编码器和解码器模型概要

14.5.5　模型训练

使用以下两个语句编译和训练模型。

```
model.compile(optimizer='Adam', loss='mse')
```

```
model.fit(vggfeatures, image_a_b_gen(x_train),
          verbose=1, epochs=100, batch_size=128)
```

使用 Adam 优化器和均方差损失进行模型训练。由于本项目使用了预训练的编码器，因此只需要训练解码器参数。在 GPU 上训练网络时，每个 epoch 的训练时间约为 1s，比上一项目的网络训练速度有了相当大的提升。

14.5.6 预测

现在，运行以下代码从测试数据集生成图像。本段代码相对简单，容易理解。

```
sample = x_test[1:6]
for image in sample:
  lab = rgb2lab(image)
  l = lab[:,:,0]
  L = gray2rgb(l)
  L = L.reshape((1,224,224,3))
  vggpred = newmodel.predict(L)
  ab = model.predict(vggpred)
  ab = ab*128
  cur = np.zeros((224, 224, 3))
  cur[:,:,0] = l
  cur[:,:,1:] = ab
  plt.subplot(1,2,1)
  plt.title("Generated Image")
  plt.imshow( lab2rgb(cur))
  plt.axis('off')
  plt.subplot(1,2,2)
  plt.title("Original Image")
  plt.imshow(image)
  plt.axis('off')
  plt.show()
```

上述代码的输出结果如图 14-13 所示。

14.5.7 未知图像预测

与上一项目一样，可以使用未知的新图像测试上述模型性能。使用与上一项目相同的示例完成测试。

```
!wget https://raw.githubusercontent.
com/Apress/artificialneural-networks-
with-tensorflow-2/main/ch14/mountain.jpg
```

使用如下代码显示原始图像。

```
img = imread("mountain.jpg")
plt.imshow(img)
```

图 14-13　模型测试结果

使用以下代码进行模型测试。

> **注意**
>
> 需要在将图像输入网络之前将图像大小更改为 224×224。

```
test = img_to_array(load_img("mountain.jpg"))
test = resize(test, (224,224), anti_aliasing=True)
test*= 1.0/255
lab = rgb2lab(test)
l = lab[:,:,0]
L = gray2rgb(l)
L = L.reshape((1,224,224,3))
vggpred = newmodel.predict(L)
ab = model.predict(vggpred)
ab = ab*128
cur = np.zeros((224, 224, 3))
cur[:,:,0] = l
cur[:,:,1:] = ab
plt.imshow( lab2rgb(cur))
plt.axis('off')
```

上述代码的输出结果如图 14-14 所示。

图 14-14 通过自动编码器迁移学习模型生成的彩色图像

14.5.8 项目源码

本项目的完整源码如清单 14-2 所示。

清单 14-2 AutoEncoder_TransferLearning 模型源码

```
import numpy as np
import pandas as pd
import cv2
import os
import sys

import matplotlib.pyplot as plt

from tqdm import tqdm
from itertools import chain
import skimage
from PIL import Image
from skimage.io import imread, imshow,
       imread_collection, concatenate_images
from skimage.transform import resize
from skimage.util import crop, pad
from skimage.morphology import label
```

源码清单
链　接：https://pan.baidu.com/s/1NV0rimQ_8kRz22xfFHN-Cw
提取码：1218

```python
from skimage.color import rgb2gray, gray2rgb,
    rgb2lab, lab2rgb
from sklearn.model_selection import train_test_split

from tensorflow.keras.applications.vgg16 import VGG16
from tensorflow.keras.preprocessing.image
    import load_img
from tensorflow.keras.preprocessing.image
    import img_to_array
from tensorflow.keras.applications.vgg16
    import preprocess_input

import tensorflow as tf
from tensorflow.keras.models
    import Model, load_model,Sequential
from tensorflow.keras.preprocessing.image
    import ImageDataGenerator
from tensorflow.keras.layers import Input, Dense,
    UpSampling2D, RepeatVector, Reshape
from tensorflow.keras.layers import Dropout, Lambda
from tensorflow.keras.layers
    import Conv2D, Conv2DTranspose
from tensorflow.keras.layers import MaxPooling2D
from tensorflow.keras.layers import concatenate
from tensorflow.keras import backend as K
#!pip install -q kaggle
#!mkdir ~/.kaggle
#!touch ~/.kaggle/kaggle.json

#api_token = {"username":"","key":""}

#import json
#with open('/root/.kaggle/kaggle.json', 'w') as file:
#    json.dump(api_token, file)

#!chmod 600 ~/.kaggle/kaggle.json
#!kaggle datasets download -d thedownhill/art-images-drawings-painting-sculpture-engraving

!wget --no-check-certificate -r 'https://drive.google.com/uc?export=download&id=1CKs7s_MZMuZFBXDchcL_AgmCxgPBTJXK' -O art-images-drawings-painting-sculpture-engraving.zip
!unzip art-images-drawings-painting-sculpture-engraving.zip

IMG_WIDTH = 224
IMG_HEIGHT = 224
TRAIN_PATH = '/content/dataset/dataset_updated/training_set/painting/'
train_ids = next(os.walk(TRAIN_PATH))[2]
```

```python
missing_count = 0
for n, id_ in tqdm(enumerate(train_ids),
                   total=len(train_ids)):
    path = TRAIN_PATH + id_+''
    try:
        img = imread(path)
    except:
        missing_count += 1

print("\n\nTotal missing: "+ str(missing_count))
X_train = np.zeros((len(train_ids)-missing_count,
        IMG_HEIGHT, IMG_WIDTH, 3), dtype=np.uint8)
missing_images = 0
for n, id_ in tqdm(enumerate(train_ids),
                   total=len(train_ids)):
    path = TRAIN_PATH + id_+''
    try:
        img = imread(path)
        img = resize(img, (IMG_HEIGHT, IMG_WIDTH),
                     mode='constant',
                     preserve_range=True)
        X_train[n-missing_images] = img
    except:
        missing_images += 1

X_train = X_train.astype('float32') / 255.

plt.imshow(X_train[5])

x_train, x_test = train_test_split
                    (X_train, test_size=1500)

datagen = ImageDataGenerator(
        shear_range=0.2,
        zoom_range=0.2,
        rotation_range=20,
        horizontal_flip=True)
def image_a_b_gen(dataset=X_train):
    # iteration for every image
    for batch in datagen.flow(dataset, batch_size=542):
        # convert from rgb to grayscale
        X_batch = rgb2gray(batch)
        # convert the rgb to Lab format
        lab_batch = rgb2lab(batch)

        X_batch = lab_batch[:,:,:,1:] /128

        return X_batch
vggmodel = tf.keras.applications.vgg16.VGG16()
newmodel = Sequential()
```

```python
num = 0
for i, layer in enumerate(vggmodel.layers):
    if i<19:
      newmodel.add(layer)
newmodel.summary()
for layer in newmodel.layers:
  layer.trainable=False

vggfeatures = []
for sample in x_train:
  sample = gray2rgb(sample)
  sample = sample.reshape((1,224,224,3))
  prediction = newmodel.predict(sample)
  prediction = prediction.reshape((7,7,512))
  vggfeatures.append(prediction)
vggfeatures = np.array(vggfeatures)
#Encoder
encoder_input = Input(shape=(7, 7, 512,))
#Decoder
decoder_output = Conv2D(256, (3,3),
                  activation='relu', padding='same')
                  (encoder_input)
decoder_output = Conv2D(128, (3,3),
                  activation='relu', padding='same')
                  (decoder_output)
decoder_output = UpSampling2D((2, 2))(decoder_output)
decoder_output = Conv2D(64, (3,3), activation='relu',
                  padding='same')(decoder_output)
decoder_output = UpSampling2D((2, 2))(decoder_output)
decoder_output = Conv2D(32, (3,3), activation='relu',
                  padding='same')(decoder_output)
decoder_output = UpSampling2D((2, 2))(decoder_output)
decoder_output = Conv2D(16, (3,3), activation='relu',
                  padding='same')(decoder_output)
decoder_output = UpSampling2D((2, 2))(decoder_output)
decoder_output = Conv2D(2, (3, 3), activation='tanh',
                  padding='same')(decoder_output)
decoder_output = UpSampling2D((2, 2))(decoder_output)
model = Model(inputs=encoder_input,
              outputs=decoder_output)
model.summary()
model.compile(optimizer='Adam', loss='mse')
model.fit(vggfeatures, image_a_b_gen(x_train),
         verbose=1, epochs=100, batch_size=128)

sample = x_test[1:6]
for image in sample:
  lab = rgb2lab(image)
  l = lab[:,:,0]
  L = gray2rgb(l)
```

```python
L = L.reshape((1,224,224,3))
vggpred = newmodel.predict(L)
ab = model.predict(vggpred)
ab = ab*128
cur = np.zeros((224, 224, 3))
cur[:,:,0] = l
cur[:,:,1:] = ab
plt.subplot(1,2,1)
plt.title("Generated Image")
plt.imshow( lab2rgb(cur))
plt.axis('off')
plt.subplot(1,2,2)
plt.title("Original Image")
plt.imshow(image)
plt.axis('off')
plt.show()
!wget https://raw.githubusercontent.com/Apress/artificialneural-
networks-with-tensorflow-2/main/ch14/mountain.jpg

img = imread("mountain.jpg")
plt.imshow(img)

test = img_to_array(load_img("mountain.jpg"))
test = resize(test, (224,224), anti_aliasing=True)
test*= 1.0/255
lab = rgb2lab(test)
l = lab[:,:,0]
L = gray2rgb(l)
L = L.reshape((1,224,224,3))
vggpred = newmodel.predict(L)
ab = model.predict(vggpred)
ab = ab*128
cur = np.zeros((224, 224, 3))
cur[:,:,0] = l
cur[:,:,1:] = ab
plt.imshow( lab2rgb(cur))
plt.axis('off')
```

总结

使用深度神经网络可以为黑白图像添加颜色。本章讲解了创建 AutoEncoder，并用它为黑白图像着色。AutoEncoder 包含一个编码器和一个解码器，编码器用来提取图像特征，解码器使用从编码器中提取的特征重新着色图像。同时，本章还讲解了如何使用预训练的图像分类器来提取图像特征，并将其用作编码器的一部分。